In Partnership with the **NJATC**

DELMAR
CENGAGE Learning™

Australia • Brazil • Japan • Korea • Mexico • Singapore • Spain • United Kingdom • United States

DELMAR
CENGAGE Learning™

Blueprint Reading for Electricians,
Second Edition
NJATC

Vice President, Technology and Trades
Professional Business Unit:
Gregory L. Clayton

Product Development Manager:
Ed Francis

Product Manager:
Stephanie Kelly

Editorial Assistant:
Nobina Chakraborti

Director of Marketing:
Beth A. Lutz

Executive Marketing Manager:
Taryn Zlatin

Marketing Manager:
Marissa Maiella

Director of Technology:
Paul Morris

Technology Project Manager:
Jim Ormsbee

Production Director:
Carolyn Miller

Production Manager:
Andrew Crouth

Content Project Manager:
Christopher Chien

Art Director:
Bethany Casey

Library of Congress Cataloging-in-Publication Data Card Number: 2008922253

ISBN-13: 978-1-4180-7310-7

ISBN-10: 1-4180-7310-5

Delmar Cengage Learning
5 Maxwell Drive
Clifton Park, NY 12065-2919
USA

Cengage Learning products are represented in Canada by Nelson Education, Ltd.

For your lifelong learning solutions, visit **delmar.cengage.com**

Visit our corporate website at **cengage.com**

Printed in Canada
1 2 3 4 5 XX 10 09 08

In Loving Memory of

※

Stan Klein

CONTENTS

INTRODUCTION

Welcome to the second edition of *Blueprint Reading for Electricians,* which has been redesigned and updated to provide knowledge of the fundamentals to electrical technologists in apprenticeship programs, vocational-technical schools and colleges, and community colleges. The text emphasizes a solid foundation of classroom theory supported by on-the-job hands-on practice. Every project, every piece of knowledge, and every new task will be based on the experience and information acquired as each technician progresses through his or her career. This book, along with the others in this series, contains a significant portion of the material that will form the basis for success in an electrical career.

This text was developed by blending up-to-date practice with long-lived theories in an effort to help technicians learn how to better perform on the job. It is written at a level that invites further discussion beyond its pages while clearly and succinctly answering the questions of how and why. Improvements to this edition were made possible by the continued commitment by the National Joint Apprenticeship and Training Committee (NJATC) in partnership with Delmar Cengage Learning to deliver the very finest in training materials for the electrical profession.

For excellence in your electrical and telecommunications curriculum, look no further. The NJATC has been *the* source for superior electrical training for thousands of qualified men and women for more than 65 years. Curriculum improvements are constant as the NJATC strives to continuously enhance the support it provides to its apprentices, journeypersons, and instructors in more than 285 training programs nationwide.

The efforts for continuous enhancement have produced the volume you see before you: this technically precise and academically superior edition of *Blueprint Reading for Electricians*. Using a distinctive blend of theory-based explanation partnered with hands-on accounts of what to do in the field and peppered with tips on professional practices, this book will lead you through the study of print reading—from basic drawing and sketching to symbology and residential, commercial, and industrial print reading.

This text has been strengthened from top to bottom with many new features and enhancements to existing content. All-new chapter features provide structure and guidance for learners. Enhanced and concrete Chapter Objectives are complemented by solid and reinforcing Chapter Summaries, Review Questions, and Practice Problems. Chapter contents are introduced at the beginning of each chapter, and then bolstered before moving on to the next chapter. Throughout each chapter, concepts are explained from their theoretical roots to their application principles, with reminders about safety, technology, professionalism, and more.

Blueprint Reading for Electricians, Second Edition, has been expanded to more fully explore a number of concepts through more robust chapters on drawing and sketching, projections and perspectives, scaling and dimensions, architectural considerations, and symbology. In addition, *Blueprint Reading for Electricians, Second Edition*, is chock full of new chapters, including framing and structures, elevations, schedules, and more. A new chapter on print-related math provides numerous opportunities to practice math applications, including conversion between feet and inches, working with improper fractions and mixed numbers, area and volume calculations, and more. Although these concepts were touched upon in the previous edition, this revised text more thoroughly delves into existing concepts and offers brand new information to further support the electrical technologist.

This book is divided into 18 chapters and has four basic groups of information. Chapters 1 through 6 cover basic drawing and sketching techniques, as well as how to recognize, create, and use the common views and projections used in construction drawings. Related math is then covered. It is important to be able to perform basic math functions to be able to transfer information from construction drawings to the job site. This section also introduces commonly used framing and construction types. Finally, this section provides an overview of construction drawings and how they relate to one another.

Chapters 7 through 10 cover the recognition and application of symbology in electrical, mechanical, hydraulic and pneumatic, and specialized symbology. Symbology is the interpretation of symbols, which are graphic representations of actual devices used by drafters on prints. The electrician must be able to read and understand all of the other craft prints and symbols so that he or she can decide where runs must be placed. HVAC and piping runs are more difficult to move than electrical and conduit runs. Therefore, the electrician runs often must work around the other crafts. The symbology used in the chapters is from the current American National Standards Institute and other standard symbols. The book notes that new symbols are being developed every day and that drawing legends should list these new symbols.

The third group of information details each of the types of construction drawings. Special attention is given to the electrical aspect of each drawing. The drawings covered include site plans, floor plans, elevations, details, and sections. This section also covers schedules and specifications.

Finally, Chapters 16 through 18 are the culmination of the print-reading text and activities. They cover residential, commercial, and industrial prints and specifications. The residential prints are supplied on CD and require no specifications other than window and door schedules on the prints and the National Electrical Code combined with local construction codes. Partial commercial and industrial specifications are supplied in the textbook. Exercises are supplied that use the specifications with the commercial and industrial prints. Along with the book specifications is a partial set of prints on CD for the commercial and industrial chapters. These prints may be viewed or printed full size for completion of the chapter review questions and other exercises. It is suggested that the instructor and class work together to develop more practice exercises with the supplied prints and specifications.

Running Glossaries are included in each chapter along with a comprehensive glossary at the end of the book.

High-Contrast Images give a clear and colorful view that "pops" off the page.

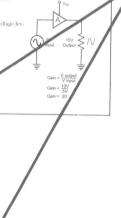

High-interest content is given in ThinkSafe!, FieldNote!, and TechTip! sidebars to make real-world connections to lessons learned in the chapters.

New photos and illustrations are located near their text references and clarify explanations.

13.6 Practical Examples

Figures 13–14 through 13–18 provide examples of installation requirements and field adjustments to signals. Each figure shows a proper installation for the given process and measurement ranges.

FIGURE 13–14 Model 3051 pressure transmitter with liquid crystal display.

FIGURE 13–15 Steam measurement. Notice the steam pots to keep the tubing full, the isolation valves, and the three-valve manifold.

Drain Valves

FIGURE 13–16 Venturi differential measurement. Observation of the tubing route, mounting location, and isolation valves indicates the type of service: liquid, gas, or steam. A gas service is shown.

FIGURE 13–17 Digital-to-analog converter from a Model 3095 multivariable transmitter that reads temperature, pressure, and differential pressure.

18 • Fiber Optics and Fiber-Optic Cable

Fibers usually come in bundles. Bundles are of two types: flexible or rigid. The flexible bundle is usually surrounded by a protective plastic coating, and at the ends of the cable the individual fibers are tied or joined together. In the rigid bundle, the individual fibers are melted together into a single rod and are shaped during the manufacturing process. Figure 18–10 shows a flexible fiber-optic bundle.

FIGURE 18–10 Fiber-optic bundles.

Coated Fiber

18.2 Optical Fiber Characteristics

To understand how a fiber-optic system operates, you need a fundamental knowledge of three areas: optics, electronics, and communications. In physics, light is treated as either electromagnetic waves or as photons (electromagnetic energy particles). For this discussion, we will concentrate on the electromagnetic wave characteristics of light. The light spectrum (light measured as a wave or electromagnetic frequency) is quite small when compared to the entire spectrum range. Figure 18–11 shows a chart of the electromagnetic spectrum. As you can see, there is only a small area of the spectrum we will consider when dealing with fiber optics, the optical spectrum from infrared to ultraviolet frequencies.

FIGURE 18–11 Electromagnetic spectrum.

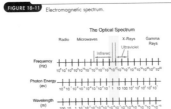

End-of-chapter problems reinforce critical concepts and relate to the worked-out examples in the chapter.

Practice Problems 159

PRACTICE PROBLEMS

Problems 1 to 6 refer to Figure 7–17.

1. Find the impedance.
2. Draw and label an impedance triangle for the series circuit as shown in Figure 7–17.
3. Find E_T.
4. Find E_R, E_C, and E_L.
5. What is the true power of the circuit?
6. What is the phase angle and power factor (PF)?

Problems 7 to 13 refer to Figure 7–18:

7. Draw and label an impedance triangle for the series circuit in Figure 7–18.
8. Find Z.
9. Draw and label a vector diagram of the circuit values for E_C, E_L, E_R, I_T, and E_T, and angle theta for the circuit in Figure 7–18.
10. Find I.
11. Find E_R, E_L, and E_C.
12. What power is used in this circuit? Prove it.
13. Find E_T and draw the vector diagram.

FIGURE 7–17 Circuit drawing for Problems 1–6

FIGURE 7–18 Circuit drawing for Problems 7–13

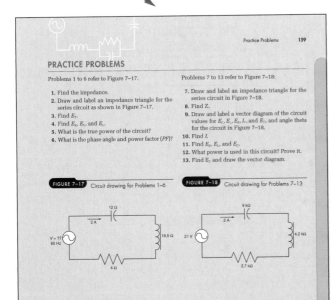

Step-by-step sample problems and solutions relate to the end-of-chapter exercises.

Example

A differential amplifier has the following input and output signals:
• Common-mode input = 1 V
• Common-mode output = 0.02 V
• Differential input = 0.1 V
• Differential output = 10 V
• Solve for $CMRR_{dB}$

Solution:
Step 1:

$$A_{V(DIFF)} = \frac{Differential\ Output}{Differential\ Input}$$

$$A_{V(DIFF)} = \frac{10\ V}{0.1\ V}$$

$$A_{V(DIFF)} = 100$$

Step 2:

$$A_{V(DIFF)\ dB} = 20 \times \log_{10} A_{V(DIFF)}$$

$$A_{V(DIFF)\ dB} = 20 \times \log_{10} 100$$

$$A_{V(DIFF)\ dB} = 40\ dB$$

Step 3:

$$A_{V(COMM)} = \frac{Common\ Mode\ Output}{Common\ Mode\ Input}$$

$$A_{V(COMM)} = \frac{0.02\ V}{1\ V}$$

$$A_{V(COMM)} = 0.02$$

Step 4:

$$A_{V(COMM)\ dB} = 20 \times \log_{10} A_{V(COMM)}$$

$$A_{V(COMM)\ dB} = 20 \log_{10} 0.02$$

$$A_{V(COMM)\ dB} = -33.98\ dB$$

Step 5:

$$CMRR = \frac{A_{V(DIFF)}}{A_{V(COMM)}}$$

$$CMRR = \frac{100}{0.02}$$

$$CMRR = 5,000$$

The Instructor Resource Kit is geared to provide instructors with all of the tools they need in one convenient package. Instructors will find that this resource provides them with a far-reaching teaching partner that includes the following:

- PowerPoint® slides (electronic and hard copy) for each book chapter that reinforce key points and feature illustrations and photos from the book,

- the Computerized Test Bank in ExamView format, which allows for test customization for evaluating student comprehension of noteworthy concepts,

- an electronic and paper version of the Instructor's Manual, with supplemental lesson plans and support,

- the image library, which includes all drawings and photos from the book for the instructor's use to supplement class discussions, and

- a transition guide (electronic and hard copy) to help instructors map the changes from the previous edition to this new, stronger edition of the book.

ABOUT THE NJATC

Should you decide on a career in the electrical industry, training provided by the International Brotherhood of Electrical Workers and the National Electrical Contractors Association (IBEW-NECA) is the most comprehensive the industry has to offer. If you are accepted into one of their local apprenticeship programs, you will be trained for one of four career specialties: journeyman lineman, residential wireman, journeyman wireman, or telecommunications installer/technician. Most importantly, you will be paid while you learn. To learn more, visit *http://www.njatc.org.*

ACKNOWLEDGMENTS

NJATC ACKNOWLEDGMENTS

Technical Writer and Editor

James L. Boyd, NJATC Senior Director

ADDITIONAL ACKNOWLEDGMENTS

This material is continually reviewed and evaluated by Curriculum Groups who are also members of the NJATC Inside Education Committee. The invaluable input provided by these individuals allows for the development of instructional material that is of the absolute highest quality. At the time of this printing the Inside Education Committee consisted of the following members: Lawrence Hidalgo, Chair; Stan Elsasser; Jonathan Gosse; Kurt Hamilton; Mitch Hegman; Carl Latona; Ed Murphy; Jim Paladino; and Bill Rusher.

PUBLISHER'S ACKNOWLEDGMENTS

Delmar Learning and the author would also like to thank the following reviewers for their valuable suggestions and expertise:

Gwen Oster, Electrical Instructor, Northwest Technical College, Bemidji, MN

Paul Westrom, Asst. Professor of Electrical Technology, New England Institute of Technology, Portsmouth, RI

Tom Collins, Instructor, Gateway Community and Technical College, Cincinnati, OH

AUTHOR'S ACKNOWLEDGMENTS

Thank you to my wife, Brandi, and my children, Lauren, Kate, and Julia, for being patient and understanding while I spent countless hours revising this text. I would like to thank the entire Delmar staff for their support, but in particular Ed Francis for giving me this opportunity and Stephanie Kelly for all of her hard work and assistance. Thank you to the following people for their contributions to this text:

- Keith Johnson of Minnesota State Community and Technical College, Moorhead, MN
- Dennis Wagner of Minnesota State Community and Technical College, Moorhead, MN
- Jon Scraper of Ulteig Engineers, Fargo, ND
- Mike Berger of Ulteig Engineers, Fargo, ND
- Troy Magnell of Ulteig Engineers, Fargo, ND
- Gary Fritz of Gary Fritz Electric Inc., Fargo, ND
- Patrick Olson of Gary Fritz Electric Inc., Fargo, ND

Rob Zachariason is an instructor at Minnesota State Community and Technical College and an instructor for the Joint Apprenticeship and Training Committee. He is currently the chair of the construction trades division at Minnesota State Community and Technical College. Courses Rob has taught include Blueprint Reading, Residential Wiring, Electrical Materials, National Electrical Code, Motors and Generators, Conduit Bending, Electrical Services, First-year NJATC Curriculum, Second-year NJATC Curriculum, and Fifth-year NJATC Curriculum.

Rob is a member of the International Brotherhood of Electrical Workers, the International Association of Electrical Inspectors, and the National Education Association. He carries a Master Electricians license in North Dakota and Minnesota.

Rob graduated from Northwest Technical College with a Diploma in Construction Electricity. He then went on to complete the National Joint Apprenticeship and Training Committee's Inside Wireman Program.

Rob worked as an electrician for 10 years before becoming a full-time instructor. His time in the field was spent with Gary Fritz Electric and Robert Gibb and Sons. While with Gary Fritz Electric he ran a service truck and worked residential, commercial, and small industrial jobs. While with Robert Gibb and Sons, he spent his time doing electrical control work and running a service truck.

1

Drawing and Sketching

OUTLINE

OVERVIEW

Throughout history, tradespeople have relied on drawings and prints to build objects and structures. To read and interpret blueprints, it is necessary to understand how drawings are produced. It is also important to understand the use of basic drawing and layout tools. Finally, basic sketching concepts are needed to converse on the job, where drafting tools and computers often are not readily available. This chapter covers these three areas to set the stage for the practice of proper blueprint reading.

OBJECTIVES

After completing this chapter, you should be able to:

- Recognize and identify the application and function of the different line types used in drawing.

- Identify and give the function of the basic drawing tools.

- List the advantages of computer-aided design over paper-and-pencil drawing.

- Sketch lines, arcs, circles, and shapes freehand.

LINE TYPES

In blueprint reading there are several types of lines. These lines are used to define and clarify what the architect or engineer is trying to convey to the electrician. It is important for the electrician to recognize and understand the meaning of each line type.

This system of line definition is called the **alphabet of lines.** Each line has a specific thickness and shape. The use of **computer-aided design (CAD)** systems has brought to light dozens of line types. However, there are actually only about 12 lines that are likely to occur in construction-trade blueprints.

To properly understand blueprints it is necessary to understand this system of line definition. The following are the lines commonly used in construction prints.

1. *Object line (visible line).* The object line is a thick, continuous line that represents all surface boundaries of a print or object. In print reading, it defines all solid objects and boundaries.

2. *Hidden line.* The hidden line is a thin series of short dashes. The function of this line is to show features or lines that are not visible in the current view. This sometimes helps clarify the shape of the object. In print reading, the hidden line can be used to show hidden features or future construction. It can also be used to identify items for removal or change.

3. *Center line.* Center lines are thin lines with a short dash every inch or so. Their function is to define centers of holes, arcs, and symmetry. In print reading, center lines can show center points, elevation lines, projections, and paths of motion.

4. *Phantom line.* Phantom lines are thin lines with pairs of two short dashes about every inch. Generally, the function of a phantom line is to show motion or alternate positions of a part or feature. In print reading, phantom lines indicate movement or boundary lines. They can also be used to indicate an existing feature (such as a column) that has a new attachment, such as a beam.

5. *Cutting-plane (viewing-plane) line.* Cutting-plane lines are thick lines that resemble phantom lines. The difference is that a cutting-plane line has arrows at each end that indicate viewing direction. The function of a cutting-plane line is to visually carve an internal section out of a drawing or feature. This is usually done to provide a larger and more readable view. The arrows indicate the viewing direction of the section that is made by the cutting-plane line. In print reading, the cutting-plane line indicates internal features or parts of drawings set aside for further and more detailed inspection.

6. *Section line.* Section lines are thin, parallel lines drawn at an angle to the main surfaces of the part. The function of section lines is to show the surfaces actually cut by the cutting-plane line. If you imagine you are cutting through the smooth part of an object with a hacksaw, the cutting-plane lines would be the scratch marks left by the saw teeth. In print reading, section lines indicate internal features. They are also used to define various materials (such as wood, steel, or concrete) used in corner sections and beam sections. In this case, the medium parallel lines are replaced by the

Alphabet of Lines
System of defining the types of lines used in technical and construction drawings.

Computer-Aided Design (CAD)
Using a computer, program, and plotter or printer to assist in the creation of drawings.

American National Standards Institute (ANSI)
An organization that coordinates the development and use of voluntary consensus standards in the United States.

American National Standards Institute (ANSI) symbol for the material of which the part is constructed.

7. *Short-break line*. The short-break line is thick and jagged in shape. The function of the short-break line is to break out a section of a part that is too long to show at the existing scale. It is also used to break out sections for further inspection. In print reading, the short-break line indicates small sections broken out of a drawing.

8. *Long-break line*. Long-break lines are thin lines interrupted by a sharp zigzag about every inch. They serve essentially the same function as a short-break line, except that they are used for larger areas. In print reading, they are used to pull out a section of a drawing for enlargement and further detailed inspection.

9. *Extension line*. Extension lines are thin, straight lines that extend the feature to be dimensioned. They usually extend from the part at 90° to the surface being dimensioned. The function of the extension line is to tell the print reader where the dimension starts and ends.

10. *Dimension line*. Dimension lines are thin, straight lines with arrowheads, dots, or slashes at each end where they contact extension lines. The actual numerical dimension is broken into the middle of this line or placed above it. The function of the dimension line is to tell the print reader the numerical distance from one point to another or the size of a given feature. Extension lines are typically used together, except in the case of leader dimensions.

11. *Leader line*. Leaders are thin lines with an angular jog (or S shape) ending with an arrowhead touching the feature. The other end of the leader ends at a dimension or note. The function of leaders is to give the print reader dimensions of irregularly shaped features or to indicate a specific drawing notation.

12. *Construction line*. Construction lines (layout lines) are very light lines used to set up the framework for a drawing. Construction lines are so light that they should not be visible at arm's length.

NOTE: Due to reproduction techniques and plotters used with some CAD systems, the line weight or thickness may not be distinguishable. In this case, the line type should be determined by its characteristics, location, and apparent function in the drawing. See Figure 1–1 for a condensed version of the preceding information.

DRAWING TOOLS

The draftsperson uses a variety of drawing tools to turn the alphabet of lines into working drawings or prints. The basic tools are a straightedge, triangles, a protractor, a divider, a compass, scales, and of course a sharp dark drafting pencil (lead holder). The straightedge may be a T square. The T square is placed on the edge of a drawing board and rides up and down the side of the board, producing parallel lines (as in Figure 1–2).

After making horizontal parallel lines it is necessary for the draftsperson to make perpendicular and angular lines. Perpendicular lines are drawn by placing the perpendicular side of a triangle on the T square and sliding the triangle right to left, as shown in Figure 1–3.

Angular lines are reproduced using the T square, triangles, or a protractor. There are three basic triangles available to the draftsperson: the 30° × 60° triangle, the 45° triangle, and the adjustable triangle. The use of

FIGURE 1–1 Alphabet of lines. Next to each line is an example of where the line may be used.

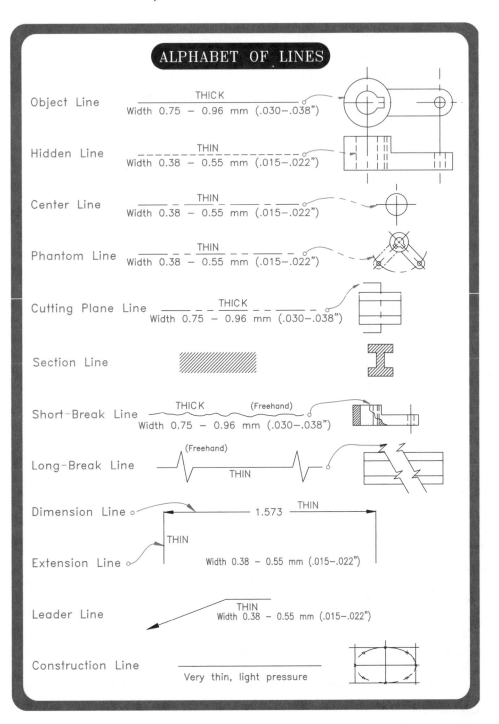

FIGURE 1-2   Drawing horizontal lines using a T-square.

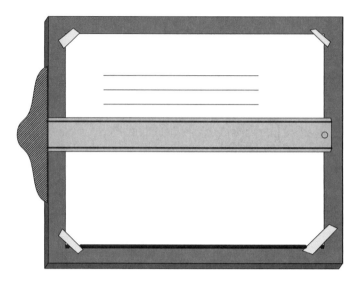

FIGURE 1-3   Drawing vertical lines using a T-square and a triangle.

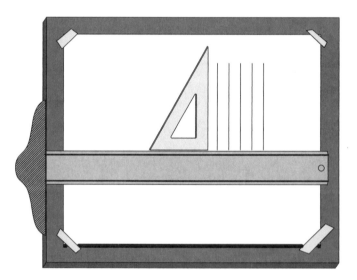

the fixed triangles individually or together can produce a variety of angles in 15° increments, as shown in Figure 1–4.

A protractor can be used to lay out any other necessary angle. With the protractor, the draftsperson first marks the angle with a pencil dot. Next, a line is drawn between the origin and the dot using any triangle or straight-edge (as illustrated in Figure 1–5).

A compass, like that shown in Figure 1–6, is used to connect straight lines with curves and radii. It can also be used to draw circles or partial circles.

A divider is used to lay out equal distances or to transfer existing dimensions. Once set to a specific dimension, they can be used to transfer

FIGURE 1-4 By using a 45° triangle and a 30 x 60 triangle, angles can be drawn in 15° increments.

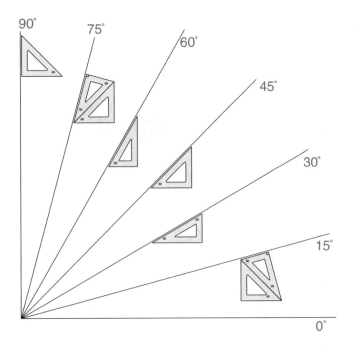

FIGURE 1-5 A protractor is used to make angles that can't me made using the combination of triangles in Figure 1-4.

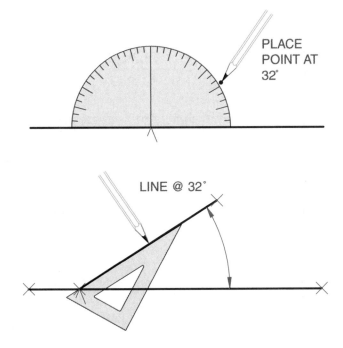

FIGURE 1-6   A compass is used to make circles or arcs.

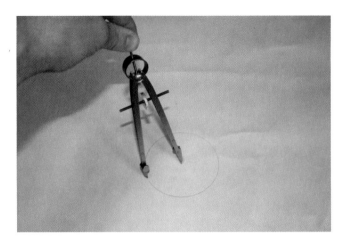

that dimension to another place on the drawing. If a draftsperson were laying out equally spaced wall studs on a drawing, dividers would be the appropriate tool (as shown in Figure 1–7).

The combination of all of these tools and a scale makes it possible to produce sharp, readable mechanical or construction drawings. Various scales are used in drawing (discussed in Chapter 4).

The next level up in drawing, after handheld instruments, is the drafting machine. The drafting machine contains a horizontal and vertical straightedge scale. It is also capable of turning to and locking at any angle in 1° increments. It is still necessary, however, to use the compass for circular shapes. The drafting machine is faster and usually more accurate than totally handheld instruments (see Figure 1–8).

FIGURE 1-7   A divider can be used to provide equal spacing.

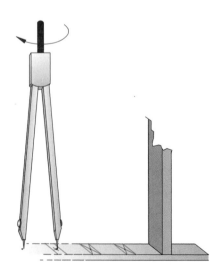

FIGURE 1-8 A drafting machine provides a horizontal and vertical straightedge. It can be rotated and locked at any angle.

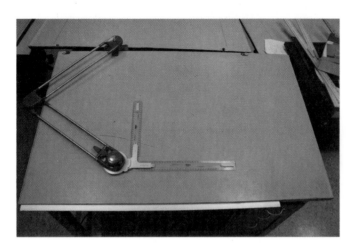

1.1 Computer-Aided Design

The final level of mechanical drawing is the CAD system (Figure 1–9). CAD is now the method most often used to create blueprints and construction drawings. CAD uses a computer program to create drawings. Once a drawing or set of prints has been completed using the computer, the information is sent to a printer or plotter. This is where the paper copy

FIGURE 1-9 Blueprints are now created on a computer loaded with CAD software. The typical workstation consists of a monitor, CPU (computer process unit), keyboard, and mouse. The blueprints are sent electronically to a plotter or printer to create the paper copy.

of the drawings is created (see Figures 1–10 and 1–11). CAD programs are very complex and take time to learn, but offer numerous advantages over paper-and-pencil drawings.

1. *Ability to easily change or revise drawings.* If a mistake is made or if something has to be revised, it is not necessary to erase or perhaps start over with the drawing. The portion of the drawing that needs to be altered is simply selected, and then changed, moved, or deleted.

2. *Precise Drawings*

   a. *Accurate.* The drawings produced by CAD will be drawn with pinpoint accuracy (provided, of course, the information was entered into the program accurately).

   b. *Consistent line weights.* When drawing with a pencil and paper it is very easy to vary the line weights, which could cause the line to have a different meaning. The computer will draw the lines perfectly every time. This ensures that the appropriate information is conveyed.

   c. *Consistent and neat lettering and symbols.* Sloppy handwriting or symbols can make a drawing confusing and difficult to comprehend. CAD lettering will be perfect every time, and a large bank of symbols makes the insertion of symbols very easy.

3. *Layers.* A CAD drawing can have layers. Layers are information that can be added or removed from a drawing with the click of a button. These layers can all have different colors, making them very easy to distinguish. Any

FIGURE 1–10   Plotters create a paper copy of a blueprint. This plotter will create color prints or drawings.

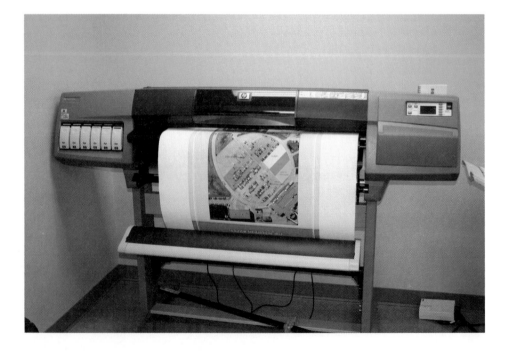

FIGURE 1–11 Printers create a paper copy of a blueprint. Printers may have the information sent directly from the workstation, or they can create a copy from another print.

number or all of the layers can be removed from the drawing. For example, imagine we have the basic floor plan of a house with nothing but the partitions (walls) in the drawing. This would be the basic drawing. We could then start adding layers. The dimensioning could be one layer, notes another layer, electrical symbols another, plumbing symbols another, and so on. If the electrician did not want to see any information except for the basic floor plan and the electrical, that alone could be printed.

4. *Easily copy information.* Rather than redrawing a plan or specific part of a plan, the information can be copied with the click of a button. For example, suppose a building has four floors that are all the same. Once one floor has been designed, it can be copied to create the other three floors.

5. *Print in colors.* The layers previously discussed can be printed in color. Printing in color makes it easier to distinguish lines and thus makes the document easier to read.

6. *Send files electronically.* Drawings can be shared or sent via e-mail or disk.

7. *Easy file storage.* Storing a set of prints on a disk or in a hard drive is much easier than having drawers of blueprints that are each several hundred pages long.

This is not to say that many preliminary drawings are not done using a pencil and paper. This is still a great way to brainstorm and work through ideas, but once the construction drawings begin CAD takes over.

PAPER TYPES AND SIZES

Drawings are typically printed on standard paper sizes. Table 1–1 outlines the standard paper sizes for architectural drawings. Most prints will be drawn on size C, D, or E paper. Detail drawings will occasionally come separately on size A or B paper. An example would be from the cabinet makers on a residential project. These drawings are often on size A paper.

TABLE 1–1   Standard paper sizes.

STANDARD PAPER SIZES	
Size	**Dimensions**
A	8½ x 11
B	11 x 17
C	17 x 22
D	22 x 34
E	34 x 44

Once in a while a general contractor will drop off a residential set of prints that are on size A or B paper. The scale of these drawings is very small, making them extremely difficult to read. A larger set of drawings can typically be acquired by contacting the person that drew up the construction drawings.

Orthographic paper (graph paper) comes in very handy for the tradesperson (Figure 1–12). Commonly used line spacing is ¼ inch. However, it is available with other spacing. This is used by many estimators and contractors when meeting with a customer. Graph paper aids in sketching a drawing or floor plan by providing faint lines that are parallel and square. Drawings can be quickly drawn nearly to scale simply by counting the number of grid lines.

Orthographic Paper
Paper with a grid of lines drawn at 90° angles.

FIGURE 1–12   Graph paper has faint horizontal and vertical lines that can be used as a guide to create sketches.

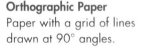

BASIC SKETCHING

As a tradesperson, it is necessary to communicate with other workers and supervisors. This communication can be verbal, written, or sketched. Drafting or sketching is the language of engineering—and of trades in general. A drafting board and tools may not be available on the job. It is therefore necessary to develop skills in freehand sketching.

Freehand sketching is the process used by tradespeople and engineers to convey ideas quickly on the job. It is therefore necessary to have a command of the drawing and sketching process. The following sections discuss some techniques and tricks to help you become better at freehand sketching.

A sharp, dark pencil and sketching pad are ideal for the process—but any paper or writing utensil can be used. The major benefit of a pencil is its erasability. First, hold the pencil firmly and in a comfortable writing position. Do not be so rigid as to reduce your ability to move. While drawing lines, make short, light strokes that you will darken later. Start from one point and draw the short sketching strokes toward the final (terminal) point. Your eye should always be on the terminal (ending) point (Figure 1–13).

Freehand Sketching
Sketching a drawing without any drawing tools other than a pencil and paper.

FieldNote!

Where Will I Use Sketching?

Imagine you are the electrical contractor or estimator and you go out to a basement remodel to give an estimate. The first thing you may need to do is sketch a quick room layout or floor plan on a sheet of paper. Then, as you walk through the basement with the owner you can write where the lights, devices, switches, and so on will go. This drawing may go back to the office for preliminary drafting, or be given to an electrician for wiring the project (see Figure 1–14). Other examples:

- Draw a sketch of a type of light fixture you think would work well for a business owner (see Figure 1–15).
- Draw a sketch to describe a part so that a new apprentice would be able to obtain it (see Figure 1–16).
- Draw how the layout of several raceways should be on a wall.

It is very helpful to be able to convey information using a quick sketch. Keep in mind that it does not have to be perfect; it simply needs to accurately get the information across.

FIGURE 1-13

When sketching a line always keep your eye on the terminal point.

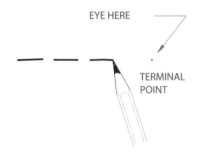

FIGURE 1-14

Sketch a contractor or estimator made while meeting with an owner about finishing off a basement.

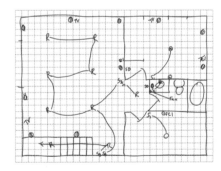

FIGURE 1-15

Sketch of an outdoor light fixture.

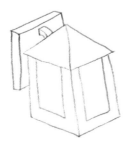

FIGURE 1-16

Sketch of an electrical metallic tubing coupling.

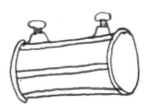

1.2 Sketching Horizontal Lines

Horizontal lines should be made with a series of short forearm movements roughly perpendicular or at 90° to the line or surface being sketched. Right-handers should sketch from left to right (Figure 1–17). Left-handers should sketch from right to left.

FIGURE 1-17 When sketching horizontally, right handers should sketch from left to right; left handers should sketch from right to left.

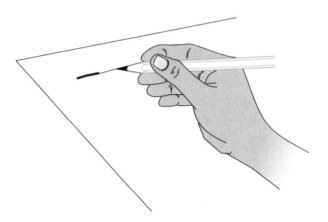

1.3 Sketching Vertical Lines

Vertical lines should be made top to bottom with short, vertical strokes. Your arm should be offset slightly from the vertical line to keep the terminal point in sight and to produce a straight line (Figure 1–18).

FIGURE 1-18 When sketching vertically, work from the top down.

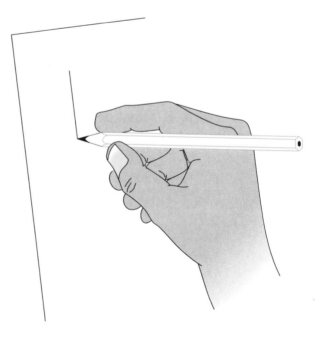

1.4 Sketching Inclined Lines

The process for sketching **inclined lines** (lines that slope or make an oblique angle with the horizon) is virtually identical to that described for horizontal and vertical lines. Turn the paper to the desired angle. Mark the start and end points, and then use short sketching lines to the end point just as in horizontal and vertical lines. Keep your eye on the end point.

1.5 Basic Sketching Steps

There are four basic steps in sketching straight lines. These are as follows:

1. Locate starting and finishing points for the desired line.
2. Position your arm by making trial movements without drawing an actual line. The sketching movements should be from left to right and top to bottom for right-handers. On horizontal lines, left-handers should sketch right to left.
3. Use a series of short lines between the starting and terminal point.

NOTE: Always keep your eye on the point where the line will end.

4. After your line is correct, darken it to a uniform weight or thickness (Figure 1–19).

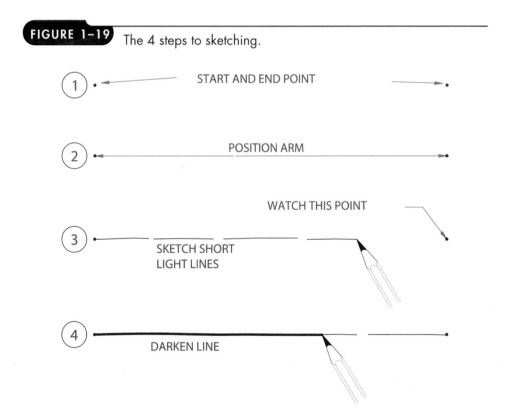

FIGURE 1–19 The 4 steps to sketching.

1 START AND END POINT

2 POSITION ARM

3 WATCH THIS POINT
   SKETCH SHORT
   LIGHT LINES

4 DARKEN LINE

PRACTICE

At this point (as in any new skill or process), it is best to practice sketching horizontal, vertical, and diagonal lines.

1. Take a sheet of plain paper and practice making parallel horizontal lines. Try making them approximately 1 inch apart.

2. On the same or another sheet of paper, try the same exercise with vertical lines.

3. On a separate sheet of paper, try making inclined lines from top to bottom and left to right.

4. On this same page, try making inclined lines from top to bottom and right to left.

NOTE: Remember to place starting and ending points on your paper for all of the previous exercises. A trick that may help at first is to use a straight-edge to draw your first line in each exercise. This will give your eye a parallel guide for the practice lines. After trying this trick, make sure you can sketch the various lines.

1.6 Sketching Angles

The process of making any sketched line is virtually the same as previously discussed. The difference is in the process of finding the starting and ending points. The starting point is at your discretion, but the end point must be determined.

In drawing angles, it is best to sketch a 90° line and a 180° line and divide them roughly into the desired angles (as shown previously in Figure 1–4). This should be done with a very light pencil line that is barely visible. After finding the desired starting and ending points of the angles, use the basic sketching process to darken them (Figure 1–20).

FIGURE 1–20 Angle estimating.

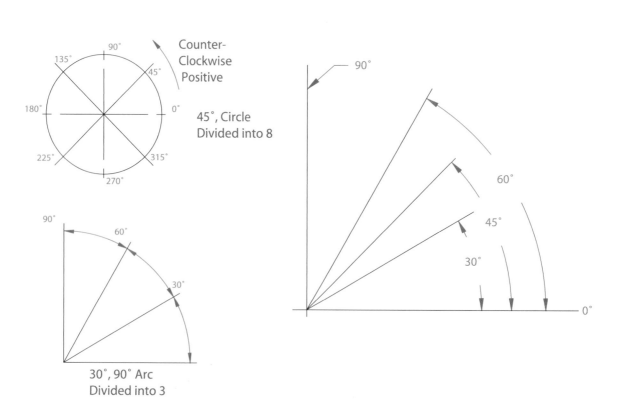

1.7 Sketching Arcs

Sketching arcs, circles, and ellipses requires the use of light layout lines to set up starting and ending points for the arcs. One proven process for making arcs is the triangle-square method. To sketch an arc using this method, perform the following steps (Figure 1–21).

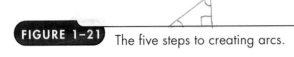

FIGURE 1–21 The five steps to creating arcs.

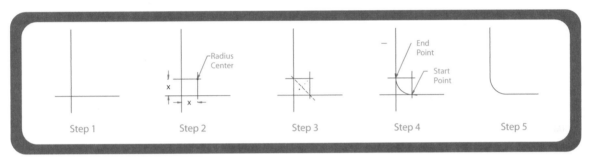

1. Sketch intersecting vertical and horizontal lines at roughly 90°.
2. Lay out the radius center by measuring equal distances for each line and drawing parallel intersecting lines. They will intersect at the arc center.
3. Make a triangle crossing the constructed square. This locates the starting and ending points of the arc. Mark a dot at the approximate center of the arc, which is the center of the triangle. This will guide the radius.
4. Sketch short strokes from the starting point to the ending point using an arcing motion, passing through the center dot.
5. Darken the arc and erase any unnecessary layout lines.

1.8 Sketching a Circle

Sketching circles is similar to sketching arcs and is accomplished by performing the following steps (see Figure 1–22).

FIGURE 1–22 The five steps to create a circle.

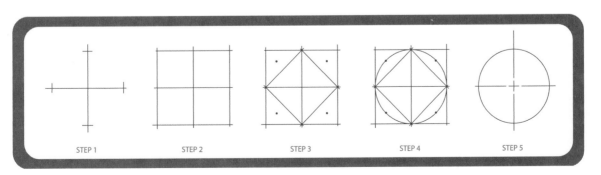

1. Sketch a horizontal and vertical line crossing at a center point. Place points at estimated equal distances from center.
2. Sketch a square crossing the end points using very light layout lines.
3. Sketch corner diagonals and locate center points, as in drawing regular arcs.
4. Sketch short arcs through the appropriate layout lines. Some people find it easier to turn the paper to the most comfortable position for each arc.
5. Darken the circle and erase unnecessary layout lines.

1.9 Sketching an Ellipse

Sketching an ellipse is virtually the same process as sketching a circle. The difference is that an ellipse is a circle elongated in one direction. Therefore, the process is as follows (see Figure 1–23).

FIGURE 1–23 The 4 steps to sketch an ellipse.

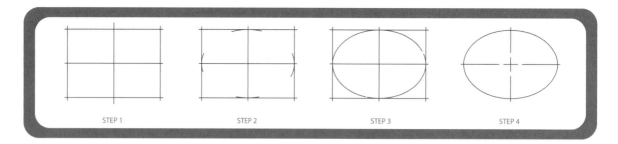

STEP 1   STEP 2   STEP 3   STEP 4

1. Use the circle setup process but construct a rectangle instead of a square. The length and width of the rectangle determine the length and width of the ellipse.
2. Sketch small arcs at the intersecting points. Remember that an ellipse is actually two large arcs and two smaller arcs converging. The arcs that touch the long sides of the rectangle are larger than those that touch the small sides.
3. Complete the ellipse using light, short strokes.
4. Darken the ellipse and erase unnecessary layout lines.

Layout and practice are the keys to becoming proficient at freehand sketching. In the layout process, a short piece of paper with evenly spaced lines can be used to keep the object to scale. If you have trouble drawing evenly spaced lines, make a paper scale. A makeshift paper scale can be made by folding a piece of paper in half again and again until it is about ½ inch wide. Next, unfold the paper and mark the fold lines with a pencil and number them. If the paper scale is too long at this point, tear off the top inch to use as a scale (Figure 1–24).

FIGURE 1-24 A paper scale can be created to help create evenly spaced lines.

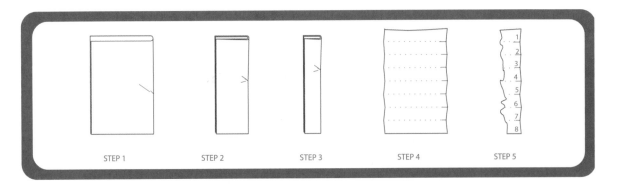

STEP 1   STEP 2   STEP 3   STEP 4   STEP 5

With these sketching techniques for straight, angular, and curved lines, it is possible to create any shape desired. All shapes are a combination of points, lines, and curves. Sketching takes practice and is easier for some people than others. At this point, you should practice sketching the arcs, circles, and ellipses described in this chapter.

Chapter 2 discusses various types of views, projections, and perspectives. After completing Chapter 2, you should practice sketching some of the basic views and projections described.

SUMMARY

There are many different types of lines associated with blueprints. Recognition of the different types of lines is necessary for the proper understanding of construction drawings.

Drawing tools have changed over the years. CAD has revolutionized the construction drawing process, saving time and money.

Sketching is a valuable tool. It is an important part of the communication process in the construction industry.

REVIEW QUESTIONS

1. Draw and label each of the 12 common line types.
2. What is used in mechanical drawing to draw horizontal and vertical lines?
3. What tool or tools are used in mechanical drawing to draw angled lines at 15° and 30° intervals?
4. What is the tool used to lay out odd angles?
5. What tool is used to lay out equal distances?
6. List five advantages CAD has over paper-and-pencil drawing.
7. Why do some estimators bring graph paper when going on an estimate?
8. List the indicating letter and dimensions of the standard paper sizes used with construction drawings.
9. What is the function of sketching?

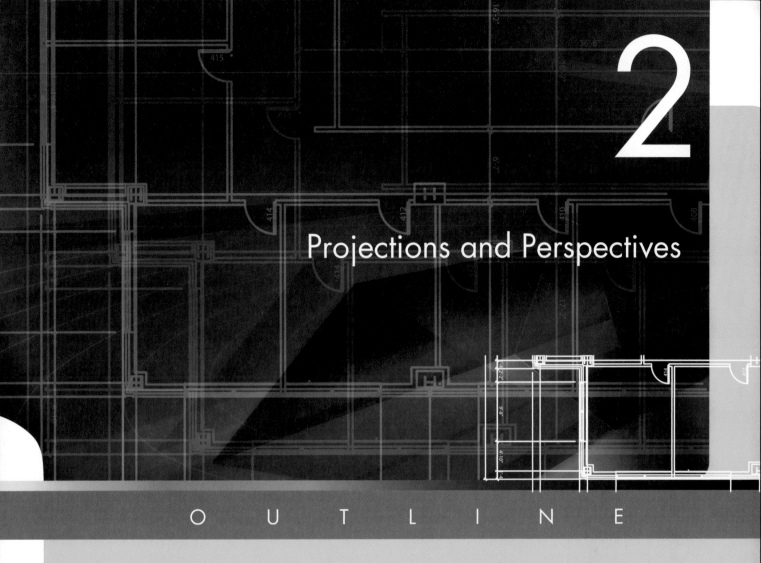

2

Projections and Perspectives

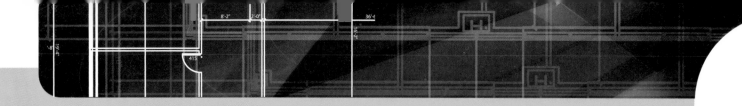

OVERVIEW

In blueprints, the mechanical drawings are in the form of projections. These projections have views, which are essentially the way we look at drawings and from what angle. Prints also have plan and elevation views, as discussed in subsequent chapters. It is important to understand projections and views to avoid costly mistakes in construction from incorrect print reading and device placement.

OBJECTIVES

After completing this chapter, you should be able to:

• Recognize and label the views on orthographic, isometric, and oblique drawings.

• Sketch or draw simple orthographic, isometric, and oblique drawings.

• Convert between orthographic, isometric, and oblique drawings.

INTRODUCTION

The purpose of drawing is to define objects in shape and size. This makes it possible to produce these objects. The purpose of trade blueprints is to show the necessary views of a building to enable the tradesperson to build, wire, or plumb that building. There are several types of projections, perspectives, and views that may be used to accomplish this task. It is extremely important that the tradesperson can visualize the relations of these views and projections. Visualization is the ability to look at a drawing and form a realistic picture of the actual object in one's mind.

ORTHOGRAPHIC PROJECTIONS

Orthographic projections consist of a possibility of two to six standard views and any number of auxiliary views. The standard views are drawn as if the object were in a transparent box. The box and the object would have six sides: front, back, right, left, top, and bottom (Figure 2–1). If this imaginary transparent box were unfolded from all edges at 90°, a flat representation of all of the object views would appear. The views would be displayed as if they were folded out on hinges and then separated (Figure 2–2).

Orthographic Projection
Using two or more two-dimensional drawings to represent a three-dimensional object.

FIGURE 2–1

Imagining an object in a transparent box can help to understand the possible views used in orthographic projection.

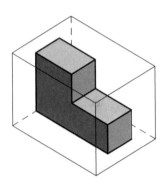

FIGURE 2–2 The six possible views of an object.

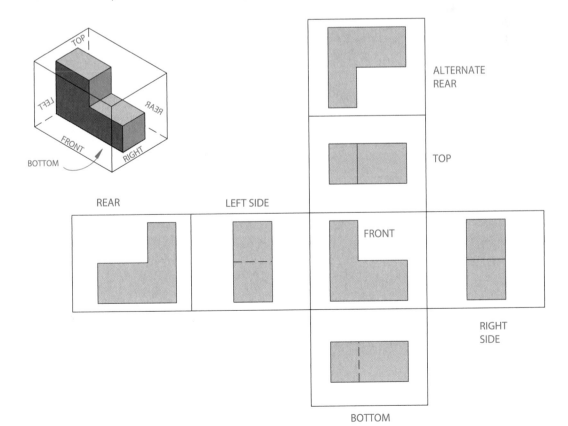

Visualization
Imagining what a finished product will look like while looking at a drawing.

Each view must be individually visualized as if the draftsperson or tradesperson were looking at it at 90°, or perpendicular to the viewing plane (Figure 2–3). **Visualization** is difficult for most beginners. Therefore, it may be useful to dissect a simple object into points, lines, and planes for clarification. Figure 2–4 shows a picture of the basic object and the three necessary views folded out into orthographic (90°) projections. Parts are marked on each view for reference, and a fold line shows how the top and right-side views are related. Dashed lines indicate layout lines.

FIGURE 2-3 An object viewed from the top and from the right side. Only the views necessary to understand an object are shown.

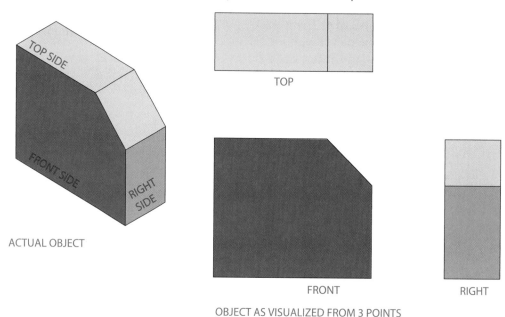

FIGURE 2-4 Numbering the points of an object can help with visualization.

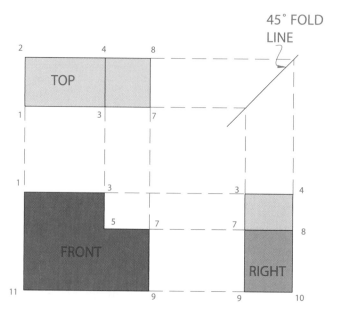

2.1 Visualization

There are several rules to follow when visualizing orthographic projections. First, look over all views shown to get a basic picture of the object. Next, study the front or main view carefully. This view is usually the most important view for shape description. Next, pick one feature in the front view and relate it to its relative position in the other views. Finally, repeat this process of relating features throughout the given views until you form a clear mental picture of the entire object.

2.2 View Selection

Although there are six basic views on an orthographic drawing, it is not necessary to use all six views every time. It is necessary to use only the minimum number of views to totally define the object's shape. In a case such as a simple cylinder, only two views would be necessary: top and front. Any further views would be unnecessary. The majority of orthographic drawings have three views; cylinders are an exception (Figure 2–5).

2.3 Auxiliary Views

In typical drawings, three or more views are necessary to thoroughly define the object's shape. Rarely are all six views needed. However, it is sometimes necessary to use special views, such as regular **auxiliary views**, sections, or partial auxiliary views. Regular auxiliary views are those objects having surfaces with some principle faces that are not parallel or perpendicular to the six regular planes of projection. They may also be cylindrical shapes not perpendicular to a regular plane (Figure 2–6).

FIGURE 2–5

Orthographic projection of a cylinder requires only two views.

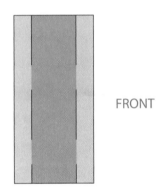

TOP

FRONT

Auxiliary View
A view of an object looking from an angle to show the actual size of a surface.

FIGURE 2–6   An auxiliary view may be used to give additional information.

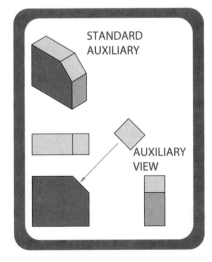

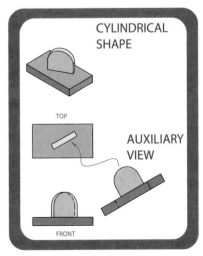

2.4 Drawing Orthographic Projections

Drawing or sketching orthographic or multiview drawings is a four-step process, as shown in Figure 2–7.

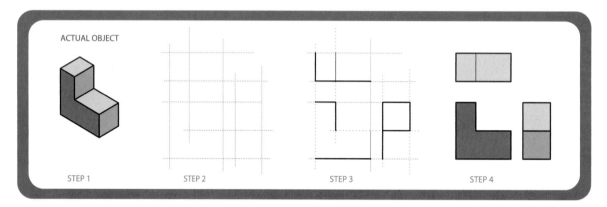

FIGURE 2–7   Drawing orthographic projection.

ACTUAL OBJECT

STEP 1   STEP 2   STEP 3   STEP 4

1. Determine which view is the most important or most detailed view. This will normally become the front view.
2. Determine how many views are necessary to totally define the object and lay out the block shapes of the necessary views in the correct relationship, as depicted back in Figure 2–2. This should be done with light lines and to the desired scale or measurements.
3. Draw necessary layout lines to complete the final actual shapes.
4. Darken the final object shapes and erase the layout lines.
5. When drawing orthographic or multiview drawings using CAD software, the layout lines can be placed on a separate layer; this can be turned off to view the final drawing.

PICTORIAL DRAWINGS

There are three types of pictorial drawings we will be studying. Pictorial drawings give a three-dimensional view that helps in the understanding of the drawings. The three types of drawings we will be looking at are isometric, oblique, and perspective.

2.5 Isometric Projections

Isometric Projection
A pictorial view of an object that shows three views of an object simultaneously. All horizontal lines are drawn at 30° off the horizontal axis, and vertical lines are drawn vertically.

Isometric projection is equal-measure projection. This means that all of the angles of projection are 120° angles that are 30° off the horizon to the right or left. In the isometric projection, the viewer sees three views simultaneously. These three views may be front, right, and top or any other combination of three connected surfaces. The viewer actually looks down the common edge between the front and right views, with a 30° downward tilt

Isometric drawings give a pictorial view of an object with all horizontal lines drawn 30° off the horizontal axis either to the left or right.

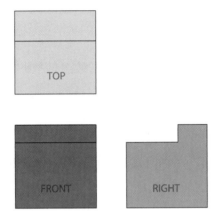

allowing the top to show (Figure 2–8). Horizontal lines found on the front or top view will be drawn 30° off the horizon to the left. Horizontal lines found on the right view and vertical lines found on the top view will be drawn 30° off the horizontal axis to the right. Vertical lines found on the front and right views will be drawn vertically in the isometric drawing.

An isometric projection is used to show a pictorial view of the object. This type of view shows the three object views tied together as they would appear to the human eye. Relationships of sides are better described in isometric projection, whereas individual features and dimensions are best described in orthographic projection. Figure 2–9 is an example of a conversion from a three-view orthographic drawing to an isometric drawing using isometric paper.

An isometric drawing of a stair on isometric paper.

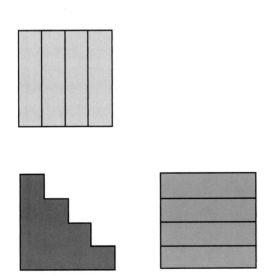

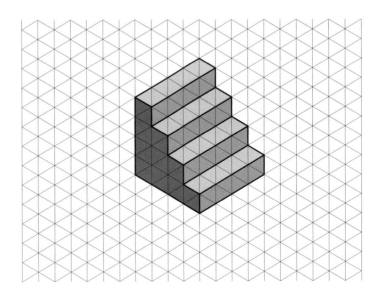

2.6 Drawing Isometric Projections

Drawing or sketching isometric projections is accomplished by the following four-step process, as shown in Figure 2–10.

FIGURE 2–10   The four steps to creating an isometric drawing.

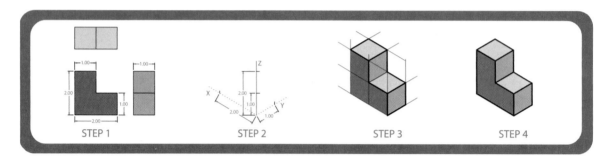

STEP 1   STEP 2   STEP 3   STEP 4

1. Determine which three adjoining views are the most important and most descriptive of the actual object.
2. Lay out the basic 120° isometric V with a vertical line straight up from the intersection of the V. Measure out the *x*, *y*, and *z* axes to the appropriate distances.
3. Block in the basic cube and the layout lines for any notches or cutaways. Draw all horizontal lines that are found on the front or top view at the angle of the *x* axis. Draw all horizontal lines that are found on the right view at the angle of the *y* axis. Draw all vertical lines from the front or right view along the *z* axis.
4. Darken the finished object lines and erase the construction lines.

NOTE: It is important to remember that all *x*-axis lines will be drawn 30° to the left, all *y*-axis lines will 30° to the right, and all *z*-axis lines vertically.

2.7 Oblique Projections

Oblique Projection
A pictorial drawing that shows three views of an object simultaneously. One side of the object is drawn true to size, as if looking straight on. The other sides are angled away, usually at 45°.

An **oblique projection** is similar to the isometric projection, but the front view is parallel to the main surface and true shape and the other two views are angled away (usually at 45°). The function of the oblique projection is basically the same as the isometric projection. It describes the object pictorially but leaves the front view's true shape and size (Figure 2–11). Figure 2–12 is an example of conversion from a three-view orthographic drawing to an oblique drawing using graph paper.

2.8 Drawing Oblique Projections

The steps in drawing oblique projections are exactly the same as those in drawing isometric projections, except for the initial axis setup. The *x* axis is set up horizontally, the *y* axis is 45° to the right, and the *z* axis is vertical (Figure 2–13).

FIGURE 2-11 Oblique drawings give a pictorial view of a drawing. It has a front view that has its true shape and size with lines projecting back at 45°.

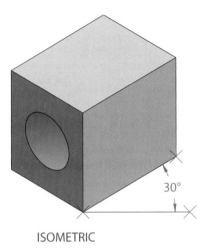

ISOMETRIC

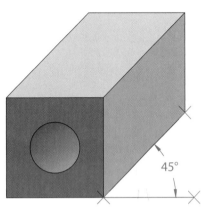

OBLIQUE

FIGURE 2-12 An oblique drawing of a stair on orthographic paper.

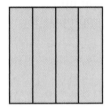
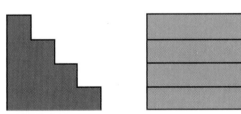
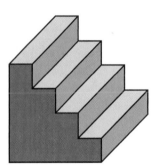

FIGURE 2-13 The four steps to creating an oblique drawing.

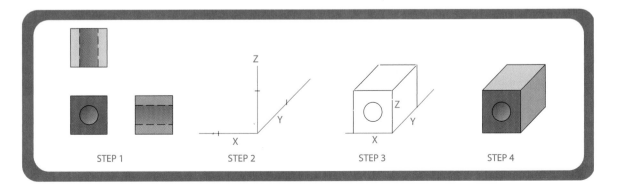

STEP 1   STEP 2   STEP 3   STEP 4

2.9 Perspectives

Perspective View
A pictorial view of an object showing three views simultaneously that is drawn as the human eye would view the object. All horizontal lines taper together as they get further away.

Perspective views are pictorial in nature. By definition, perspectives are the process of representing spatial relations of plane or curved surfaces of objects as they appear to the human eye. The reason Perspectives are used primarily as a sales presentation tool. The difference between perspectives and isometric or oblique projections is that the object lines taper away from the point of view rather than being parallel. The human eye brings parallel lines together as they go away from the viewing point. This is evident if you look down railroad tracks. The tracks seem to converge, but they actually do not.

Similarly, if you look down the side of a long building the far end of the building looks shorter than the near end. A perspective view tapers the angular lines as they move away from the foreground of the drawing to trick the eye into thinking the object is real (Figures 2–14 and 2-15). It is very unlikely that you will ever have to draw perspective, but you will see them in print sets as pictorial views of the overall building project.

FIGURE 2–14   A perspective drawing of a box. All horizontal lines if extended would end at a vanishing point.

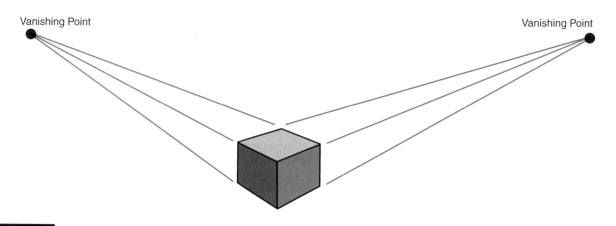

Vanishing Point   Vanishing Point

FIGURE 2–15   A perspective drawing of a building.

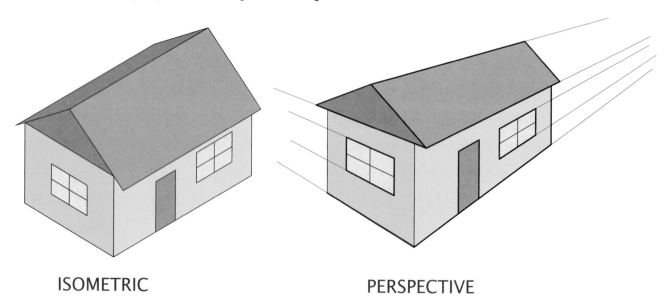

ISOMETRIC   PERSPECTIVE

ORTHOGRAPHIC AND ISOMETRIC PAPER

At first, sketching on orthographic paper may help the beginner improve his or her skills (Figure 2–16). Orthographic paper can be used for orthographic projection as well as oblique drawings. **Isometric graph paper** is available to assist the beginner with practicing isometric sketching. (See Figure 2–17.)

Isometric Graph Paper
Paper with a grid of lines drawn 30° off the horizontal axis.

FIGURE 2-16   Orthographic paper.

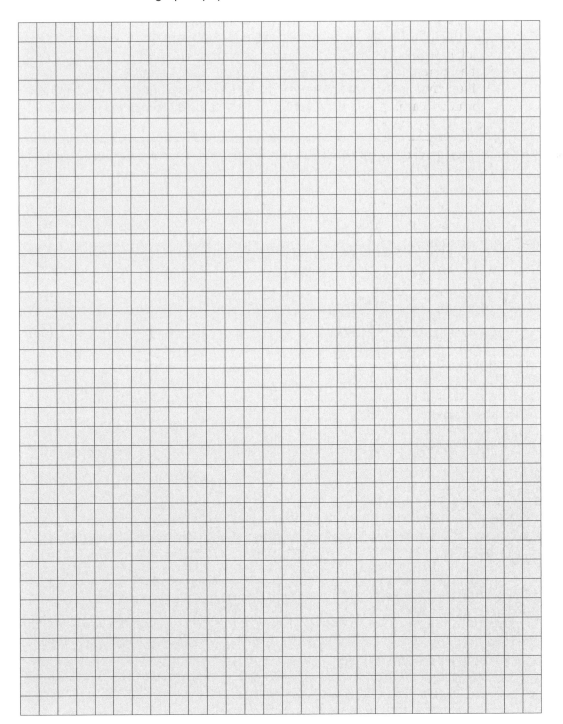

FIGURE 2-17   Isometric paper.

PRACTICE

Now that you have an understanding of the basic projections and the pictorial drawings, it would be wise to practice sketching them. The practice problems at the end of this chapter will give you the opportunity to practice sketching orthographic projection, isometric, and oblique drawings. Once a tradesperson has the ability to visualize and make these drawings, sketching shapes, ideas, and locations will become easy.

SUMMARY

The ability to visualize a set of construction drawings is absolutely necessary. All required information to build a project must be given in the drawings. However, a person has to be able to apply this information.

Orthographic projection consists of two or more views of an object. The front view typically gives the most important information for shape description.

Pictorial views are used to aid in the visualization of objects. The three types of pictorial drawings typically used in blueprints are isometric, oblique, and perspective.

Being able to sketch simple isometric and oblique drawings is an important part of communicating information on the job.

REVIEW QUESTIONS

1. What is an orthographic projection?
2. What is an isometric projection?
3. What is an oblique projection?
4. What is a perspective?

PRACTICE PROBLEMS

Change these isometric drawings to orthographic (three-view) drawings.

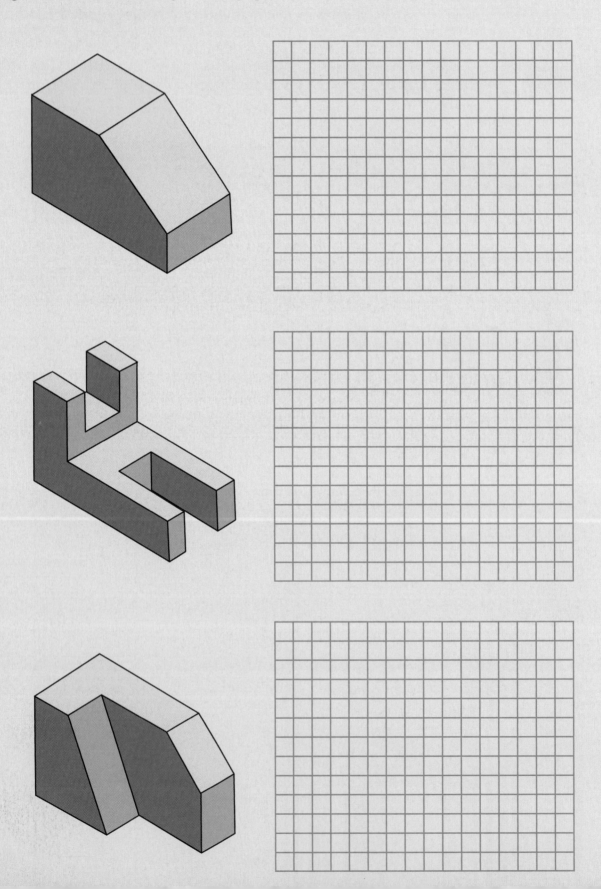

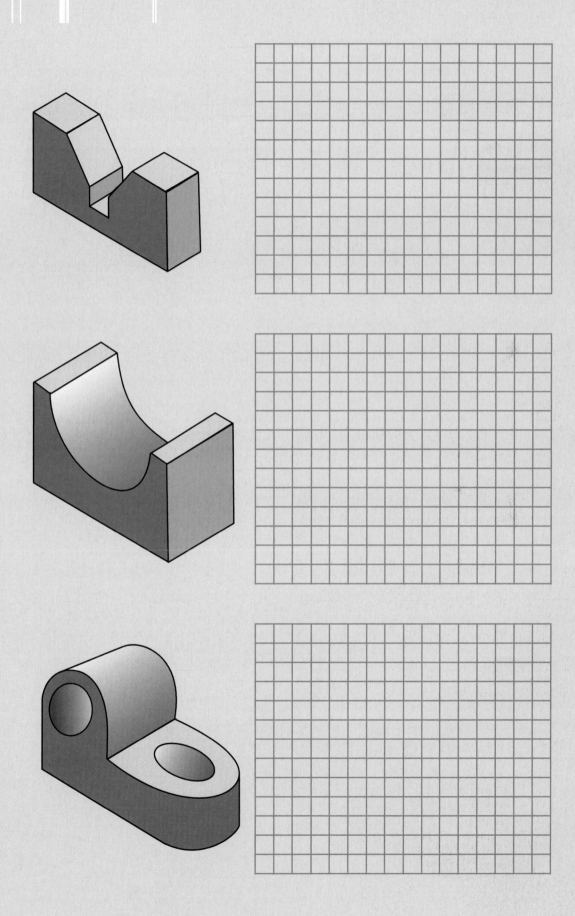

Change these orthographic drawings to isometric drawings.

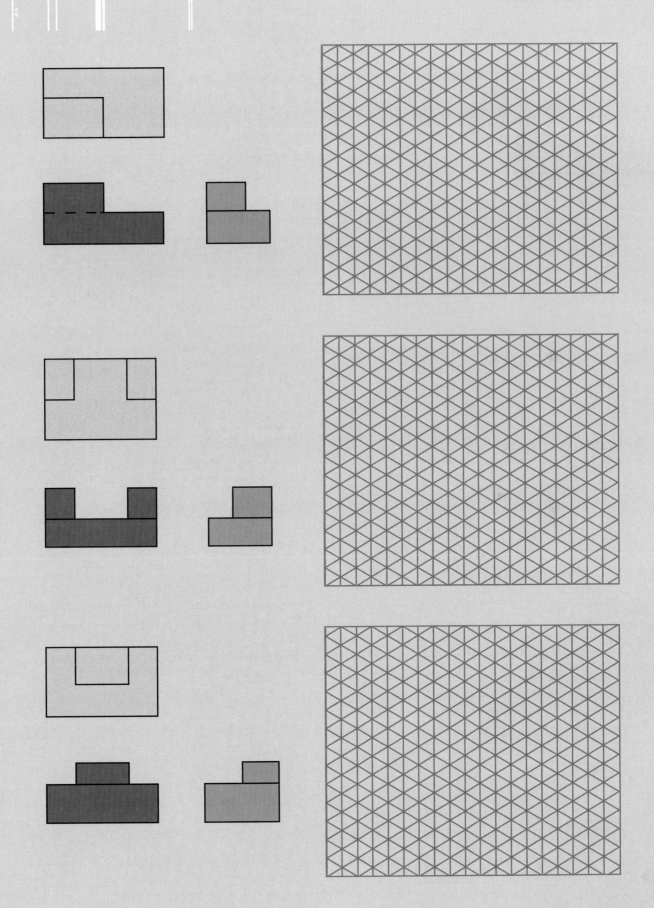

Change these orthographic drawings to oblique drawings.

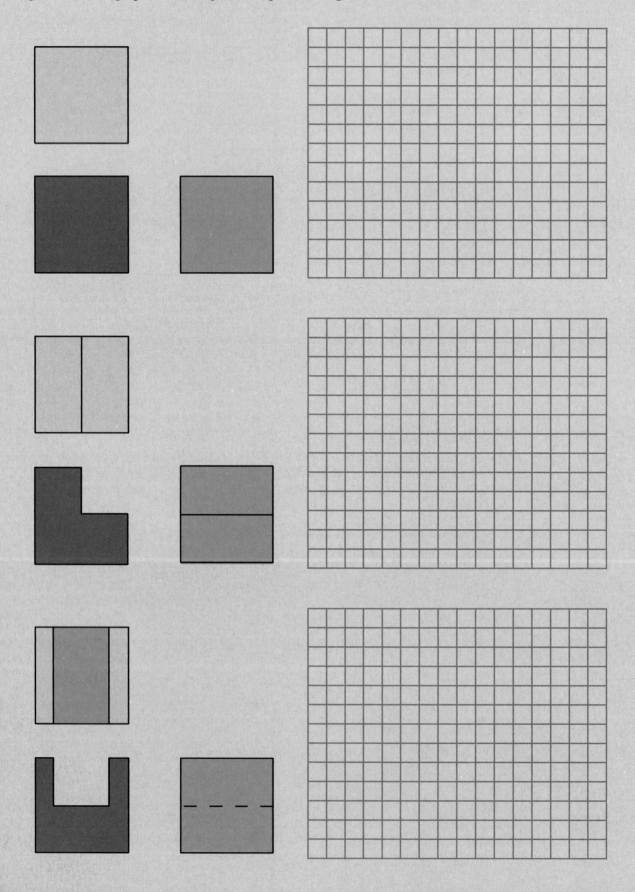

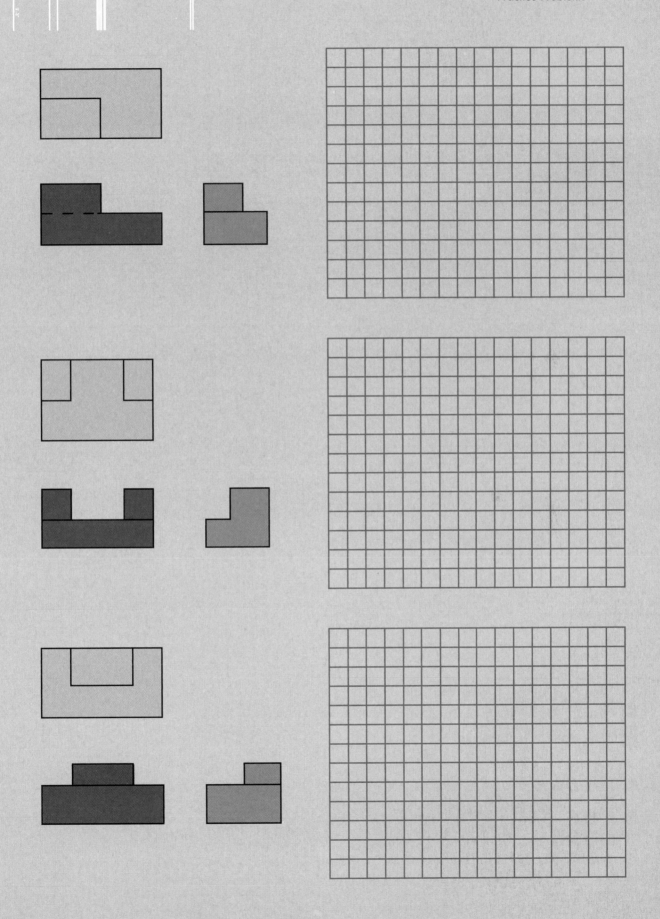

3

Related Math

OVERVIEW

Mathematics plays an important role in all aspects of the construction industry. A good foundation in math is necessary to be able to understand the information on blueprints, and to be able to transfer that information to the job site. The day-to-day calculations an electrician will do involve feet and inches, fractions, area, and volume. The purpose of this unit is to provide a review of some of the math functions needed to read and work with blueprints.

OBJECTIVES

After completing this chapter, you should be able to:

• Add and subtract feet and inches.

• Add, subtract, multiply, and divide fractions.

• Calculate area.

• Calculate volume.

FEET AND INCHES

Adding and subtracting feet and inches is the key to being able to do the rest of the math this chapter will be covering. An electrician performs the calculations every day. To find the location of items on a blueprint, you have to add a string of dimensions that are all in feet and inches. To find the correct spacing of lighting fixtures in a room, you will have to divide the room dimensions, which will involve feet and inches (Figure 3–1).

FIGURE 3–1   A tape measure is one of the most commonly used tools in construction.

When using feet and inches, one must remember that there are 12 equal divisions in a foot. Each of those divisions is 1 inch. This makes adding and subtracting feet and inches a bit trickier than adding decimal numbers.

3.1 Converting Feet to Inches

The electrician needs to be able to convert feet to inches and inches to feet. To convert feet to inches, you multiply the number of feet you have by 12. You do this because there are 12 inches in a foot.

Convert 5 feet to inches:

5 (feet) × 12 (inches in a foot) = 60 inches

If the dimension you want to convert has both feet and inches in it, it becomes a two-step process. The first step is to convert the feet to inches by multiplying the number of feet by 12. The second step is to add the number of inches from the first step to the number of inches in the original number.

Convert 6 feet 4 inches to inches:

Step 1. 6 (feet) × 12 (inches in a foot) = 72 inches

Step 2. 72 inches (from step 1) + 4 inches (inches from original dimension) = 76 inches

3.2 Converting Inches to Feet

To convert inches to feet, you divide the number of inches by 12. You divide by 12 because there are 12 inches in a foot.

Convert 120 inches to feet:

120 inches ÷ 12 (inches in a foot) = 10 feet

Most of the time the number of inches will not divide perfectly to arrive at a whole number (such as 10 in the previous example). If 12 does not evenly divide into the number of inches, the remainder left after dividing by 12 is the number of inches you have in addition to the number of feet.

Convert 50 inches to feet and inches:

$$12\,\overline{)50}\quad 4$$
$$\underline{48}$$
$$2$$

Remainder 2 = 4 feet 2 inches = 4'-2"

50 ÷ 12 = 4, with a remainder of 2 = 4 feet 2 inches

3.3 Practice Exercises: Converting Feet and Inches

1. Convert 12 feet to inches.
2. Convert 4 feet to inches.
3. Convert 20 feet to inches.
4. Convert 4 feet 6 inches to inches.
5. Convert 13 feet 7 inches to inches.
6. Convert 16 feet 9 inches to inches.
7. Convert 72 inches to feet and inches.
8. Convert 132 inches to feet and inches.
9. Convert 192 inches to feet and inches.
10. Convert 16 inches to feet and inches.
11. Convert 45 inches to feet and inches.
12. Convert 165 inches to feet and inches.

FIGURE 3-2 Floor plan of motor controls lab.

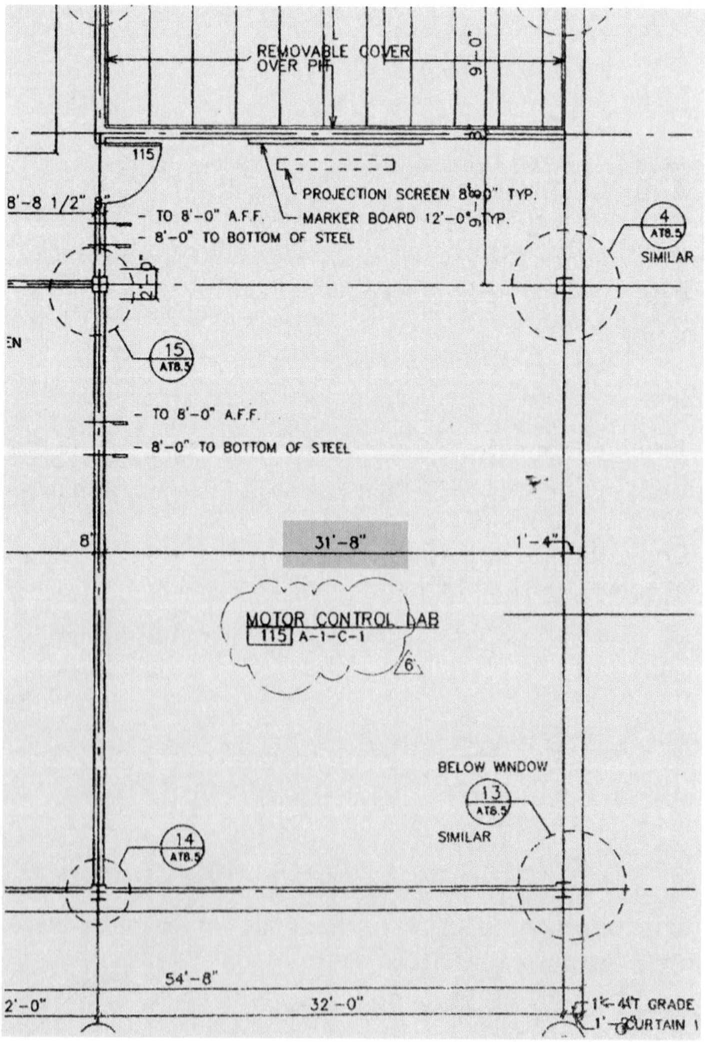

FieldNote!

The motor control lab in Figure 3–2 needs to be divided into four equal parts to install rows of light fixtures. The first step is to convert 31 feet 8 inches into inches.

31 (feet) × 12 (inches in a foot) = 372 inches

372 (inches) + 8 (inches) = 380 inches

The next step is to divide the number of inches (380) by 4 to create the four equal parts.

380 ÷ 4 = 95 inches

You could use 95 inches to find the spacing. However, you could also convert 95 inches back into feet and inches.

$$12\,\overline{)93}\quad 7$$
$$\underline{84}$$
$$11$$

7 remainder 11 or 7'-11"

The room can be equally divided into four 7-foot by 11-inch sections.

3.4 Adding Feet and Inches

There are two methods used to add measurements containing feet and inches.

Method 1

Method 1 involves adding the numbers together, adding the inches together, and then combining the two answers.

Step 1. Add the number of feet together.
Step 2. Add the number of inches together (step 2 is the answer if the number of inches does not exceed 12).
Step 3. Convert the number of inches from step 2 to feet (if the number exceeds 12).
Step 4. Add the number of feet and inches from step 3 to the number of feet from step 1.

4 feet 7 inches + 3 feet 10 inches

Step 1. Add the two feet measurements together.
Step 2. Add the inches measurements together.
Step 3. Here, 17 is more than 12 and so it must be converted.
Step 4. Add 7 feet to 1 foot 5 inches.
Answer: 8 feet 5 inches

Method 2

Method 2 involves converting the measurements to inches before adding.
Step 1. Convert the two numbers to inches.
Step 2. Add the numbers together.
Step 3. Convert the number back to feet and inches (if necessary).

4 feet 7 inches + 3 feet 10 inches

Step 1. Convert the numbers from feet and inches to inches.
Step 2. Add the numbers together.
Step 3. Convert the answer in inches back to feet and inches.
Answer: 8 feet 5 inches

3.5 Subtracting Feet and Inches

There are two methods used to subtract measurements containing feet and inches.

Method 1

Subtracting feet and inches using method 1 is accomplished with a three-step process.

Step 1. Check to see that the number of inches being subtracted is smaller than the number of inches being subtracted from. If it is not, subtract 1 from the number of feet and add 12 to the number of inches.

Step 2. Subtract the inches.

Step 3. Subtract the feet.

6 feet 2 inches − 4 feet 10 inches

$$
\underbrace{\begin{array}{r} \overset{5}{\cancel{6}}\text{'-}\overset{14}{2}\overset{}{\cancel{}}12 \\ -\ 4\text{'-}10\text{''} \\ \hline \end{array}}_{\text{Step 1}}
\qquad
\underbrace{\begin{array}{r} 5\text{'-}14\text{''} \\ -\ 4\text{'-}10\text{''} \\ \hline 4\text{''} \end{array}}_{\text{Step 2}}
\qquad
\underbrace{\begin{array}{r} 5\text{'-}14\text{''} \\ -\ 4\text{'-}10\text{''} \\ \hline 1\text{'-}4\text{''} \end{array}}_{\text{Step 3}} \quad Answer
$$

Step 1. Subtract 1 from 6 feet and add 12 to 2 inches.
The new equation is 5 feet 14 inches − 4 feet 10 inches.

Step 2. Subtract the inches: 14 − 10 = 4 inches.

Step 3. Subtract the feet: 5 − 4 = 1 foot.

Answer: 1 foot 4 inches

Method 2

Method 2 involves converting the measurements to inches before subtracting.

Step 1. Convert the two numbers to inches.

Step 2. Subtract the numbers.

Step 3. Convert the number back to feet and inches (if necessary).

6 feet 2 inches − 4 feet 10 inches

$$
6\text{'-}2\text{''} - 4\text{'-}10\text{''} = \quad \underset{\text{Step 1}}{74\text{''} - 58\text{''}} \quad \underset{\text{Step 2}}{74\text{''} - 58\text{''} = 16\text{''}} \quad \underset{\text{Step 3}}{\begin{array}{r} 1\text{'-}4\text{''} \\ 12\overline{)16} \\ \underline{12} \\ 4\text{''} \end{array}} \quad \underset{\text{Answer}}{1\text{'-}4\text{''}}
$$

Step 1. Convert the two measurements to inches.

Step 2. Subtract the numbers.

Step 3. Convert the numbers back to feet and inches.

Answer: 1 foot 4 inches

3.6 Practice Exercises: Adding and Subtracting Feet and Inches

1. 8 feet + 4 feet 3 inches
2. 2 feet + 7 inches
3. 17 feet 1 inch + 10 feet 7 inches
4. 3 feet 2 inches + 3 feet 9 inches
5. 7 feet 9 inches + 13 feet 7 inches
6. 12 feet 4 inches + 21 feet 11 inches

7. 8 feet 6 inches − 6 feet
8. 23 feet 9 inches − 4 inches
9. 13 feet 8 inches − 10 feet 2 inches
10. 24 feet 9 inches − 13 feet 2 inches
11. 11 feet 3 inches − 6 feet 7 inches
12. 35 feet 4 inches − 19 feet 7 inches

FIGURE 3–3   Floor plan of mechanical room.

FieldNote!

The overall dimension of the mechanical room in Figure 3–3 is needed. To find this, the string of dimensions must be added together.

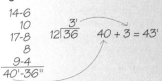

The room is 43 feet long.

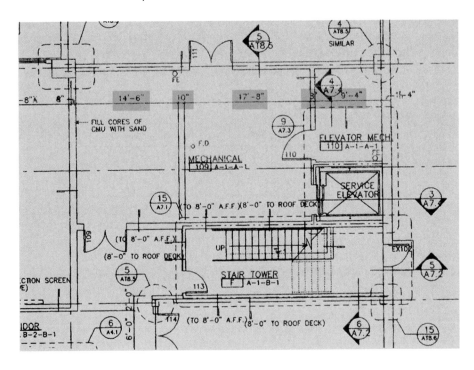

FRACTIONS

Fractions are an important part of reading prints and performing layout. Reading fractions of an inch on a tape measure is an everyday occurrence in the construction industry. A **fraction** is a way of representing a part of a whole amount. $\frac{1}{16}$ is 1 part of the 16 possible parts, as shown in Figure 3–4.

Fraction
A way of representing a part of a whole amount using a numerator and a denominator.

FIGURE 3–4   Pie chart and bar graph representing 1 part of 16 (1/16).

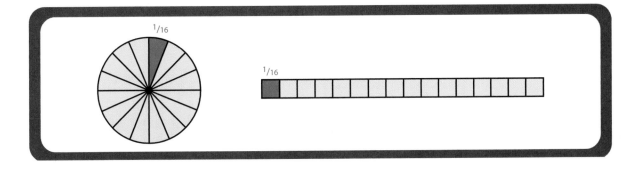

Numerator
The top number of a fraction.

Denominator
The bottom number of a fraction.

$$\frac{Numerator}{Denominator} \qquad \frac{1}{16}$$

The **numerator** is the top number of a fraction. The numerator represents how many parts you have.

The **denominator** is the bottom number of a fraction. The denominator represents how many equal parts the whole amount has been divided into.

3.7 Reducing a Fraction

A fraction should always be reduced to its lowest possible form. This is done by dividing both the numerator and denominator by any number that will divide into both and still leave a whole number (Figure 3–5).

FIGURE 3–5   The bar graphs show that 4 parts of 16 (4/16) is the same as 1 part of 4 (1/4).

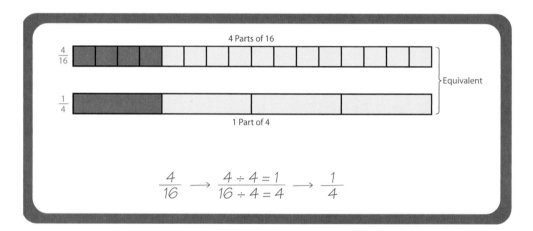

3.8 Converting Improper Fractions and Mixed Numbers

Improper Fraction
A fraction that has a numerator that is larger than the denominator.

An **improper fraction** has a numerator that is larger than the denominator. This must be converted into a **mixed number.** A mixed number has a whole number as well as a fraction.

Improper Fraction to Mixed Number

Mixed Number
A number that contains a whole number as well as a fraction.

$$\frac{21}{16} \qquad 1\frac{5}{16}$$
Improper   *Mixed*
Fraction   *Number*

To convert an improper fraction to a mixed number, the numerator is divided by the denominator. The number of times the denominator divides into the numerator becomes the whole number, and the remainder becomes the new numerator.

$$\frac{21}{16} \qquad 16\overline{\smash{\big)}21} \qquad 1\frac{5}{16}$$
$$\phantom{\frac{21}{16} \qquad} \underline{16}$$
$$\phantom{\frac{21}{16} \qquad 16\overline{)}} 5$$

Mixed Number to Improper Fraction

To convert a mixed number to an improper fraction, multiply the whole number by the denominator and add that result to the numerator. That result is the numerator, which is over the original denominator.

Convert $1\frac{3}{8}$ to an improper fraction:

$$1\frac{3}{8} \quad \underset{\text{Step 1}}{8 \times 1 = 8} \longrightarrow \underset{\text{Step 2}}{8 + 3 = 11} \longrightarrow \frac{11}{8} \;\; \text{Answer}$$

3.9 Practice Exercises: Converting Improper Fractions and Mixed Numbers

Convert the following improper fractions to mixed numbers:

1. $\dfrac{15}{4}$

2. $\dfrac{33}{2}$

3. $\dfrac{17}{16}$

4. $\dfrac{11}{8}$

5. $\dfrac{13}{2}$

Convert the following mixed numbers to improper fractions:

1. $5\dfrac{1}{4}$

2. $12\dfrac{1}{16}$

3. $10\dfrac{3}{4}$

4. $15\dfrac{1}{2}$

5. $1\dfrac{1}{16}$

3.10 Adding Fractions

When adding fractions, the key is to have the same denominator. Once you have the same denominator, you simply add the numerators together and this gives how many parts of the denominator you have. Fractions can be added using the following three steps.

Step 1. Convert the fractions to have a common denominator.

Step 2. Add the numerators together.

Step 3. Reduce the fraction to its lowest possible form.

$$\frac{1}{8} + \frac{3}{8}$$

$$\underset{\text{Add}}{\frac{1}{8} + \frac{3}{8} = \frac{4}{8}} \longrightarrow \underset{\text{Reduce}}{\frac{4 \div 4 = 1}{8 \div 4 = 2}} \longrightarrow \frac{1}{2} \;\; \text{Answer}$$

Step 1. No action is required because the fractions have a common denominator.

Step 2. Add the numerators of the fractions together.

Step 3. Reduce the fraction to its lowest possible form by dividing the numerator and the denominator by 4.

$$\frac{3}{8} + \frac{1}{4}$$

$$\overbrace{\frac{3}{8} + \frac{1 \times 2 = 2}{4 \times 2 = 8}}^{\text{Convert to Common Denominator}} \longrightarrow \frac{3}{8} + \frac{2}{8} \overbrace{= \frac{5}{8}}^{\text{Add}} \text{ Answer}$$

Step 1. Multiply the numerator and denominator of $\frac{1}{4}$ by 2 to have a common denominator with $\frac{3}{8}$.

Step 2. Add the numerators together.

Step 3. No action is needed because the fraction is in the lowest possible form.

3.11 Subtracting Fractions

When subtracting fractions, the two fractions must have a common denominator. Once the fractions have a common denominator, the numerators may be subtracted. This will give you how many parts of the denominator you have left. Fractions can be subtracted using the following three steps.

Step 1. Convert the fractions to have a common denominator.

Step 2. Subtract the numerators.

Step 3. Reduce the fraction to its lowest possible form.

$$\frac{5}{8} - \frac{1}{8}$$

$$\overbrace{\frac{5}{8} - \frac{1}{8} = \frac{4}{8}}^{\text{Subtract}} \longrightarrow \overbrace{\frac{4 \div 4 = 1}{8 \div 4 = 2}}^{\text{Reduce}} \longrightarrow \frac{1}{2} \text{ Answer}$$

Step 1. No action is necessary because the fractions have the same common denominator.

Step 2. Subtract the numerators.

Step 3. Divide the numerator and denominator by 4 to have the fraction in its lowest possible form.

$$\frac{3}{4} - \frac{1}{8}$$

$$\frac{3}{4} - \frac{1}{8} = \longrightarrow \overbrace{\frac{3 \times 2 = 6}{4 \times 2 = 8}}^{\text{Convert to common denominator}} \longrightarrow \frac{6}{8} - \frac{1}{8} \overbrace{= \frac{5}{8}}^{\text{Subtract}} \text{ Answer}$$

Step 1. Multiply the numerator and denominator of $\frac{3}{4}$ by 2 to have a common denominator with $\frac{1}{8}$.

Step 2. Subtract the numerators.

Step 3. No action is necessary because the fraction is in its lowest possible form.

Subtracting Mixed Numbers

Subtracting mixed numbers may involve more steps than subtracting regular fractions. This is due to that fact that the fraction of the number being subtracted from may have a numerator that is smaller than the number being subtracted. When this is the case, 1 must be borrowed from the whole number and the value of the denominator added to the numerator.

Step 1. Convert the fractions to have a common denominator.

Step 2. Borrow from the whole number to give to the numerator.

Step 3. Subtract the numerators of the fractions.

Step 4. Subtract the whole numbers.

Step 5. Reduce the fraction to its lowest possible form.

$$7\frac{1}{8} - \frac{3}{8}$$

Step 1. No action is necessary because the fractions have the same denominator.

Step 2. Borrow 1 from the whole number so that the numerators can be subtracted. This is done by taking 1 away from 7 (whole number) and adding 8 (value of the denominator) to 1 (numerator of the fraction).

Step 3. Subtract the numerators of the fractions.

Step 4. Subtract the whole numbers.

Step 5. Reduce the fraction to its lowest possible form by dividing the numerator and denominator by 2.

3.12 Practice Exercises: Adding and Subtracting Fractions and Mixed Numbers

1. $\dfrac{1}{16} + \dfrac{5}{16} =$

2. $\dfrac{3}{8} + \dfrac{7}{8} =$

3. $\dfrac{1}{2} + \dfrac{3}{8} =$

4. $\dfrac{3}{4} + \dfrac{7}{16} =$

5. $1\dfrac{1}{2} + \dfrac{3}{4} =$

6. $5\dfrac{3}{4} + 6\dfrac{1}{16} =$

7. $\dfrac{3}{4} - \dfrac{1}{4} =$

8. $\dfrac{5}{16} - \dfrac{1}{2} =$

9. $1\dfrac{3}{8} - \dfrac{1}{8} =$

10. $3\dfrac{3}{4} - \dfrac{7}{8} =$

11. $10\dfrac{7}{8} - 3\dfrac{1}{2} =$

12. $3\dfrac{1}{2} + 4\dfrac{1}{2} - 6\dfrac{1}{4} =$

3.13 Multiplying Fractions

Fractions can be multiplied using the following five steps. Remember that anytime a fraction is multiplied by a whole number the whole number is placed over 1 (example: $8 = \frac{8}{1}$).

Step 1. Convert mixed numbers into an improper fraction.
Step 2. Multiply the numerators together.
Step 3. Multiply the denominators together.
Step 4. Convert the fraction to a mixed number.
Step 5. Reduce the fraction to its lowest form.

$\dfrac{1}{4} \times \dfrac{3}{4}$

$$\dfrac{1}{4} \times \dfrac{3}{4} = \dfrac{3}{16}$$

Step 1. No action is necessary because the numbers are not improper fractions.
Step 2. Multiply the numerators together.
Step 3. Multiply the denominators together.
Step 4. No action is necessary because the numerator is smaller than the denominator.
Step 5. No action is necessary because the fraction is already in its lowest form.

$2\dfrac{3}{4} \times \dfrac{1}{4}$

$$2\dfrac{3}{4} = \dfrac{11}{4} \longrightarrow \dfrac{11}{4} \times \dfrac{1}{4} = \longrightarrow \dfrac{11}{16} \ Answer$$

Convert to an improper fraction   Multiply

Step 1. Convert $2\frac{3}{4}$ from a mixed number to an improper fraction.
Step 2. Multiply the numerators together.
Step 3. Multiply the denominators together.

Step 4. No action is necessary because the numerator is smaller than the denominator.

Step 5. No action is necessary because the fraction is already in its lowest form.

3.14 Dividing Fractions

To divide a fraction, the number following the division sign is inverted and then the fractions are multiplied.

$$\frac{3}{4} \div \frac{1}{2} = \frac{3}{4} \times \frac{2}{1}$$

Fractions can be divided using the following six steps.

Step 1. Convert mixed numbers into an improper fraction (if necessary).
Step 2. Invert the fraction that follows the division sign.
Step 3. Multiply the numerators together.
Step 4. Multiply the denominators together.
Step 5. Convert the fraction to a mixed number (if necessary).
Step 6. Reduce the fraction to its lowest form (if necessary).

$$12\frac{1}{2} \div 4$$

Convert mixed numbers to an improper fraction   Invert and multiply   Reduce

$$12\frac{1}{2} \div 4 \longrightarrow \frac{25}{2} \div \frac{4}{1} \longrightarrow \frac{25}{2} \times \frac{1}{4} \longrightarrow \frac{25}{8} \longrightarrow 3\frac{1}{8}$$

3.15 Practice Exercises: Multiplying and Dividing Fractions and Mixed Numbers

1. $\frac{1}{2} \times 3 =$

2. $\frac{3}{4} \times \frac{1}{2} =$

3. $10\frac{3}{4} \times 5 =$

4. $13\frac{3}{4} \times \frac{1}{4} =$

5. $\frac{7}{8} \div 4 =$

6. $\frac{3}{4} \div \frac{1}{2} =$

7. $1\frac{3}{4} \div 4 =$

8. $20\frac{1}{2} \div \frac{3}{4} =$

9. $12\frac{3}{16} \div 1\frac{1}{2} =$

10. $10\frac{1}{2} \div 4 =$

CALCULATING AREA

Area
The number of square units it takes to cover a space.

Finding the *area* of a square or rectangular object is not very difficult. It is simply multiplying the length and the width: Area = Length × Width. Table 3–1 outlines some of the common area formulas and conversions.

Finding the area of a room when the room has feet and inches involved is a bit trickier. There are two ways to do this.

TABLE 3-1   Important area formulas and conversions

IMPORTANT AREA FORMULAS AND CONVERSIONS	
Area of a square or rectangle = Length × Width	$A = L \times W$
Area of a triangle = Length × Width × ½	$A = \frac{1}{2} L \times W$
Area of a circle = π × radius²	$A = \pi \times r^2$
1 square foot = 144 square inches	
1 square yard = 9 square feet	
1 square yard = 1,296 square inches	

Method 1

Method 1 involves converting each dimension into inches. First convert the length and width into inches. Multiply the number of feet by 12 and add that number to the number of inches. Then multiply the length (in inches) by the width (in inches). This gives the area in square inches. Now it must be converted into square feet. To do this you divide the number by 144 (there are 144 inches in a square foot, 12 × 12 = 144). To convert the square feet into square yards, you divide the number of square feet by 9 (there are 9 square feet in a square yard, 3 × 3 = 9).

Step 1. Convert the length to inches.
Step 2. Convert the width to inches.
Step 3. Multiply the length times the width (this is the area in square inches).
Step 4. Convert the number of square inches into square feet or yards.

Find the area of a room that is 10 feet 6 inches by 12 feet 4 inches

Step 1. (10 × 12) + 6 = 126 inches.
Step 2. (12 × 12) + 4 = 148 inches.
Step 3. 126 inches × 148 inches = 18,684 square inches
Step 4. 18,684 ÷ 144 = 129.5 square feet

Method 2

Find the decimal equivalent of the inches. Take the number of inches and divide it by 12. This is the decimal equivalent of the number of inches. Add this to the number of feet. Do this to both dimensions if necessary. Then multiply the length and width.

Step 1. Find the decimal equivalent of the number of inches for the length. Add this to the number of feet.
Step 2. Find the decimal equivalent of the number of inches for the width. Add this to the number of feet.

Step 3. Multiply the length by the width.
Find the area of a room that is 10 feet 6 inches by 12 feet 4 inches.
Step 1. (6 ÷ 12) + 10 = 10.5 feet
Step 2. (4 ÷ 12) + 12 = 12.3334 feet
Step 3. 10.5 × 12.3334 = 129.5 square feet

3.16 Practice Exercises: Calculating Area

1. Find the area in square feet of a room that is 10 feet by 12 feet.
2. Find the area of a building in square feet that is 123 feet by 10 feet 6 inches.
3. Find the area in square yards of a bedroom that is 13 feet 9 inches by 10 feet 6 inches.
4. Find the area in square yards of a garage that is 25 feet 6 inches by 22 feet.

CALCULATING VOLUME

Finding the **volume** of a cube is multiplying the length times the width times the height.
    Volume = Length × Width × Height
    Table 3–2 outlines common volume formulas and conversions.

Volume
The amount of space occupied by a three-dimensional object or region of space expressed in cubic units.

TABLE 3–2   Important volume formulas and conversions

IMPORTANT VOLUME FORMULAS AND CONVERSIONS	
Volume of a cube = Length × Width × Height	V = L × W × H
Volume of a cylinder = π × radius² × Height	V = π × r² × H
1 cubic foot = 1,728 cubic inches	
1 cubic yard = 27 cubic feet	
1 cubic yard = 46,656 cubic inches	

FieldNote!

The square footage of a house needs to be known to calculate the general lighting load (see Figure 3–6). The house has three levels that are 32 feet 6 inches by 45 feet 9 inches. Convert the dimensions into their decimal equivalent.
32 feet 6 inches = 32.5 feet
45 feet 9 inches = 45.75 feet
Multiply the length by the width.
32.5 feet × 45.75 feet =
    1,486.875, or 1,487 square feet per floor
Multiply the square footage per floor times the number of floors.
1,487 × 3 = 4,461
The area of the house is 4,461 square feet.

FIGURE 3–6   The square footage of a residence is used to calculate the general lighting load and the minimum number of circuits required in a house.

Although finding volume in itself is not that difficult, finding the volume when the dimensions include both feet and inches is a bit trickier. There are two ways to do this.

Method 1

Method 1 involves converting each dimension into inches. First convert the length, width, and height into inches. Multiply the number of feet by 12 and add that number to the number of inches for all three numbers. Then multiply the length (in inches) by the width (in inches) by the height (in inches). This gives the volume in inches. Now it must be converted into cubic feet. To do this you divide the number by 1,728 (there are 1,728 cubic inches in a cubic foot). Thus, $12 \times 12 \times 12 = 1,728$. To convert the cubic feet into cubic yards, you divide the number of cubic feet by 27 (there are 27 cubic feet in a yard). Thus, $3 \times 3 \times 3 = 27$.

Step 1. Convert the length to inches.
Step 2. Convert the width to inches.
Step 3. Convert the height to inches.
Step 4. Multiply the length, the width, and the height (this is the volume in cubic inches).
Step 5. Convert the cubic inches from step 4 into cubic feet or yards.

Find the volume of a concrete slab that is 8 feet 6 inches long by 6 feet 6 inches wide by 8 inches high.

Step 1. $(8 \times 12) + 6 = 102$ inches
Step 2. $(6 \times 12) + 6 + 78$ inches
Step 3. 8 inches
Step 4. $102 \times 78 \times 8 = 63,648$ cubic inches
Step 5. $63,648 \div 1,728 = 36.83$ cubic feet

Method 2

Method 2 involves finding the decimal equivalent of the inches. Take the number of inches in the equation and divide it by 12. This is the decimal equivalent of the number of inches. Add this to the number of feet. Do this to all three dimensions if necessary. Then multiply the length, width, and height.

Step 1. Find the decimal equivalent of the number of inches for the length. Add this to the number of feet.
Step 2. Find the decimal equivalent of the number of inches for the width. Add this to the number of feet.
Step 3. Find the decimal equivalent of the number of inches for the height. Add this to the number of feet.
Step 4. Multiply the length, the width, and the height.

Find the volume of a concrete slab that is 8 feet 6 inches long by 6 feet 6 inches wide by 8 inches high.

Step 1. $(6 \div 12) + 8 = 8.5$
Step 2. $(6 \div 12) + 6 = 6.5$
Step 3. $(8 \div 12) = .6667$
Step 4. $8.5 \times 6.5 \times .6667 = 36.83$ cubic feet

FIGURE 3-7 When pouring light bases the volume of concrete required must be calculated.

FieldNote!

We need to calculate the amount of concrete needed to pour eight light bases like the one shown in Figure 3–7. Each base is a cylinder that is 6 feet deep and 2 feet in diameter. The concrete will be purchased in cubic yards. First find the volume of a cylinder by using the formula Area (circle) = $\pi \times r^2 \times H$.
3.14 $\times$ 1 $\times$ 6 = 18.84 cubic feet per cylinder
Multiply the volume of each cylinder by 8 (there are eight bases).
18.84 $\times$ 8 = 150.72 cubic feet
150.72 cubic feet has to be converted into cubic yards by dividing the number by 27 (number of cubic feet in a yard).
150.72 $\div$ 27 = 5.58 cubic yards of concrete

3.17 Practice Exercises: Calculating Volume

1. Find the volume in cubic feet of a concrete slab that is 13 feet long, 10 feet wide, and 1 foot thick.
2. Find the volume in cubic feet of a concrete slab that is 12 feet 6 inches long, 9 feet 6 inches wide, and 9 inches thick.
3. Find the volume in cubic yards of a bed of sand that is 40 feet long, 15 feet wide, and 2 feet thick.
4. Find the volume in cubic yards of a bed of sand that is 24 feet 6 inches long, 8 feet 9 inches wide, and 18 inches thick.
5. Find the volume in cubic feet of a round light pole base that is 2 feet in diameter and 10 feet deep.
6. Find the volume in cubic yards of a round light pole base that is 18 inches in diameter and 6 feet deep.

SUMMARY

When working with feet and inches, remember that inches are a base 12 number—meaning that there are 12 inches in a foot. Most jobs today still use feet and inches for measurements. It is absolutely necessary for a tradesperson to be able to add, subtract, multiply, and divide feet and inches measurements.

Many of the dimensions on a construction site involve fractions. An electrician adds, subtracts, multiplies, and divides fractions many times daily. This skill should be second nature.

When ordering materials and planning jobs, area and volume calculations are performed. The ability to calculate area and volume using numbers that include feet and inches is essential.

4

Scaling and Dimensions

OVERVIEW

It is important to understand the concept of scales and scaling. There are a variety of scales used in manufacturing and construction. The architectural scale is used most commonly in construction. Scales make it possible to draw a 50,000-square-foot building on a 24-inch by 30-inch piece of print paper. This smaller proportional drawing is a reduced-size model of the larger building. It is important to understand and use scales and dimensions accurately, as a slight error in scaling or measurement may result in misplacement of electrical devices and therefore costly rework.

OBJECTIVES

After completing this chapter, you should be able to:

• Recognize and select the proper scale for the appropriate task.

• Scale a drawing accurately.

• Differentiate between extension lines and dimension lines.

• Understand the rules and concepts of print dimensioning.

SCALING

The term **scale** has many definitions. In print reading, *scale* has the following two common definitions.

- An instrument, with evenly spaced graduations, that is used for measuring.
- A mathematical size relationship between the actual object and a drawing of the object.

It is necessary to have a blueprint drawn to scale for a couple of reasons. First, to construct a building from a drawing the drawing must be an accurate representation. The drawing must have the correct proportions. Second, it would be impossible to draw a building print to actual scale. A reduced scale must be used. Some drawings of small objects, patterns, and electronic circuit path drawings are drawn to full scale. You will rarely ever see a full-scale blueprint of a construction project.

The first thing the architect or engineer must do is to determine which type of scale and what ratio or size of scale to use. There are numerous types and ratios of print scales. The type of scale is determined by the finished product and the level of detail required. An architectural or electrical blueprint will be produced with an architect's scale. A plot plan or site plan will be drawn with a civil engineer's scale. A drawing of a mechanical part may be produced with a mechanical engineer's scale. Other scales are used for other specific functions.

4.1 Common Scales

There are several common types of scales used in home, industry, and construction. An engineer or architect may use any of the common scales. The determination as to which is the appropriate scale is made by the type of drawing. An engineer or architect will choose the best scale for the job at hand. Scales get their names from the scale type or whether an architect or engineer commonly uses them. Figure 4–1 depicts five of the most common scales.

Most people are familiar with the standard inch and metric scales (or rulers). They are used around the house and in school for measuring objects and distances. On a construction job, you use an extended version of the ruler—a folding ruler or tape measure. The inch and metric scales measure actual size and imply no ratio of increase or decrease (see Figure 4–2).

The decimal scale is the same as the inch scale, but it divides the inches into 0.1 (one-tenth of an inch) and 0.01 (one-hundredth of an inch) instead of into fractions. Engineers will use this scale if the object was designed in decimal inches (see Figure 4–3).

The mechanical engineer's scale can be flat or triangular in shape. If it is triangular, the scales may be mixed English, metric, or decimal. The actual mechanical engineer's scales are "size" scales. This means that they are related to the actual size by a fraction, such as ¼ inch being equal to one actual inch or ½ inch being equal to one actual inch. The small divisions at the scale's end are for the uneven fractions of the inches. There

Scale

1. An instrument, with evenly spaced graduations, that is used for measuring.
2. A mathematical size relationship between the actual object and a drawing of the object.

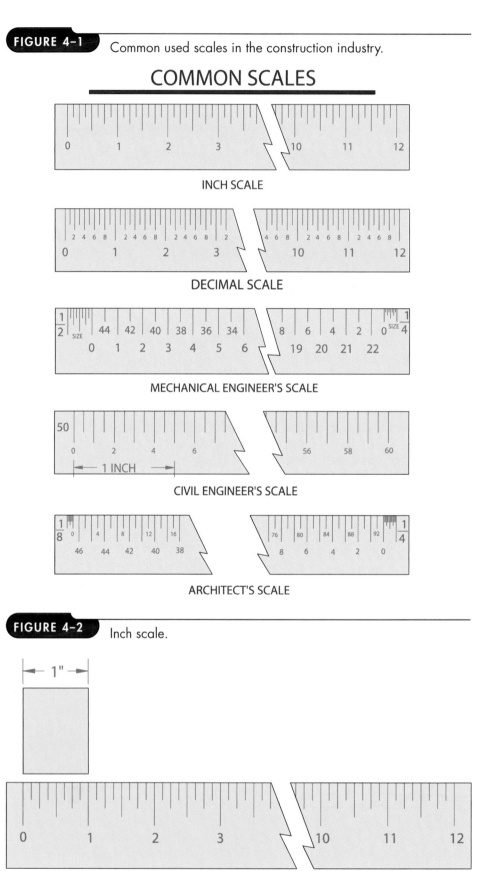

FIGURE 4-1 Common used scales in the construction industry.

COMMON SCALES

INCH SCALE

DECIMAL SCALE

MECHANICAL ENGINEER'S SCALE

CIVIL ENGINEER'S SCALE

ARCHITECT'S SCALE

FIGURE 4-2 Inch scale.

INCH SCALE

FIGURE 4-3　Decimal scale.

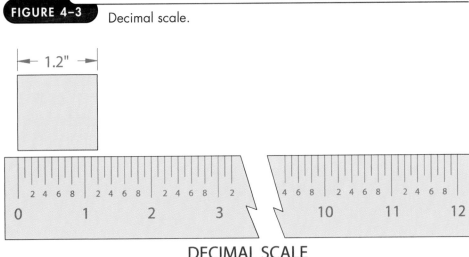

DECIMAL SCALE

may be 8 or 16 small divisions for ¼- or ½-size scales, respectively. These scales may be on the same line if they are comparable scales, such as ½ inch. You read the ½-inch scale left to right, and the ¼-inch scale right to left (see Figure 4–4).

FIGURE 4-4　Mechanical engineer's scale.

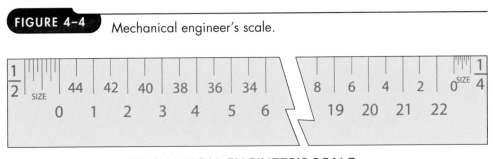

MECHANICAL ENGINEER'S SCALE

The civil engineer's scale is used to lay out large areas such as plots. The scales are 1 inch equals 10, 20, 30, 40, 50, or 60 feet. The size of the drawing is read directly from the scale in feet. Figure 4–5 depicts a "1 inch equals 50 feet" scale. It is also important to understand that civil engineers use tenths of a foot instead of inches. A distance may be dimensioned as 40.5 feet, which is not the same as 40 feet 5 inches. The .5 rep-

FIGURE 4-5　Civil engineer's scale.

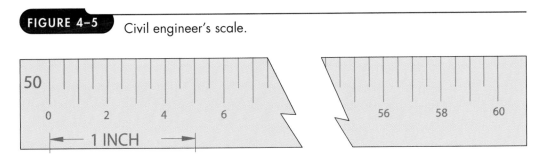

CIVIL ENGINEER'S SCALE

resents 5 tenths of a foot, and thus the measurement in feet and inches would be 40 feet 6 inches. It would be advisable to purchase a tape measure with tenths of a foot rather than trying to convert all dimensions (see Figure 4–5).

The architect's scale is the most common scale used in blueprint production for the building trades industry. This scale is a scaled-down relation, such as ¼ inch equals 1 foot. Using the architect's scale makes it possible to draw a very large house or building on a relatively small piece of paper. The architect's scale may be flat or triangular in shape and may contain various other scales. The scales may be mixed, as with mechanical engineer's scales.

The common architectural scales are ¹⁄₁₆ inch equals 1 foot, ⅛ inch equals 1 foot, ¼ inch equals 1 foot, ½ inch equals 1 foot, ¾ inch equals 1 foot, 1 inch equals 1 foot, 1½ inches equals 1 foot, and 3 inches equals 1 foot. The 1½- or 3-inch scales are for details. The small divisions on the scale's end are always divided into multiples of 12. The small divisions relate to actual inches in size. Architect's scales often involve dual reading (in pairs), like the mechanical engineer's scale. Figure 4–6 shows the ¼-inch and ⅛-inch scales, depending on which direction the scale is read. Architectural scales often have two scales that are marked on the same plane of the scale. An example of this is shown in Figure 4–6, which shows a scale with ¼-inch and ⅛-inch scales. If read from the left, the ⅛-inch scale is used, and if read from the right, the ¼-inch scale is used.

FIGURE 4–6   Architect's scale.

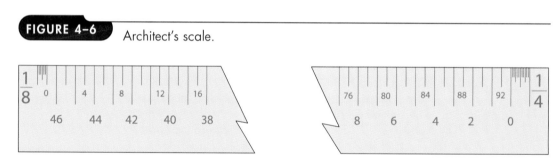

ARCHITECT'S SCALE

4.2 Scaling Drawings

Scaling
Using the appropriate scale to determine the actual size of a distance or feature.

The term **scaling** refers to using the appropriate architectural or engineering scale to measure a drawing and determine the actual size of a certain feature. For example, on the "¼ inch equals 1 foot" scale a 1-inch-long feature on the print would represent 4 feet in actual distance.

It is said that you should never scale a drawing, and that you should always go by the given dimensions. This statement is essentially true for three reasons.

- Even if the drawing is drawn to scale, reproduction techniques (such as blue-line printing, photocopying, offset printing, and plotting) may cause a slight discrepancy in reproduction size.
- The more scale reduction the drawing has, the more line width affects the scaled measurement. For instance, on a "1⁄16 inch equals 1 foot" measurement the width of the draftsperson's line may equal several inches in actual distance.
- Not all drawings are drawn or reproduced to an accurate scale.

In electrical print reading, it is often necessary to scale drawings. On residential prints, scales are not extremely important. This is due to the fact that the device will normally be mounted to the nearest available stud. *National Electrical Code® (NEC®)* and local codes often mandate the layout of residential devices and make their placement more of an "on-the-job" process. In commercial print reading, scale is more important. If accurate fixture placement is necessary, a detailed section will be supplied. This enlarged section will show precise locations and mounting concerns.

If you must scale a print, do it properly. The first step is to check the drawing for the scale being used. Either the title block will indicate the scale of the drawing (Figure 4–7) or the scale will be written below the drawing description (Figure 4–8). Second, make sure the architectural scale is the

FIGURE 4-7   A title block will typically indicate the scale of a drawing.

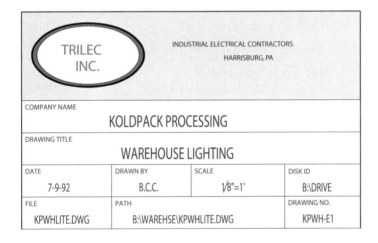

FIGURE 4-8   The scale of a drawing is often written below the drawing.

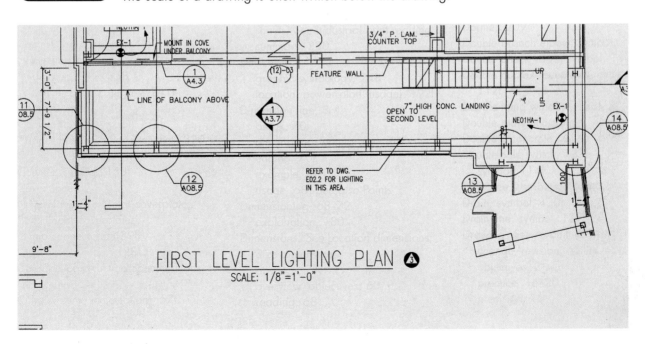

same as that indicated on the print and then check the print reproduction scale with a known dimension. Use at least one vertical and one horizontal dimension. The scaled dimension should equal the dimension written on the print (see Figure 4–9).

FIGURE 4-9   It is important to verify the scale of a drawing is correct with a known measurement.

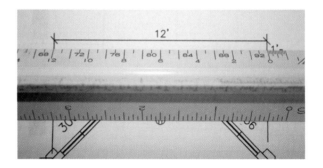

Finally, use proper scaling procedure. Start the measurement from the 0 (zero) point on the scale, not the end of the scale or the first notch. (The small series of notches before the 0 on the scale divide the scale into smaller divisions.) Then place the 0 at the start of the feature being measured. Look at the number at the other end of the feature being measured. If this number is not a whole number, slide the scale to the previous whole number. This measurement is stated by the main scale number and the number of inches (for example, 8 feet 6 inches, as in Figure 4–10).

FIGURE 4-10   A scale can be used to find unknown distances.

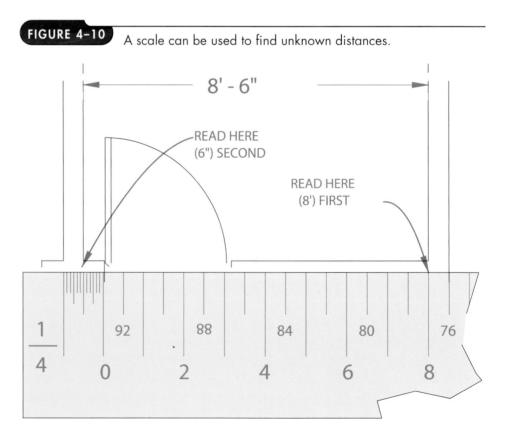

Scaling on the job may be necessary from time to time. Because typically there is no scale on the job this may be accomplished with a common tape measure. Using a tape measure requires some conversion that is not necessary with an architect's scale. If the print scale is ⅛ inch equals 1 foot, 1 inch on the tape is equal to 8 feet in actual print distance. For example, the GFCI (Ground Fault Circuit Interrupter) outlet on the partial print shown in Figure 4–11 is 10 feet from the corner wall as measured by the tape.

FieldNote!

A **curvimeter** (plan wheel, map measure), shown in the following figure, is often used by electrical estimators or contractors while estimating a project. It is typically used to find the length of a conduit run or lengths of conductors/cables, or for calculating square footage. To measure a distance with a curvimeter, you simply roll it along the plan. A reset button will reset it to zero after each use. The curvimeter shown in the following figure is an analog version containing "¼ inch equals 1 foot" and "⅛ inch equals 1 foot" scales. Digital versions of curvimeters, often called plan wheels, give you the option of selecting the scale to be used.

| **FIGURE 4-11** | A tape measure can be used to find an unknown dimension when a scale isn't available. |

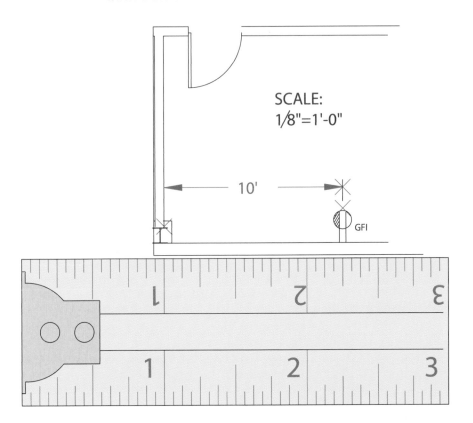

Curvimeter
An instrument used for measuring distances on plans and maps. It has a wheel that rolls along the print, recording the distance traveled.

NOTE: The measurement is taken from the 1-inch mark to eliminate measurement inaccuracy due to the hook play and thickness of the tape. This is not a very accurate way of obtaining information and should never be used to find locations that must be exact. Always look to dimensions for information when they are available.

Practice makes perfect when scaling, but never forget to check the drawing for reproduction scale loss. In addition, most important is the fact that no matter what the scale tells you the written dimension is law. You will never go wrong to trust a written dimension. Errors may be made when scaling. If a scaling error is made, it may cost time and money.

DIMENSIONING

4.3 Reading Dimensions

Dimensions are the preferred way of obtaining information from a print. A dimension will always supersede any measurement taken using a scale or tape measure. If a dimension seems to be wrong, contact the architect or engineer who drew the prints to ascertain whether a correction is necessary. There are three key pieces of information to know when it comes to dimensioning.

- Extension lines
- Dimension lines
- Types of dimensions

Extension Lines

Extension Lines

Lines on a blueprint that transfer a point on a plan out to get it away from the drawing, where it can be dimensioned.

Extension lines transfer a point on a plan out to get it away from the drawing, where it can be dimensioned. Extension lines will stop short of the drawing and not touch it. One extension line may be used for more than one dimension, as shown in Figure 4–12.

FIGURE 4-12   Extension lines are used to extend the points being dimensioned out from the drawing.

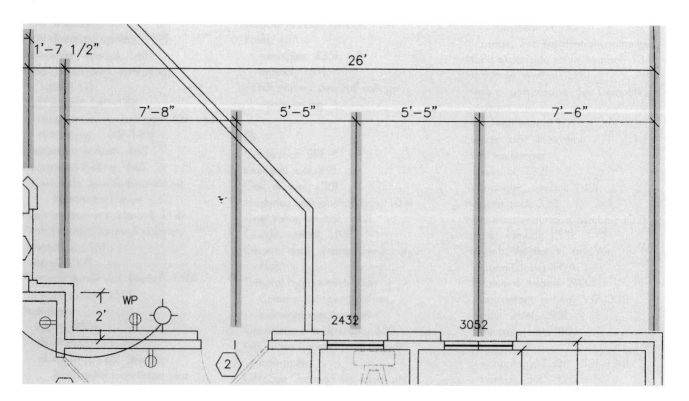

Dimension Lines

Dimension lines indicate between which two points a measurement has been taken (see Figure 4–13). The measurement may be from one extension line to another or between points on the drawing. The end of the dimension line will typically have an arrowhead, slash, or dot (see Figure 4–14).

FIGURE 4–13   Dimension lines indicate which two points are being dimensioned.

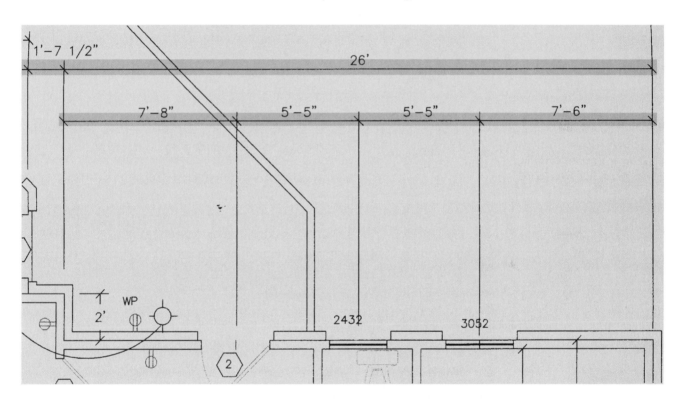

FIGURE 4–14   Dimension line ends indicate where a dimension line ends. There are several ways to draw a dimension line end.

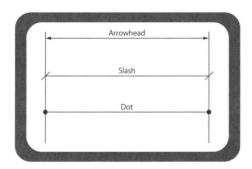

FIGURE 4-15   Dimensions may be written over the dimension line or in a break in the dimension line.

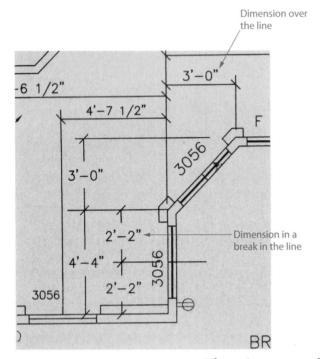

Dimension over the line

Dimension in a break in the line

Size Dimensions
Dimensions that define the length, width, and height of a feature.

Location Dimensions
Dimensions that define the distance a feature is from a known reference point.

Types of Dimensions

There are two types of dimensions on construction prints: size and location. **Size dimensions** define the length, width, and height of a feature. **Location dimensions** define the distance the feature is from a known reference point. This is essential for proper device placement. The dimensions are typically placed over the dimension line, or in a break in the dimension line (see Figure 4–15).

Where there is insufficient room to write a dimension, a leader may be used. At one side of the leader there will be an arrow pointing to the location being dimensioned, and at the other end of the leader will be the dimension (see Figure 4–16).

When looking at electrical prints, a consideration is whether the device is dimensioned to the center or to the edge. Normally, a device is mounted to its center unless it is dimensioned otherwise (such as to an edge). The edge dimension will have a distinct extension line (see Figure 4–17). Device mounting height is also a consideration. Several things may determine the height of mounting.

- The most current national or local electrical code, which should always be consulted.
- The print set may have elevation drawings or details for locating devices.
- Notes on the drawing, schedules, or the specifications may indicate the mounting heights (see Figure 4–18).
- If none of these sources contains mounting height information, consult the project architect or engineer.

There is a concept *called standard practice* (or *company practice*). If the code, print, or specifications do not determine the mounting height of the devices, the manufacturer's mounting sheet or a standard/company practice may be used.

FIGURE 4-16   When there isn't room to write a dimension, a leader is used to move the dimension away from the drawing. The arrow of the leader will point to the location being dimensioned.

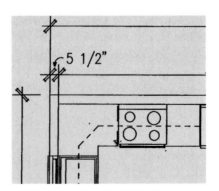

FIGURE 4-17 A device may be dimensioned to the edge, center, or may have no dimensions at all. If there is no dimension indicated, a scale can be used to find the location.

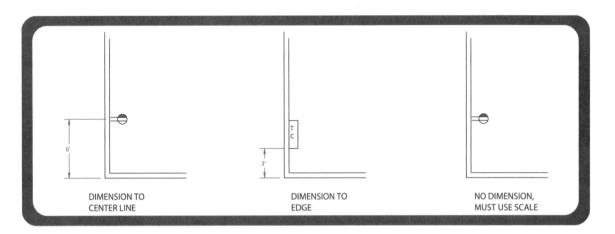

DIMENSION TO
CENTER LINE

DIMENSION TO
EDGE

NO DIMENSION,
MUST USE SCALE

FIGURE 4-18 A legend may contain notes indicating the height of receptacles.

	LEGEND		
⊖	120 V DUPLEX RECEPTACLE: MOUNT 18" AFF (UNLESS NOTED), 20 AMP.	$S^{3}_{K}{}^{4}_{P}$	SWITCH, MOUNT 48" AFF, SINGLE POLE (UNLESS NOTED), 3 OR 4 WAY, "P" – INDICATES PILOT LIGHT EQUIPPED; "K" – INDICATES KEY OPERATED.
⊖	120 V DUPLEX RECEPTACLE: MOUNT 42" AFF OR 8" ABOVE COUNTER, 20 AMP.	⊄/⊐	DISCONNECT SWITCH, FUSED/NON–FUSED, SIZE AS NOTED.
⊟	250 V, 1 PHASE, 3 WIRE RECEPTACLE; MOUNT 18" AFF (UNLESS NOTED), AMP AS NOTED.	Ⓘ	MOTOR (HP, VOLTAGE AND USE AS NOTED); NUMBER INDICATES HORSEPOWER.
⊟	250 V, 3 PHASE, 4 WIRE RECEPTACLE; MOUNT 18" AFF (UNLESS NOTED), AMP AS NOTED.	⊠	COMBINATION DISCONNECT/MAGNETIC MOTOR STARTER.
⊖	COMBINATION 120/208 V, 1 PHASE, 3 WIRE RECEPTACLE; MOUNT 18" AFF (UNLESS NOTED), 20 AMPERES.	⊘	MANUAL MOTOR STARTER.
		⊠	MAGNETIC MOTOR STARTER.
⊙	FLUSH FLOOR OUTLET, "E" – INDICATES ELECTRIC, PROVIDE WITH TRIM PLATE TO SUIT FLOOR TYPE, ("T" – INDICATES TELEPHONE), ("C" – INDICATES COMPUTER), ("M" – INDICATES MICROPHONE).	⊏T⊐	DRY TYPE TRANSFORMER, VOLTAGE, PHASE AND USE AS NOTED.
		▬	ELECTRIC PANEL, TOP 6'-0" AFF
⊖	120 V DUPLEX RECEPTACLE	TTS	TELEPHONE TERMINAL SPACE (3/4" X 4'-0"

SUMMARY

Drawings are drawn to scale to obtain an accurate representation of a building that will fit on a sheet of paper. There are several different types of scales that may be used, depending on what type of drawing it is. The commonly used scales for construction drawings are mechanical engineer's, civil engineer's, and architect's.

The most commonly used scale for construction drawings is the architectural (architect's) scale. A typical triangular architectural scale will have 11 different scales on it.

When scaling a drawing using a scale or tape measure, always verify the scale of the drawing with a known measurement. The scale on the drawing may be mismarked, or may have been altered during reproduction.

Extension lines transfer a point on a drawing out away from the drawing to be dimensioned. Dimension lines indicate between which two points a measurement has been taken. Leaders are used in an area that is too small in which to write a dimension.

There are two types of dimensions: location and size. Location dimensions indicate where a feature or point is located. Size dimensions indicate the length, width, or height of a feature.

Written dimensions always supersede any measurement taken using a scale or tape measure. If a written dimension is available, that measurement should always be used.

REVIEW QUESTIONS

1. What are the unit divisions on a decimal scale?
2. What are the divisions on an inch scale or tape measure?
3. Which scale is most commonly used on electrical prints?
4. What determines the scale per foot used by the drafter or on an electrical print?
5. Should you ever scale a drawing on the job and disregard the dimension? Why or why not?
6. What two checks must be made to determine print-scale accuracy?
7. Where are the most common places to find the drawing scale?
8. When scaling a drawing, should you trust the end of the ruler or tape measure as an accurate 0 (zero) starting point?
9. What are the three types of dimension line ends that may be used?
10. What two types of dimensions exist on construction prints?
11. When would a leader be used for dimensioning?
12. Describe the difference between dimension lines and extension lines.

SCALING PRACTICE

NOTE: The use of an architect's (architectural) scale will be required to complete this exercise. Refer to the figure on page 73 to answer the following questions. The drawing is drawn to "¼ inch equals 1 foot" scale. Verify that the scale is correct using one of the known dimensions.

1. Does a known dimension on the print verify that the print is drawn to "¼ inch equals 1 foot" scale?
2. What is the width of door 5?
3. What are the room dimensions of the living room?
4. What are the room dimensions of the lavatory?
5. How wide is the opening in the wall for door 10?
6. How wide is the bow window in the living room?
7. How deep are the foyer closets?
8. What are the dimensions of the cabinet in the lavatory?
9. How wide is the opening in the wall for door 11?
10. How wide is door 1?

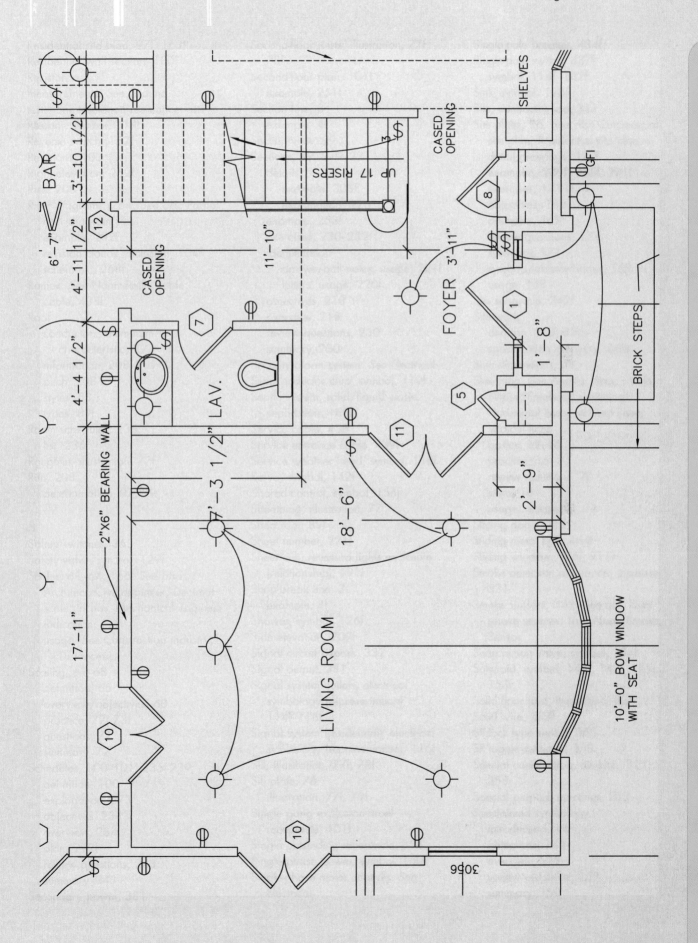

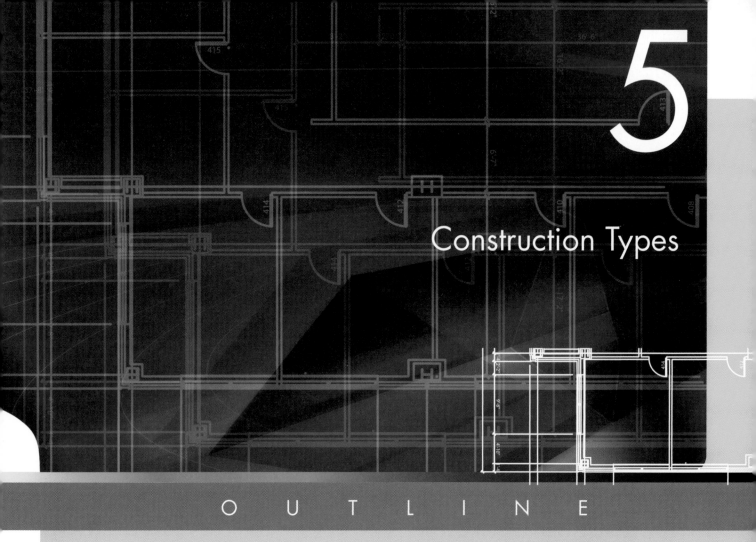

5

Construction Types

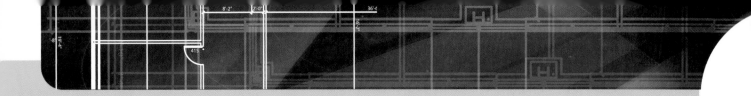

OVERVIEW

There are many types of construction with which an electrician will work. It is important that an electrician have knowledge of the different ways of constructing buildings to have a thorough understanding of what role he/she will have in the construction process. Knowing the terminology used in the construction process is imperative in understanding instructions and being able to effectively communicate on the job. This chapter describes the types of construction commonly used and identifies some of the terminology associated with each type of construction method.

OBJECTIVES

After completing this chapter, you should be able to:

• Identify the different styles of wood-framed construction.

• Identify the different styles of steel construction.

• Identify the different styles of concrete construction.

• Differentiate different styles of roofs.

• Differentiate styles of residential construction.

WOOD CONSTRUCTION

Wood construction is most commonly used in residential and light commercial construction. The three main types of wood construction used are platform, balloon, and post-and-beam.

5.1 Platform Framing

Platform framing, also known as western framing, is the most commonly used framing method. When constructing a house using platform framing, the first level is constructed—with the floor joists and the subfloor for the next level being installed. This creates a platform for the workers to work from to construct the next floor. The process continues for each level. Figure 5–1 contains the terminology typically used with platform framing.

The stud spacing for platform framing and balloon framing is typically 16 or 24 inches on center. This allows the seams of 4 × 8 sheets of wall sheathing or sheetrock to always end on a stud. The board fastened to the foundation wall is called the **sill plate.** The sill plate is a pressure-treated

FieldNote!

The terms used to describe wood studs (basically, the dimensions of the lumber) do not represent their actual sizes. For example, a 2 × 4 is actually 1-½ by 3-½ inches. Table 5–1 outlines the terms used to describe various sizes of stud and their actual sizes. Note that actual sizes may vary from the ones shown. Factors that affect this are the source of the lumber used, the age of the construction, and the specific requirements of the probject.

TABLE 5–1 Construction Lumber Sizes	
Term	Actual Size (Inches)
2 × 4	1-½ × 3-½
2 × 6	1-½ × 5-½
2 × 8	1-½ × 7-¼
2 × 10	1-½ × 9-¼
2 × 12	1-½ × 11-¼

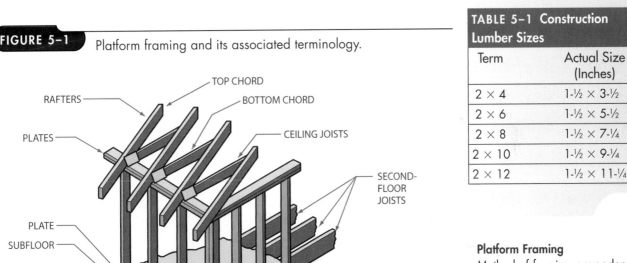

FIGURE 5–1 Platform framing and its associated terminology.

TOP CHORD
RAFTERS
BOTTOM CHORD
PLATES
CEILING JOISTS
SECOND-FLOOR JOISTS
PLATE
SUBFLOOR
HEADER JOIST
STUDS
PLATES
HEADER JOIST
SILL
FIRST-FLOOR JOISTS
SHEATHING
FOUNDATION
SUBFLOOR

Platform Framing
Method of framing a wooden structure in which the wall studs are one floor in height. After one level has been completed, a platform is created that the framers can stand on to complete the next level.

Sill Plate
A piece of lumber (attached to foundation walls and concrete floors) that is pressure treated to prevent rot from the moisture in the concrete.

FieldNote!

When an electrician wires a building that has been constructed with wood framing types such as platform or balloon, he will have to **"rough in"** the job (see the following figure). During rough-in, the electrician will drill the holes in the studs, mount the boxes, and insert the cable or raceway in the wall. This is done after the wooden structure has been built but before the insulators or sheetrockers get started.

Rough-In
The stage in the construction process where electricians can install the boxes, cables, and raceways in the wall before the finished wall material is installed.

Balloon Framing
A method of framing in which the exterior wall studs run from the sill plate to the top plate of the second floor.

Let-In Ribbon
In balloon framing, a board notched into the framing members to support the floor joists of the second floor.

piece of lumber that will prevent the moisture from the concrete or block from rotting the wood.

5.2 Balloon Framing

Balloon framing, also known as eastern framing, involves exterior wall studs that are continuous from the sill plate to the top plate of the second floor (see Figure 5–2). Balloon framing requires long pieces of wood that are not always available and are costly. With this method of framing, a board called a **"let-in ribbon"** is notched into the exterior wall framing to support the floor joists of the second floor. A fire stop is typically installed between the exterior studs at the second-floor level to slow a fire from traveling up the studs. Fire stops often make "fishing" wires and cables impractical. Figure 5–2 contains the terminology used with balloon framing.

FIGURE 5–2 Balloon framing and its associated terminology.

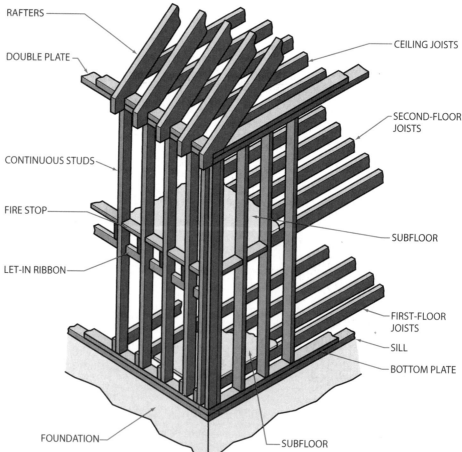

5.3 Post-and-Beam Framing

Wood post-and-beam framing involves framing members that are larger and spaced farther apart than the studs used with platform and balloon framing. The **posts** are the vertical framing members, and the beams are the horizontal or angled framing members. This method of construction is used for storage buildings, churches, and some residential construction. The **beams** in churches and residential construction are often finished and left exposed. Planks are used for roof and floor decking to be able to span the long distances between beams. Figure 5–3 contains the terminology used with post-and-beam framing.

Wood Post-and-Beam Framing
A type of construction in which large wooden members are used for the horizontal and vertical framing.

Post
The vertical structural supporting member.

Beam
The horizontal structural supporting member.

Floor Joist
A horizontal framing member that supports a floor.

FIGURE 5–3   Wood post-and-beam framing and its associated terminology.

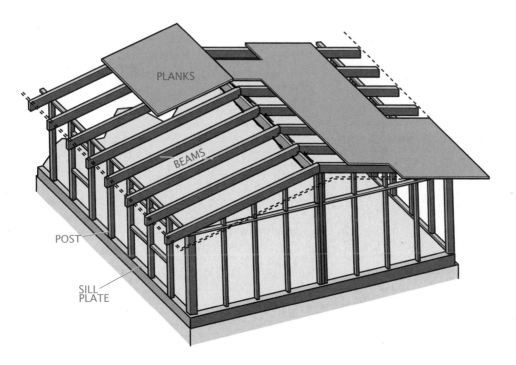

5.4 Floor Joists for Wood Construction

There are three main types of **floor joists** used in wood construction: solid wood floor joists, wood I-beam (TJI®) floor joists, and trussed floor joists (web floor joists). Solid wood floor joists (see Figure 5–4) will be found in older homes, but are rarely used in new construction. TJI® floor joists (see Figure 5–5) have a top and bottom cord of laminated wood with oriented strand board or plywood between. This type of floor joist typically has pre-stamped holes that just need to be knocked out for running electrical wires through.

FIGURE 5–4   Solid floor joist.

FIGURE 5–5   Wood I-beam TJI® floor joist.

Truss (web) floor joists (see Figure 5–6) are the most commonly used floor joist in residential construction. One of the big advantages is that the plumbing and ductwork can be installed within the truss area. This eliminates ductwork and plumbing hanging below the ceiling.

Floor joist spacing may be 12, 16, 19.2, or 24 inches on center. A standard tape measure will have every multiple of 12, 16, and 19.2 inches highlighted in some manner to make them stand out. This provides a quick reference when laying out spacing for wall or floor joists.

FIGURE 5-6 Truss (web) floor joist.

STEEL CONSTRUCTION

Steel construction is typically a column-and-beam type of construction, which is used for commercial and industrial jobs. Steel construction is most often used in conjunction with one or more other types of construction methods, such as metal stud framing, wood stud framing, concrete block, precast concrete, and so on.

5.5 Steel Column-and-Beam

Steel column-and-beam construction is often used to provide the structural support for commercial and industrial buildings (see Figure 5–7). The steel structural members are typically large, can span long distances, and support heavy loads. The vertical members are called *columns*, and the horizontal members are called *beams*. Figure 5–8 contains the common terminology associated with steel column-and-beam construction.

Steel Column-and-Beam Construction
A type of construction in which large steel members are used for the horizontal and vertical framing members.

FIGURE 5-8 Metal column-and-beam framing and its associated terminology.

FIGURE 5-7 Metal column-and-beam framing is commonly used in commercial construction.

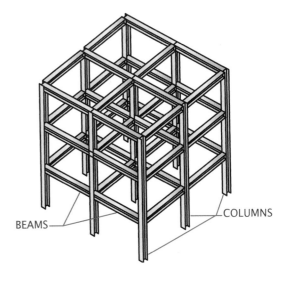

BEAMS

COLUMNS

FieldNote!

A stud punch is used to cut holes in metal studs when the factory-made openings in the stud are not in the place needed (see the following figure). Plastic bushings [per the *National Electrical Code®* *(NEC®)*] must be inserted into the notch or punched hole to protect a cable (such as non-metallic sheathed cable) from sharp edges. The bushings will also make it easier to pull cables through the holes.

Metal Stud Framing
A framing method in which metal studs are used to create walls.

FIGURE 5–9   Metal stud framing is commonly used in commercial construction.

5.6 Metal Stud Framing

Metal stud framing (Figure 5–9) is typically used for partition walls in commercial and industrial occupancies. Metal stud framing consists of a top track, a bottom track, and studs (see Figure 5–10). The metal studs come in different thicknesses. Load-bearing and exterior walls are a thicker stud, whereas the interior non–load-bearing partition walls are a lighter gauge. The studs come pre-punched with openings where electrical raceways or cables can be run. Although metal stud framing has been typically used on commercial installations, its use in the residential market has been increasing.

5.7 Steel Joists

Metal truss-type joists (bar joists) are often used to provide the structure for a roof or an upper floor (see Figure 5–11). Metal truss joists are capable of spanning long

FIGURE 5–10

Metal stud framing and its associated terminology.

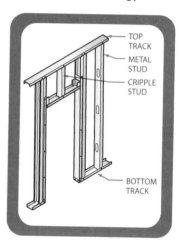

FIGURE 5–11   Metal truss joists.

distances. C-channel joists (see Figure 5–12) are often used to support floor-ing when long spans are not needed. They come with pre-punched holes that raceways and cables can be run through.

Metal C-shape joists.

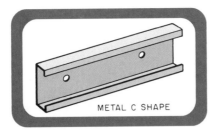

CONCRETE CONSTRUCTION

Concrete is used in almost all construction to one extent or another. In the residential market, it is often used for a foundation and flooring—whereas for commercial and industrial construction it may be used for the walls and floors and as support columns and beams. It can be installed many different ways that can be divided into three main categories: **concrete masonry units** (CMUs; i.e., concrete blocks), poured-in-place concrete, and precast concrete.

Concrete Masonry Unit
A block of concrete that is mortared in place to create walls.

5.8 Concrete Masonry Units

CMUs can be seen in all types of construction (see Figure 5–13). CMUs can be used for exterior walls, interior partition walls, and load-bearing walls. The exterior walls are often filled with an insulating material to prevent heat loss. If there is electrical that must be installed in the block wall, the electrician will have to be on site working with the brick-layers to get the conduits and boxes placed properly as the walls are constructed.

FIGURE 5–13 Concrete block (CMU) wall.

5.9 Poured-in-Place Concrete

Poured-in-Place Concrete
Concrete mixed and poured at the job site.

Poured-in-place concrete will show up on almost every job in one form or another. It might be just a sidewalk and driveway, or it might be the walls and floors of an entire building. To pour concrete, there must be forms installed to hold the concrete in place as it sets up (see Figure 5–14). Typical wall forms consist of facing paired wood panels, the space between which constitutes the thickness of the finished wall.

FIGURE 5–14 Concrete in poured concrete walls is poured into forms that will be removed when the concrete has set.

Some contractors are now using Styrofoam forms. This type of form gets put together like a puzzle to create the foundation wall. After the form is assembled, concrete is poured in the void between the panels (see Figure 5–15). These forms are becoming very popular due to the fact that they have a higher insulating value. The form has strips of plastic built into the form for fastening the finished wall material, eliminating the need for a framed wall. The sheetrock or other finished material is simply screwed into the plastic strips in the form.

Poured concrete floors that are at ground level or in a basement will have a bed of sand under the concrete. Poured concrete floors (e.g., parking

FIGURE 5–15 Styrofoam concrete forms will stay in place after the concrete has been poured. The finished wall surface will be attached to the forms.

ramp, floor in a multi-level building, and so on) that have an open space below them will either be formed up from below to support the concrete while it sets or poured on corrugated steel panels that will stay in place. Any electrical raceways or boxes to be in the floor have to be installed before the concrete is poured (see Figure 5–16). Burial depths for electrical cables and raceways, as required by the *NEC*®, are often dictated by the presence and thickness of concrete poured above them.

FIGURE 5–16 Electricians often place raceways in the sand before the concrete is poured.

5.10 Precast Concrete

Precast concrete may be used for the walls, floor, or roof. The precast panels are made in a factory to meet the specific needs of a particular job. The precast concrete products often have pre-stressed steel strands to add strength. Care must be taken not to drill through or cut any of the steel strands to avoid ruining the integrity of the panel.

Precast wall panels, sometimes called architectural panels, may have a finished wall surface (such decorative rock) as part of the panel (see Figure 5–17). The wall panels are brought to the job and tipped up into

Precast Concrete
Concrete panels poured at a plant and shipped to the construction site.

FIGURE 5–17 Precast concrete walls can be ordered with a decorative finish built into the wall.

FIGURE 5-18 Precast concrete walls are poured in a plant and set in place on the job.

place (see Figure 5–18). The wall panels will have a structural layer on the inside surface, hardboard insulation in their middle, and a decorative layer on the exterior.

Precast floor panels typically have pre-stressed cables and can span fairly long distances. They are often cored, meaning that they have holes running from one end to the other (see Figure 5–19). Electricians will sometimes use these holes to run cables. Precast panels used for the roof structure are typically a single-T or double-T type. These types of panels have pre-stressed cables and can span long distances (see Figure 5–20).

FIGURE 5-19 Precast concrete floor panels.

FIGURE 5-20 Double-T concrete ceiling joists.

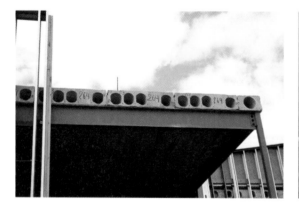

ROOF STYLES

There several styles of roofs used in construction. The flat roof is most often used in commercial and industrial construction. The roofing style for residential construction varies greatly and changes with trends. Figure 5–21 shows the commonly used roofing styles. It is not uncommon to see more than one roof style incorporated into a building.

FieldNote!

A flat roof is not actually flat. It will have a small pitch that will lead to a drain. The drain may be on the exterior wall, where there is a downspout, or it may be in the middle of the roof (with a pipe running through the building to drain the water). A flat roof is the most commonly used roof for commercial and industrial jobs (see the following figure).

FIGURE 5–21   Commonly used roof types.

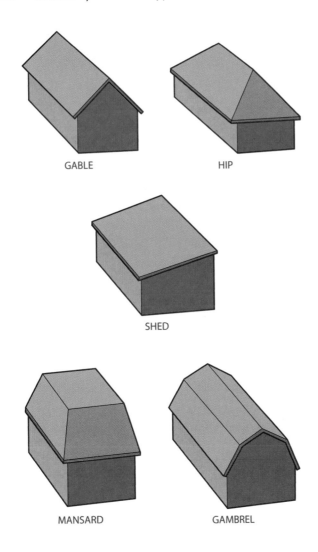

GABLE

HIP

SHED

MANSARD

GAMBREL

SUMMARY

The three main types of wood construction are platform, balloon, and post-and-beam. Platform framing is the most commonly used wood framing method.

The two types of steel construction are column-and-beam and metal framing (metal studs). Metal column-and-beam framing is typically used in conjunction with another construction method.

The three types of concrete construction are CMUs (concrete block), poured-in-place concrete, and precast concrete.

To be able to communicate on the job, one must understand the terminology associated with the various construction types. The electrician must also understand the construction process and the role he/she will play to be in the right place at the right time.

REVIEW QUESTIONS

1. What are the actual dimensions of a 2 × 6 piece of lumber?

2. What type of wood construction method has studs that run from the sill plate to the double top plate of the second floor?

3. What is the term used to represent the roof decking used in wood post-and-beam construction?

4. What are the three types of floor joists used in wood construction?

5. What are the two types of steel construction typically used on construction sites?

6. Which type of metal floor joists can span a longer distance: C-channel or truss?

7. What does CMU stand for?

8. Care must be taken when drilling in precast concrete to watch for _____.

9. True or false?: If drilled or cut, the integrity of a panel will be jeopardized.

10. What role might an electrician play with respect to bricklayers?

11. List four types of roofs used in construction.

PRACTICE PROBLEMS

1. Label the areas indicated on the following drawings.

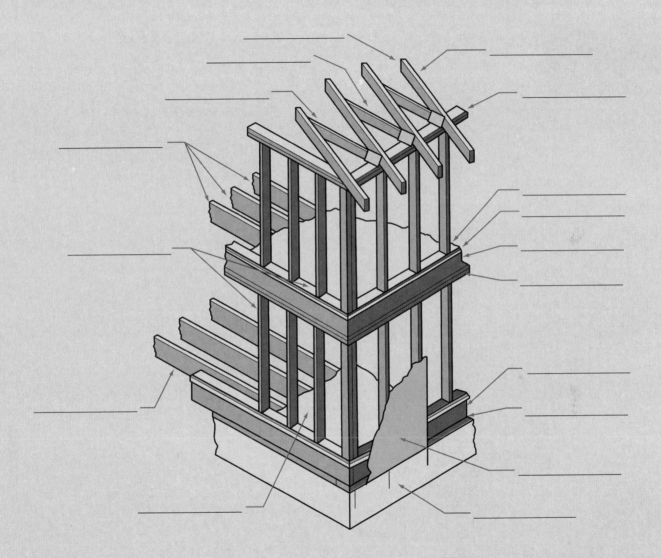

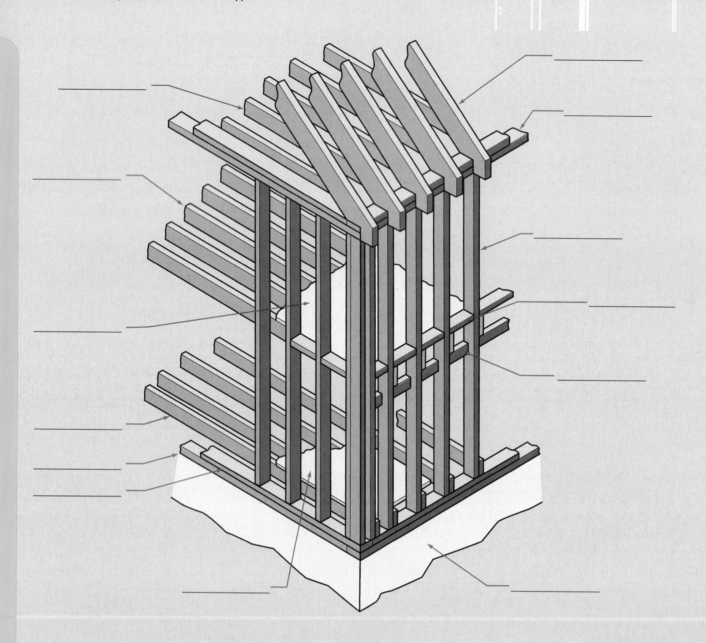

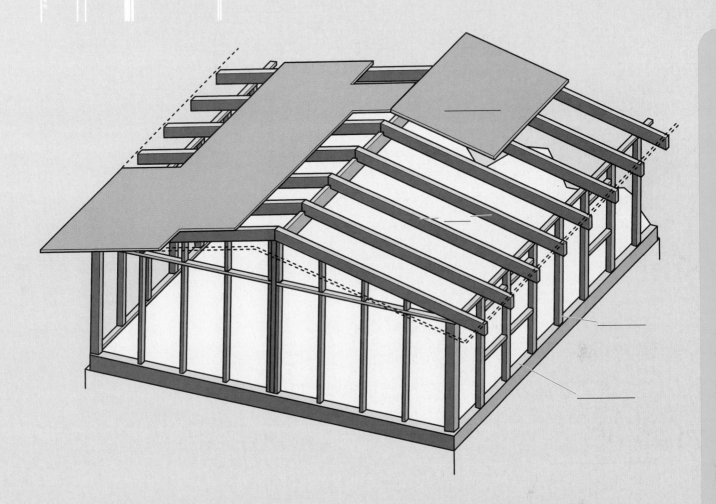

6

Architectural Considerations

OUTLINE

OVERVIEW

A set of construction drawings will have many different types of prints. It is important to understand the information contained in the prints, and the relationship they have to one another. It is also important to be able to visualize a building—to be able to imagine how the finished product will be from just looking at a set of construction drawings. This chapter provides an overview of the types of prints found in a set of construction drawings and the relationship they have to one another and the job.

OBJECTIVES

After completing this chapter, you should be able to:

• Describe the function of plans and elevations.

• Describe the function of sections and details.

• Describe the function of schedules and specifications.

• Explain the impact addenda and revisions have on the construction process.

• Identify the information found on a title block.

• Describe the division of prints by discipline.

ARCHITECTURAL CONSIDERATIONS

There are several architectural considerations that must be observed when reading a print. Obviously, electrical workers must understand electrical floor plans and schedules in detail. However, their understanding of prints must go further. Electrical workers must understand and be able to interpret all of the types of related craft symbology and construction drawings that will be detailed in the forthcoming chapters. The electrical worker must make runs around all of the other crafts (including heating and air conditioning, plumbing, piping, and structural) during construction. It is of utmost importance to understand the entire print set.

6.1 Site Plans

Site plans give the electrician a relation of the building to the surrounding area of the building site, and show a detailed contour (or "lay of the land"). To an electrical worker, the use of these site plans is primarily for building orientation and location of exterior lighting, such as parking lots and entryways. Figure 6–1 illustrates a typical topographical site plan for a residential house.

Site Plan
A scaled drawing that shows the natural features of a property as well as any man-made features.

FIGURE 6–1   Residential site plan.

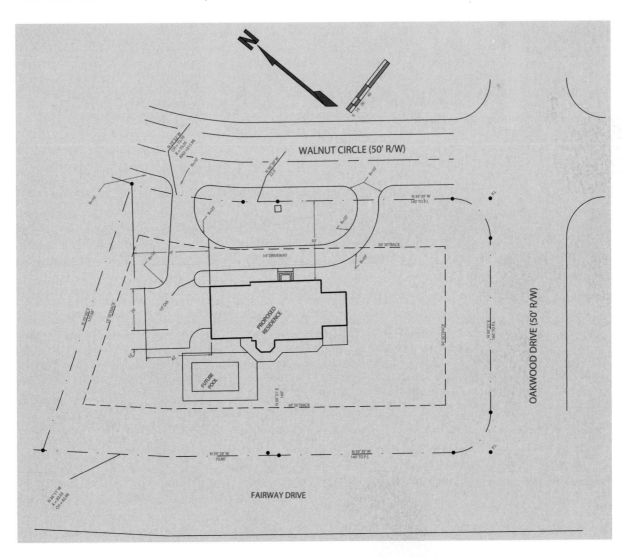

6.2 Elevations

Interior Elevation
A scaled view of an interior wall that often contains height and device locations.

Exterior Elevation
A scaled view of an exterior wall that often contains height and device locations.

There are two types of elevations: interior and exterior. They are orthographic projections of how a wall or side of a structure looks. **Interior elevations** show counter heights and device locations. **Exterior elevations** show relations of lighting to windows and doors and to the structure. Exterior elevations show floor-level relations that tell the print reader where the various floor plans originate. This is done by using cutting-plane lines on the elevations. Exterior elevations also assist in locating exterior lighting fixtures and devices not standardized by code. Figure 6–2 shows a typical exterior elevation.

FIGURE 6-2 Residential elevation.

FRONT ELEVATION

6.3 Floor Plans

Floor Plan
A scaled view looking downward at one level or floor of a building.

The **floor plan** is the plan most commonly used by the electrician. The floor plan looks as if the print reader took the roof off the structure and is looking straight down at the floors and cut-off walls. Cutting-plane lines found on elevations determine the cutting plane for each floor plan. The level of the cutting-plane line is usually such that windows and doors are displayed. Floor plans show shape and relationship of rooms and, most important, locations of devices. These devices are depicted by symbols and are located with dimensions. Sometimes it is necessary to scale the drawings for the location of devices. Figure 6–3 shows a plan of a residence.

FIGURE 6-3   First-floor plan.

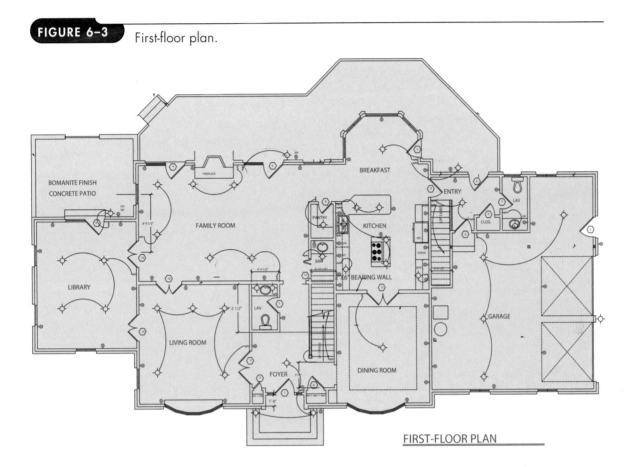

FIRST-FLOOR PLAN

6.4 Run Determination

One key to good electrical work is determining efficient runs. The architect or engineer locates the fixtures, devices, and panels on the print, but the electrician must often determine the run. Run determination takes practice and skill in reading the plan and elevation views simultaneously. In residential lighting, if the electrician fails to check on the elevation, a floor-length window or door may entirely change the choice of run direction.

When reading commercial and industrial prints, the understanding of mechanical plans, electrical plans, elevations, and their relationships is far more important than in residential prints. In commercial or industrial construction, there are often larger mechanical devices such as boilers, pipes, and duct work that get in the way of electrical runs. Electrical workers may get last choice of run location. Therefore, they must use care to read and understand all related craft prints to determine the most logical and efficient runs.

6.5 Sections

Section View
A scaled drawing giving a view of an object or part of a building that has been cut away to see the inside features.

On construction prints, it is often difficult to determine materials and sizes of wall and framing members due to the drawing scale. A **section view** is used to show a cutaway part of a drawing. This section is usually an increased-size view of the structure or detail. These sections are used to clarify points of construction not readily apparent in the standard views. Common sections are roof, wall, and foundation. The purpose of a wall section is to show the materials used in construction of that wall. This will aid the electrical worker in determining how to pass through that wall if necessary. Plaster ring depths and box mounting requirements might also be clarified. Figure 6–4 shows an example of a residential wall section.

FIGURE 6–4

Residential wall section.

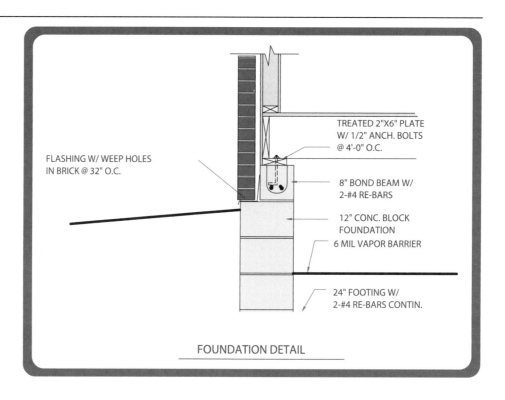

FLASHING W/ WEEP HOLES
IN BRICK @ 32" O.C.

TREATED 2"X6" PLATE
W/ 1/2" ANCH. BOLTS
@ 4'-0" O.C.

8" BOND BEAM W/
2-#4 RE-BARS

12" CONC. BLOCK
FOUNDATION

6 MIL VAPOR BARRIER

24" FOOTING W/
2-#4 RE-BARS CONTIN.

FOUNDATION DETAIL

6.6 Details

Details are plan, elevation, or section views drawn to a large scale for clarification. The scale usually runs from ¾ inch equals 1 foot to 3 inches equals 1 foot. The detail shows exactly how the feature is made or where it is located. The detail is either completely dimensioned or drawn to an accurate scale designated to reproduce the desired feature with accuracy not possible from a print with a smaller scale. Figure 6–5 shows an example of an explosion-proof receptacle detail.

Detail
A scaled drawing drawn to a larger scale to show exactly how a feature is made or where it is located.

FIGURE 6–5   Explosion-proof receptacle detail.

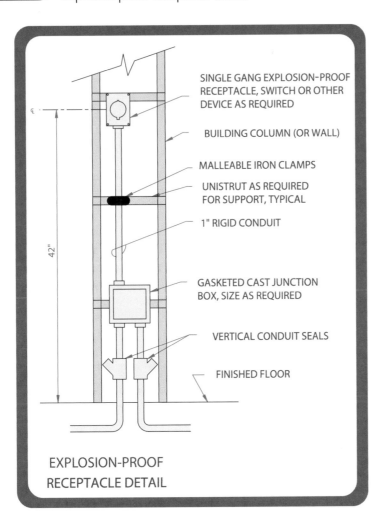

SINGLE GANG EXPLOSION-PROOF
RECEPTACLE, SWITCH OR OTHER
DEVICE AS REQUIRED

BUILDING COLUMN (OR WALL)

MALLEABLE IRON CLAMPS

UNISTRUT AS REQUIRED
FOR SUPPORT, TYPICAL

1" RIGID CONDUIT

GASKETED CAST JUNCTION
BOX, SIZE AS REQUIRED

VERTICAL CONDUIT SEALS

FINISHED FLOOR

42"

EXPLOSION-PROOF
RECEPTACLE DETAIL

Schedule
A chart that contains detailed information about what size and type of materials are to be used or how a job is to be completed.

6.7 Schedules

There are various **schedules** included in a complete set of construction drawings. Schedules contain detailed information about what size and type of materials are to be used on a job. A door schedule (Figure 6–6) is an example of a schedule containing types and sizes of materials to be used. Schedules might also contain information about how a job is to be finished. A panel schedule (Figure 6–7) contains information about how the circuitry is to be laid out. A schedule may be included on a print with another view, or there may be sheets dedicated to schedules. Various types of prints contain various types of schedules. Schedules are valuable in estimating, ordering, and organizing the entire construction project.

6.8 Specifications

Specifications are a list of detailed job requirements, under which all work must be performed. The specifications may be

Specifications
A list of detailed job requirements, under which all work must be performed.

FIGURE 6-6   Door schedule.

Note: This is a clip from the second–floor plan.

SECOND-FLOOR PLAN
SCALE; 1/4" = 1'-0"

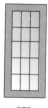 

"F"   "G"

DOOR SCHEDULE

NUMBER	SIZE	TYPE	DESCRIPTION
1	3-0/6-8	A	ENTRANCE W/SIDELITES
2	3-0/6-8	B	EXT. METAL
3	3-0/6-8	C	EXT. METAL/INSUL. GL.
4	6-0/6-8	D	GARDEN DOOR
5	1-3/6-8	E	INT. 6 PANEL WD.
6	2-0/6-8	E	INT. 6 PANEL WD.
7	2-4/6-8	E	INT. 6 PANEL WD.
8	2-6/6-8	E	INT. 6 PANEL WD.
9	2-8/6-8	E	INT. 6 PANEL WD.
10	PAIR 2-0/6-8	F	FRENCH DOORS
11	PAIR 2-6/6-8	F	FRENCH DOORS
12	3-0 BI-FOLD	G	RAISED PANEL WD.
13	4-0 BI-FOLD	G	RAISED PANEL WD.
14	5-0 BI-FOLD	G	RAISED PANEL WD.

FIGURE 6-7   Panel schedule.

PANELBOARD XX2 SCHEDULE

VOLTAGE: 277/480 VOLT, 3 PH
MAINS: 40 AMP 3-POLE MAIN BREAKER   LOCATION: EAST ELEC. ROOM
MOUNTING: SURFACE
FEEDER: 1"C - 4#8, 1#10 GRD.   30 POLES

CKT NO.	BREAKER	PHASE - A	B	C (KVA)	CIRCUIT IDENTIFICATION	NOTES
		9.6	9.4	8.8		
1	20/1	2.4			LIGHTING	
2	20/1		2.2		LIGHTING	
3	20/1			1.8	LIGHTING	
4	20/1	1.2			LIGHTING	
5	20/1		1.2		LIGHTING	
6	20/1			1.0	LIGHTING	
7	20/1	1.0			SPARE	
8	20/1		1.0		SPARE	
9	20/1			1.0	SPARE	
10	20/1				PROVISIONS	
11	20/1				PROVISIONS	
12	20/1				PROVISIONS	
13	30/3	5.0	5.0	5.0	PANEL X1	VIA 15 KVA TRANSF.
14	20/1				PROVISIONS	
15	20/1				PROVISIONS	
16	20/1				PROVISIONS	
17	20/1				PROVISIONS	
18	20/1				PROVISIONS	
19	20/1				PROVISIONS	
20	20/1				PROVISIONS	

PANELBOARD X2 SCHEDULE

VOLTAGE: 120/208 VOLT, 3 PH, 4 W
MAINS: 40 AMP 3-POLE MAIN BREAKER   LOCATION: EAST MECH/
MOUNTING: SURFACE   ELECT. ROOM
FEEDER: 1"C-4#8, 1#10 GRD.   30 POLES

CKT NO.	BREAKER	PHASE - A	B	C (KVA)	CIRCUIT IDENTIFICATION	NOTES
		2.0	2.0	2.0		
1	20/1	1.0			CCTV RECEPTACLE	
2	20/1		1.0		SPARE	
3	20/1			1.0	ELECTRTIC GATES	
4	20/1	1.0			SPARE	
5	20/1		1.0		SPARE	
6	20/1			1.0	SPARE	
7	20/1				PROVISIONS	
8	20/1				PROVISIONS	
9	20/1				PROVISIONS	
10	20/1				PROVISIONS	
11	20/1				PROVISIONS	
12	20/1				PROVISIONS	
13	20/1				PROVISIONS	
14	20/1				PROVISIONS	
15	20/1				PROVISIONS	
16	20/1				PROVISIONS	
17	20/1				PROVISIONS	
18	20/1				PROVISIONS	
19	20/1				PROVISIONS	
20	20/1				PROVISIONS	
21	20/1				PROVISIONS	

only a few pages long to hundreds of pages long, depending on the job. The specifications are a very important part of a set of construction drawings. They give an abundance of information about what products are to be used, and how they are to be installed. A set of specifications is typically not written on the large sheets of construction drawings but printed on 8-½ × 11 sheets of paper and bound. Figure 6–8 shows an example of a page from the electrical division of a set of specifications.

FIGURE 6–8   Specifications contain important information about a construction project. This is a page from the industrial specifications in Appendix B.

```
                        SECTION 16180
                  Overcurrent Protective Devices


Part 1.     General
Not used


Part 2.     Products
2.1         INDIVIDUALLY MOUNTED CIRCUIT BREAKERS: Molded case type,
with the frame and ampacities noted on the drawings; mounted in NEMA 1
or flush-mounted enclosures as noted on the drawings
2.2         FUSES FOR POWER DISTRIBUTION PANELBOARDS AND SAFETY
SWITCHES: Bussman or A/E-approved equal with the following
capabilities:
            A. HI-CAP Type KRP-C: rated 601 A and larger; time delay
               type holding 500% of the rated current for a minimum of
               4 seconds and clearing 20 times the rated current in
               0.01 second or less; 200,000 A, RMS, symmetrical minimum
               interrupting rating; UL listed as Class RK1
            B. Low-Peak Type LPS-RK: rated 0 to 600 A, 600 V; dual-
               element time delay type with separate overload and
               short-circuit elements, holding 500% of the rated
               current for a minimum of 10 seconds; 200,000A, RMS,
               symmetrical minimum interrupting rating; UL listed as
               Class RK1
Part 3.     Execution
3.1         Provide and install fuses and circuit breakers as noted on
the drawings.
3.2         Do not install fuses until equipment is ready to be
energized.
3.3         Provide 2 spare sets of fuses for each size used.
3.4         Provide Bussman Catalog No. SFC, or equal, spare fuse
cabinet. Install near the main switchboard/panelboard as shown on the
drawings or as directed by the A/E.


                        END OF SECTION
```

6.9 Addenda and Revisions

Addendum

A change to prints or specifications before bids are opened.

An **addendum** is a change to the prints or specifications before the bids are opened, which is treated as a new contract item. It is not uncommon to have an addendum come out the day before or the final day the bids are due. It is up to the contractor to be sure the project was bid with the latest information. The electrical worker needs to be aware of any addenda, changes, or print revisions because these might affect work and installations in progress. When any drawing changes are made, it is normal practice to issue a revised print set.

Revision

A dated change to blueprints or specifications.

Revisions are dated changes in prints or specifications. Revisions on the print will often have a cloud around the changes with a number or letter that indicates the date of that particular revision (see Figure 6–9). The electrical worker must be aware of and in possession of the most currently dated print set. This will ensure that all work has been installed correctly and within current guidelines.

Change Order

A change relating to a construction project that occurs after bids have been opened. Changes often cause an increase or decrease in job cost.

Changes **(change orders)** occur after bids are opened. Changes may cause an increase or decrease in job cost. It is very important to use the proper procedure with a change order, which typically requires the signature of the architect/engineer or owner. The signature ensures that the change was authorized. Without a signed change order, the extra work that was done may not be paid for.

As-Built Drawings

A set of drawings that indicates exactly how the job was completed.

A detailed record of material, installation, or print revisions necessary during the construction process should be maintained. Drawings of record, working drawings, and in-process drawings ("as-builts") should be updated daily and conveyed to the engineer or owner at the end of the job. **As-built drawings** are a set of drawings that indicates exactly how the job was completed. Most jobs will not be done exactly as they were initially planned. Pipes may have to move due to an obstruction, boxes may have to be relocated for various reasons, or change orders may add to or remove from the initial plan.

It is important to have a record of how the job was actually built—not only for the engineers or architects but for the contractor. If a person has to go back to that job to repair, add, or change something, there is an accurate representation of what actually was done to help in the planning process. Often the contractor will not receive final payment until working drawings or "as-builts" are completed and handed in. It is therefore imperative that daily in-process updates be kept current. It is a good practice to initial your personal print set for reference.

FIGURE 6-9 Riser diagram with revision clouds. Revision clouds will have a letter identifying the date of the revision.

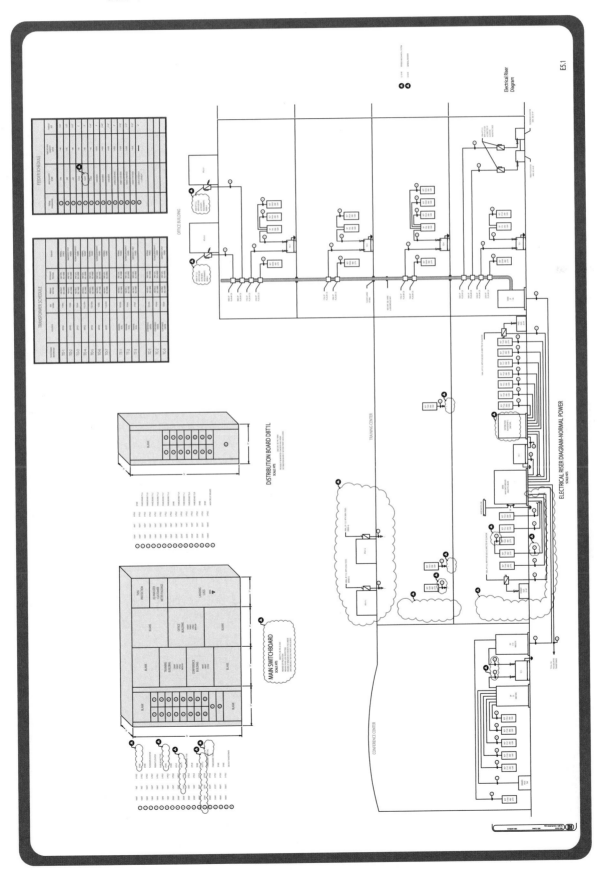

6.10 Title Block and Print Layout

Title Block
An area of a blueprint dedicated to providing additional information about the drawing, designer, job, and so on.

The **title block** has very important information relating to the job and to the construction drawings (see Figure 6–10). The first thing you should always read when looking at a sheet is the title block. Information typically found on the title block will include the following:

- Name of the project
- Project location
- Name of the architect or engineer
- Date the plans were completed
- Revision date
- Scale of the sheet
- Name of the sheet
- Sheet number
- Total number of sheets

FIGURE 6–10   A title block has information about the drawings, owner, designer, as well as the building or project.

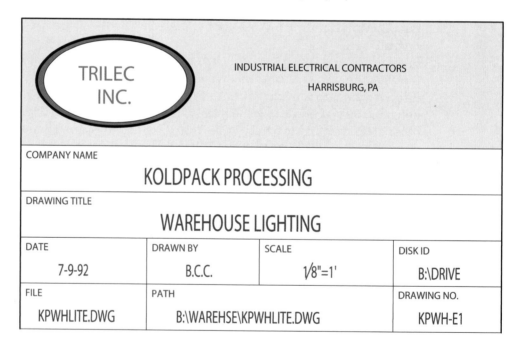

TRILEC INC.	INDUSTRIAL ELECTRICAL CONTRACTORS HARRISBURG, PA		
COMPANY NAME **KOLDPACK PROCESSING**			
DRAWING TITLE **WAREHOUSE LIGHTING**			
DATE 7-9-92	DRAWN BY B.C.C.	SCALE 1/8"=1'	DISK ID B:\DRIVE
FILE KPWHLITE.DWG	PATH B:\WAREHSE\KPWHLITE.DWG		DRAWING NO. KPWH-E1

Most residential construction drawings will have their sheets numbered from 1 to the number of sheets in the set. Thus, if there are six sheets in a set of residential prints they will be numbered from 1 to 6. Commercial and industrial construction drawings are numbered a bit differently. Each sheet will be numbered by the discipline of the drawing, as well as by the number of the sheet in relation to that category. For example, a sheet numbered

E5 would mean sheet 5 of the electrical drawings. The common denotations for disciplines of drawings are as follows:

- A: Architectural
- C: Civil
- E: Electrical
- F: Fire protection
- M: Mechanical
- L: Landscape
- P: Plumbing
- S: Structural

6.11 Visualization of Construction Drawings

It is very important to be able to visualize what a project looks like from simply looking at the construction drawings. This is a skill that will come with time and practice. Each person will develop their own way of visualizing a project, but for beginners there are a few steps to follow. It is helpful to start with a set residential print that does not have very many pages.

Step 1. Look at the site plan. This will give you an idea of the overall shape of the building, and its relationship to the property it is on.

Step 2. Look at the exterior elevation drawings. This will indicate what type of building it is and the roof style—starting the visualization process.

Step 3. Beginning with the main floor plan, start to mentally walk through the project. Imagine walking in the door and going through each room. Pay close attention to windows, ceiling heights, and locations of cabinets, lighting, devices, and any other details available on the drawings. Be sure to imagine going up and down all flights of stairs, which means referring to other prints, and work your way through the entire project.

After completely walking through a set of construction drawings in your mind, you will begin to see what the finished project will look like. This will also help with planning the work to be done. While walking through the project in your mind, you will see aspects of the project that may require special attention or consideration.

Chapters 7 through 10 of this text depict and discuss electrical and related symbology. It is important to be able to read and understand the various symbols used on blueprints. Chapters 11 through 15 cover in detail each of the various types of prints found in a set of construction drawings.

SUMMARY

It is very important to understand the information that will be found on each type of print, as well as the relationship the prints have with one another.

The types of drawings typically found on a print include the following:

- Site plans
- Elevations
- Floor plans
- Sections
- Details
- Schedules
- Specifications

Addenda are changes to a project before bids have been opened. It is important to have all addenda, or the project may not be bid accurately.

Revised prints are drawings with updated changes that are dated to indicate the date of the revision.

As-built drawings will indicate how a project was put together. It is important to keep an accurate record of how a project was built.

Visualization of a project is an important skill that anyone who reads construction drawings should have.

REVIEW QUESTIONS

1. What is the function of a site plan?
2. What is the function of elevation views?
3. What is the function of floor plans?
4. What is the function of schedules?
5. What are print specifications?
6. What are addenda and revisions?
7. List six pieces of information that may be found on the title block.
8. What steps are taken to help visualize a project?

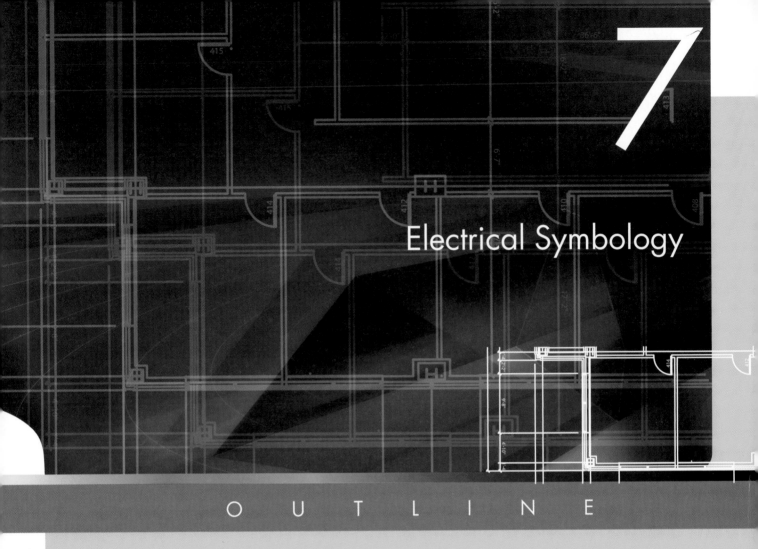

7

Electrical Symbology

OUTLINE

OVERVIEW

Chapters 7 through 10 cover symbology. Symbology is a shorthand semipictorial notation system used to designate a variety of objects on a print. These symbols are drafting representations of the objects they portray, not pictures. The symbology chapters of this text show you how to locate, recognize, and interpret a variety of craft symbols on residential, commercial, and industrial jobs. Understanding and reading symbols is a must on residential, commercial, and industrial prints because there is no way to put actual pictures or written explanations of the devices in the small space available on a print.

OBJECTIVES

After completing this chapter, you should be able to:

• Understand the function, location, and use of symbols.

• Recognize and identify the common electrical and electronic symbols.

SYMBOLOGY

Symbology is the study of making or interpreting symbols. Symbols are simple graphic or alphanumeric representations of an object or device. There is often an attempt to make the symbol represent the actual object in some way. Symbols are used to save space, simplify notation, and standardize notation. Figure 7–1 illustrates two common symbols for electrical devices.

FIGURE 7–1 Symbology.

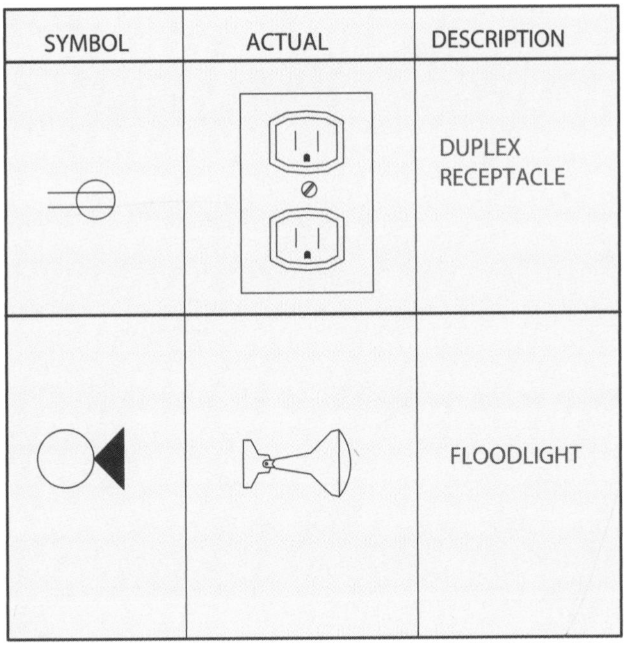

SYMBOL	ACTUAL	DESCRIPTION
		DUPLEX RECEPTACLE
		FLOODLIGHT

Symbols are sometimes arbitrary and have no obvious relation to the device they represent. The American National Standards Institute (ANSI) is one organization responsible for standardizing all symbols. This text is written using the most current ANSI standards for graphic symbology. The ANSI standardized symbols are not, however, required to be used. It is possible that an electrical worker may encounter symbols that do not meet ANSI standards or that are job specific. If this is the case, the engineer or architect should have a legend or symbol library on the print (see Figure 7–2). This legend or library is for use as a reference. The symbols in this legend or library are produced by using templates or computer-aided design (CAD) symbol libraries.

It is necessary for the electrical worker to understand various symbology from all related crafts. Tables of standard graphic symbology are available in Chapters 7 through 10 of this text. They may also be found in a library or directly from ANSI. Other sources of information on graphic symbols are trade journals and manufacturers' information sheets. It is not necessary to memorize all graphic symbols for all crafts. It is, however, necessary to know the common symbols of the electrical trade and to know how and where to find the meanings of the related craft symbols.

FIGURE 7-2 Symbol schedule (Courtesy of Ulteig Engineers).

ELECTRICAL SYMBOLS SCHEDULE

THIS IS A COMPREHENSIVE SYMBOL SCHEDULE. NOT ALL SYMBOLS ARE APPLICABLE TO THESE DRAWINGS

DESCRIPTION	MOUNTING HEIGHT TO CENTER OF DEVICE UNLESS OTHERWISE INDICATED
CEILING SURFACE MOUNTED LIGHT FIXTURE, CAPITAL LETTER INDICATES FIXTURE TYPE, SMALL LETTER INDICATES SWITCHING	
RECESSED CEILING MOUNTED LIGHT FIXTURE	
WALL MOUNTED INCAND. OR H.I.D. LIGHT FIXTURE, SURFACE OR, RECESSED MOUNTED	
CEILING JUNCTION BOX	
WALL JUNCTION BOX	18 INCHES
CEILING EXIT SIGN, SHADED SIDE INDICATES LIGHTED FACE, ARROWS INSTALLED AS SHOWN	
WALL EXIT SIGN	6 INCHES ABOVE DOOR FRAME TO BOTTOM
SUSPENDED FIXTURE	
SURFACE FLUORESCENT	
RECESSED FLUORESCENT	
FIXTURES ON EMERGENCY CIRCUIT OR CIRCUIT THAT OPERATES CONTINUOUSLY	
DASHED LINES INDICATE EXISTING FIXTURES, DEVICES, OR EQUIPMENT	
EMERGENCY BATTERY LIGHT	7 FOOT 6 INCHES
EMERGENCY REMOTE LAMP	7 FOOT 6 INCHES
SWITCH: SINGLE POLE (HORSEPOWER RATED WHEN USED AS MOTOR DISCONNECT)	48 INCHES
DOUBLE POLE	48 INCHES
3-WAY	48 INCHES
4-WAY	48 INCHES
KEY OPERATED	48 INCHES
WITH PILOT LIGHT (CAN BE USED WITH OTHER SWITCH TYPES)	48 INCHES
FLUORESCENT DIMMER	48 INCHES
MOMENTARY CONTACT	48 INCHES
DIMMER (600 WATT UNLESS OTHERWISE NOTED)	48 INCHES
LIGHTING INTENSITY SELECTOR	48 INCHES
LOW VOLTAGE SWITCH	48 INCHES
ELECTRIC HEAT	
DUPLEX RECEPTACLE	18 INCHES

DESCRIPTION	MOUNTING HEIGHT TO CENTER OF DEVICE UNLESS OTHERWISE INDICATED
VARIABLE FREQUENCY CONTROLLER	
THERMAL SWITCH (MOTOR OVERLOAD TYPE)	48 INCHES
THERMOSTAT - PROVIDE BY DIVISION 16	48 INCHES
THERMOSTAT - FURNISHED BY DIVISION 15, INSTALLED BY DIVISION 16	48 INCHES
HASH MARKS INDICATE ITEM NOTED TO BE REMOVED	
SPEAKER	
SUSPENDED SPEAKER (SEE SPECIFICATION)	
WALL MOUNTED SPEAKER	90 INCHES TO BOTTOM
CLOCK	94 INCHES TO BOTTOM
MICROPHONE OUTLET	18 INCHES
VOLUME CONTROL	48 INCHES
SECURITY DEVICES: REQUEST TO EXIT	3 INCHES ABOVE DOOR FRAME
DOOR POSITION SWITCH, RECESSED	
DOOR POSITION SWITCH, SURFACE MOUNT	
DOOR POSITION SWITCH	
CARD READER	
CEILING MOUNTED MOTION SENSOR SWITCH	
WALL MOUNTED SMALL AREA MOTION SENSOR SWITCH	48 INCHES
WALL MOUNTED LARGE AREA MOTION SENSOR SWITCH	84 INCHES
MOTION SENSOR SWITCH POWER PACK	
AUXILIARY INPUT OUTLET	18 INCHES
COMPUTER OUTLET (NUMBER INDICATES QUANTITY OF CABLES AND JACKS, NO NUMBER INDICATES ONE)	18 INCHES
COMBINATION COMPUTER/TELEPHONE OUTLET IN SINGLE GANG OPENING (NUMBER INDICATES QUANTITY OF CABLES AND JACKS,NO NUMBER INDICATES ONE)	18 INCHES
TELEPHONE OUTLET ("W" INDICATES WALL MOUNTED AT 48 INCHES)	18 INCHES
DATA PATCH PANEL	
COMPUTER/TELEPHONE OUTLET, PROVIDE ROUGH-IN AND CONDUIT ONLY	18 INCHES
TELEVISION OUTLET	18 INCHES

Electrical Symbols Legend

Symbol	Description	Mounting Height
	DOUBLE DUPLEX RECEPTACLE	18 INCHES
	SPECIAL PURPOSE RECEPTACLE LETTER INDICATES TYPE, SEE SPECIFICATION	18 INCHES
	MULTI-OUTLET ASSEMBLY - M.O.A., PROVIDE DEVICES AS SHOWN ON PLANS	
	DUPLEX RECEPTACLE, HALF SWITCHED	18 INCHES
	FLOOR OUTLET, ADDITIONAL SYMBOL INDICATES TYPE	
	TIME SWITCH	
	PHOTO CELL	
	TRANSFORMER	
	UNDER FLOOR DUCT	
	FIRE ALARM: MANUAL STATION	48 INCHES
R	RATE OF RISE DETECTOR	
	FIXED TEMPERATURE DETECTOR	
DPD	DUCT PHOTOELECTRIC DETECTOR	
P	PHOTOELECTRIC DETECTOR	
C	COMBINATION FIRE/SMOKE DAMPER BY DIVISION 15	
S	HORN ("S" INDICATES WITH STROBE LIGHT)	BOTTOM OF LENS 80 INCHES
S	SPEAKER HORN ("S" INDICATES WITH STROBE LIGHT)	BOTTOM OF LENS 80 INCHES
S	CHIME ("S" INDICATES WITH STROBE LIGHT)	
	FIRE ALARM MODULE/RELAY	
	STROBE LIGHT	BOTTOM OF LENS 80 INCHES
	MAGNETIC DOOR HOLD OPEN	
I	INTEGRAL MAGNETIC DOOR CLOSER (PROVIDED BY DIVISION 8)	
2	TAMPER SWITCH (NUMBER INDICATES NUMBER OF DEVICES AT THIS LOCATION)	
FP	FIRE PUMP	
	FLOW SWITCH	
	LIGHTING AND APPLIANCE PANELBOARD	
	SWITCHBOARD OR MOTOR CONTROL CENTER AS NOTED	
	CONTROL CABINET	
	SPECIAL EQUIPMENT CABINET AS NOTED	

Symbol	Description	Mounting Height
	INTERCOM OUTLET	18 INCHES
M	INTERCOM MASTER	
S	INTERCOM STAFF STATION	42 INCHES
	BUZZER	
	PUSH BUTTON	
	POWER DOOR HANDICAP PUSH PAD	
	BELL	
1	CONTACTOR (NUMBER INDICATES NUMBER OF POLES)	
WP	WEATHERPROOF	
AC	ABOVE COUNTER	
EWC	ELECTRIC WATER COOLER	
	CONDUIT CONCEALED IN WALL OR CEILING, QUANTITY OF CONDUCTORS NOT SHOWN, PROVIDE AS REQUIRED FOR DEVICE/CIRCUIT NUMBERS SHOWN	
	CONDUIT UP	
	CONDUIT DOWN	
	CONDUIT CONCEALED IN FLOOR	
#10 / L1-2,4,6	HOME RUN TO PANELBOARD, QUANTITY OF CONDUCTORS REQUIRED NOT INDICATED, PROVIDE QUANTITY AS REQUIRED FOR CIRCUIT NUMBERS SHOWN. SWITCHING ARRANGEMENT, OR NUMBER OF HOME RUNS SHOWN. #10 INDICATES WIRE SIZE, NO NUMBERS INDICATES #12, 3/4 INCH CONDUIT MINIMUM	
	SURFACE CONDUIT OR SURFACE RACEWAY	
1	NOTE IDENTIFICATION	
	WIRE BASKET TYPE CABLE TRAY	
	COMMUNICATIONS CABLE SUPPORT HANGER	
2	POWER POLE	
	MOTOR (NUMBER REFERS TO MOTOR AND EQUIPMENT SCHEDULE SEE SCHEDULE FOR WIRING AND CONTROL REQUIREMENTS)	
	DISCONNECT ("F" INDICATES FUSED WHEN SHOWN)	
	MAGNETIC STARTER	
	COMBINATION STARTER-DISCONNECT	

NOTES:
1. FOR DEVICES SHOWN AS 48 INCHES MOUNTING HEIGHT, WHEN INSTALLED IN MASONRY BLOCK WALLS, MOUNTING AT 48 INCHES TO THE TOP OF THE OUTLET BOX IS ACCEPTABLE.

ELECTRICAL SYMBOLOGY

Electrical symbology is the graphic representation of electrical devices and materials used in residential, commercial, and industrial construction and maintenance. In some cases, the graphic symbol represents the device. Some symbols look like simple pictures of the device. These are called pictographs. In other cases, the symbol was picked for no obvious reason. The following pages and figures present the most current ANSI graphic standard symbols for electrical wiring. They are separated into logical categories for presentation.

7.1 Lighting and Receptacle Outlets

Figure 7–3 shows the commonly used symbols for lighting and receptacle outlets. A letter is often used inside or next to the symbol to indicate the type of outlet being represented.

FIGURE 7–3

Electrical 1.

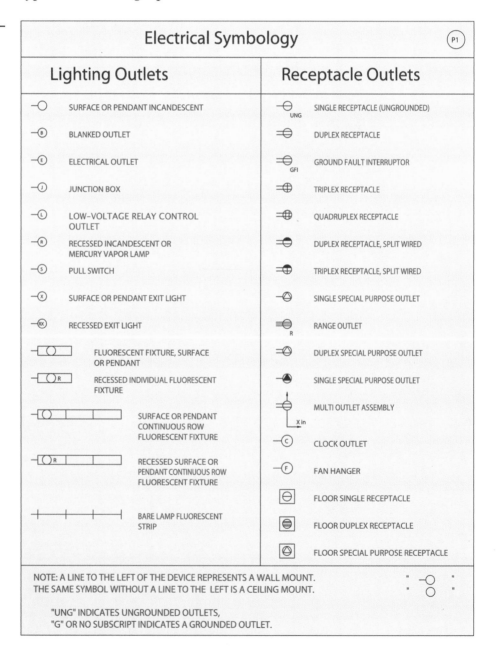

Lighting outlet symbols that have a line extending to a wall indicate a wall mount. A lighting outlet without the line represents a ceiling mount. In commercial and industrial prints, capital or lowercase letters will often be written next to the lighting outlet symbols (see Figure 7–4). The capital letters are used to identify the type of fixture and will match up with the lighting schedule. The lowercase letters correspond to a switch to indicate which switch is controlling the light.

FIGURE 7–4 A capital letter written next to the light is used to identify the light. It will correspond to a letter on the lighting schedule. The lower case letters indicate which switch controls the light. Some fluorescent light fixtures have two lower case letters next to the light. Part of the light controlled by one switch and the rest of the light controlled by another switch. (Courtesy of Ulteig Engineers.)

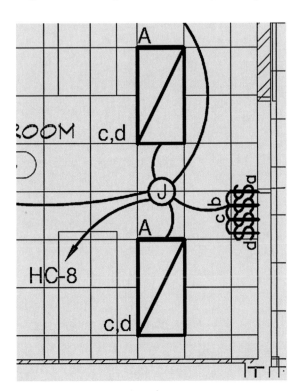

7.2 Switch and Signal System Outlets

Switches are represented by an S (see Figure 7–5). A subscript letter or number is often written next to the S to indicate the type of switch. On commercial and industrial prints, a lowercase letter is often written next to the switch that will correspond to a light to indicate which light the switch is controlling (see Figure 7–4). Figures 7–5 through 7–7 contain signal system outlets. The symbols used for signal system outlets often vary, and therefore it is very important to verify the correct symbols with the legend.

FIGURE 7–5

Electrical 2.

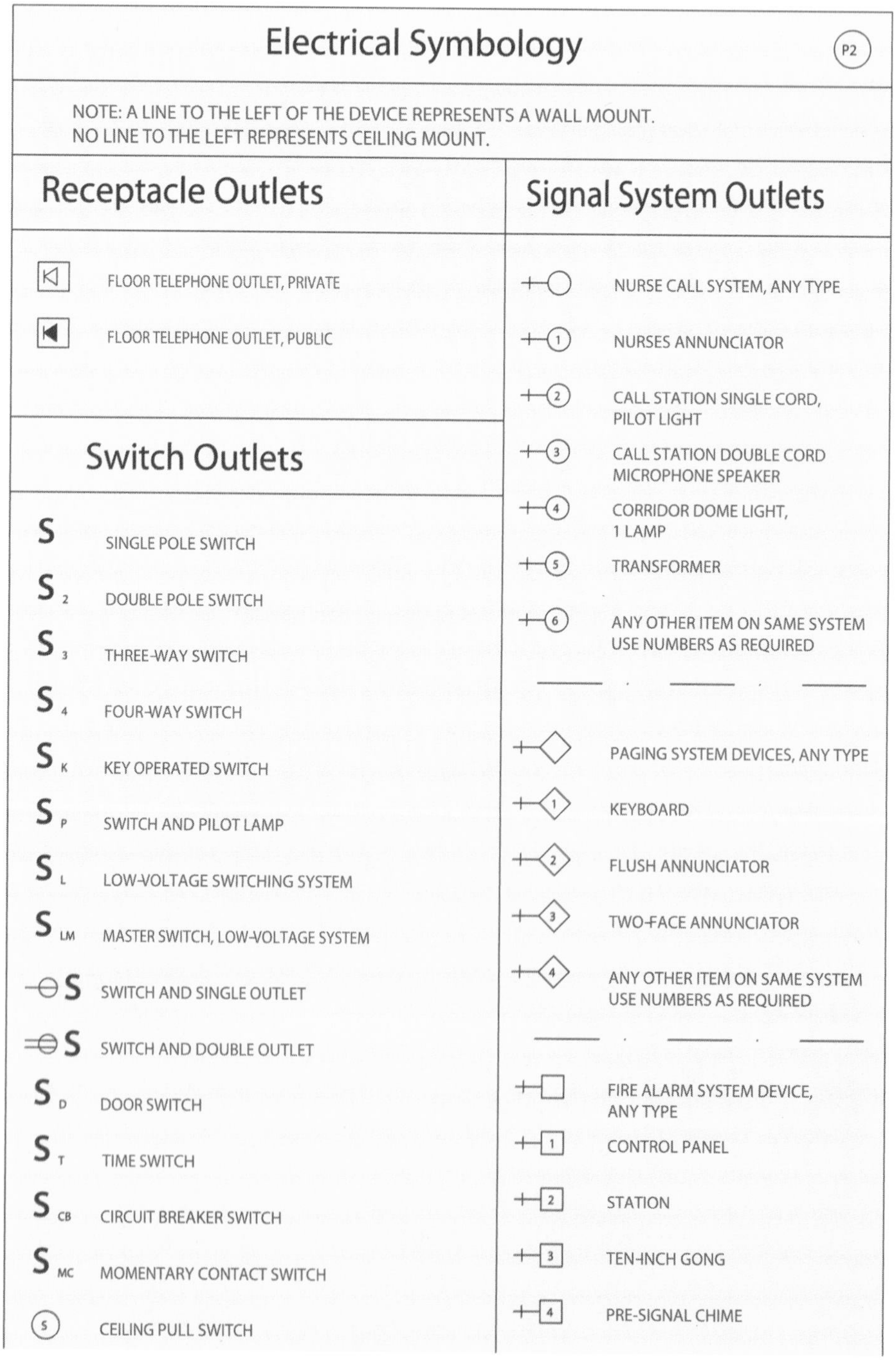

Electrical Symbology (P2)

NOTE: A LINE TO THE LEFT OF THE DEVICE REPRESENTS A WALL MOUNT.
NO LINE TO THE LEFT REPRESENTS CEILING MOUNT.

Receptacle Outlets

◁	FLOOR TELEPHONE OUTLET, PRIVATE
◀	FLOOR TELEPHONE OUTLET, PUBLIC

Switch Outlets

S	SINGLE POLE SWITCH
S₂	DOUBLE POLE SWITCH
S₃	THREE-WAY SWITCH
S₄	FOUR-WAY SWITCH
Sₖ	KEY OPERATED SWITCH
Sₚ	SWITCH AND PILOT LAMP
Sₗ	LOW-VOLTAGE SWITCHING SYSTEM
S_LM	MASTER SWITCH, LOW-VOLTAGE SYSTEM
⊖S	SWITCH AND SINGLE OUTLET
⊜S	SWITCH AND DOUBLE OUTLET
S_D	DOOR SWITCH
S_T	TIME SWITCH
S_CB	CIRCUIT BREAKER SWITCH
S_MC	MOMENTARY CONTACT SWITCH
(S)	CEILING PULL SWITCH

Signal System Outlets

⊸○	NURSE CALL SYSTEM, ANY TYPE
⊸①	NURSES ANNUNCIATOR
⊸②	CALL STATION SINGLE CORD, PILOT LIGHT
⊸③	CALL STATION DOUBLE CORD MICROPHONE SPEAKER
⊸④	CORRIDOR DOME LIGHT, 1 LAMP
⊸⑤	TRANSFORMER
⊸⑥	ANY OTHER ITEM ON SAME SYSTEM USE NUMBERS AS REQUIRED
⊸◇	PAGING SYSTEM DEVICES, ANY TYPE
⊸◇1	KEYBOARD
⊸◇2	FLUSH ANNUNCIATOR
⊸◇3	TWO-FACE ANNUNCIATOR
⊸◇4	ANY OTHER ITEM ON SAME SYSTEM USE NUMBERS AS REQUIRED
⊸▢	FIRE ALARM SYSTEM DEVICE, ANY TYPE
⊸▢1	CONTROL PANEL
⊸▢2	STATION
⊸▢3	TEN-INCH GONG
⊸▢4	PRE-SIGNAL CHIME

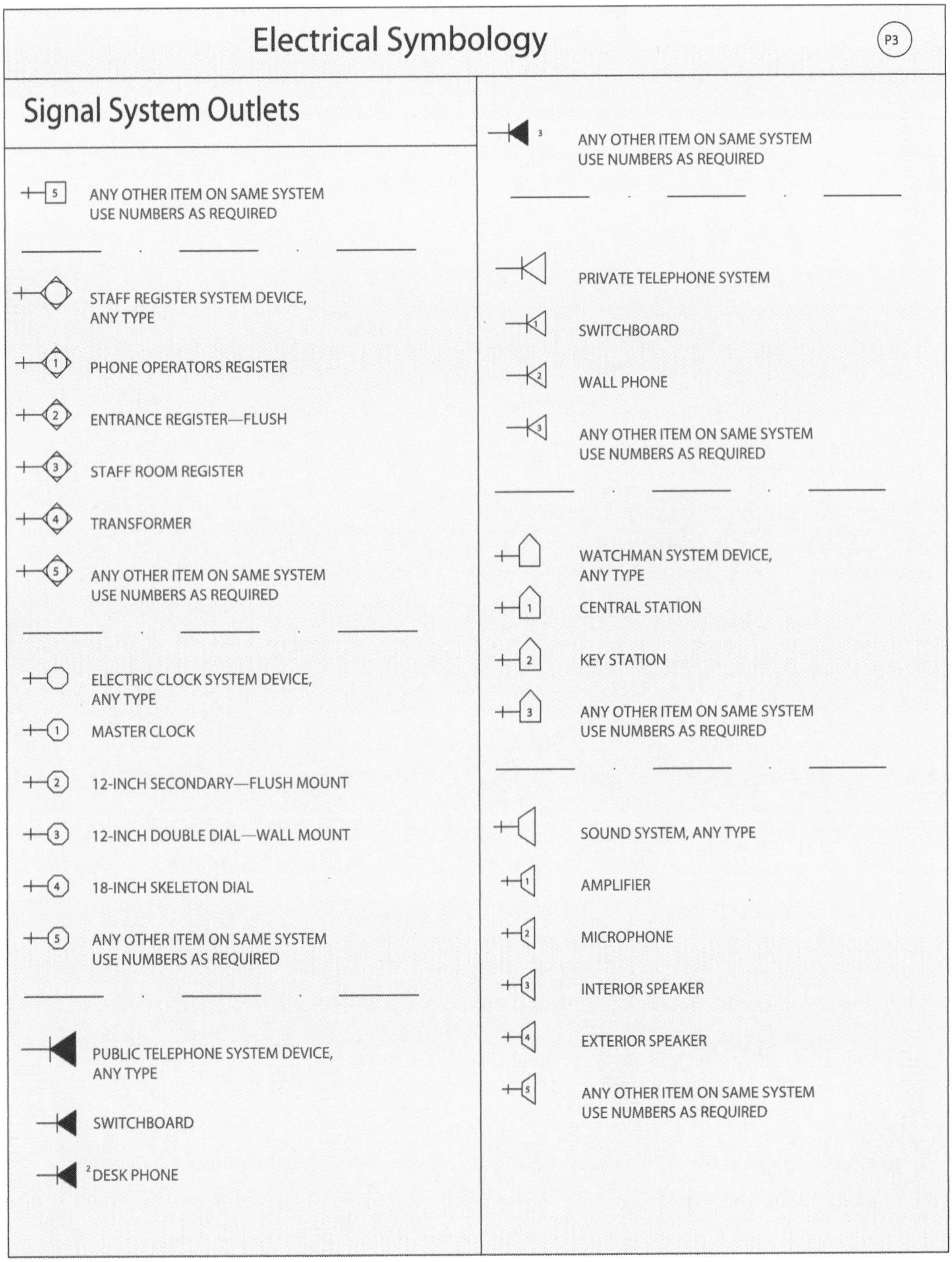

FIGURE 7–6   Electrical 3.

Electrical Symbology   (P3)

Signal System Outlets

— ⊡ 5 ANY OTHER ITEM ON SAME SYSTEM
USE NUMBERS AS REQUIRED

— ◇ STAFF REGISTER SYSTEM DEVICE,
ANY TYPE

— ◇1 PHONE OPERATORS REGISTER

— ◇2 ENTRANCE REGISTER—FLUSH

— ◇3 STAFF ROOM REGISTER

— ◇4 TRANSFORMER

— ◇5 ANY OTHER ITEM ON SAME SYSTEM
USE NUMBERS AS REQUIRED

— ◯ ELECTRIC CLOCK SYSTEM DEVICE,
ANY TYPE

— ◯1 MASTER CLOCK

— ◯2 12-INCH SECONDARY—FLUSH MOUNT

— ◯3 12-INCH DOUBLE DIAL—WALL MOUNT

— ◯4 18-INCH SKELETON DIAL

— ◯5 ANY OTHER ITEM ON SAME SYSTEM
USE NUMBERS AS REQUIRED

— ◀ PUBLIC TELEPHONE SYSTEM DEVICE,
ANY TYPE

— ◀ SWITCHBOARD

— ◀2 DESK PHONE

— ◀3 ANY OTHER ITEM ON SAME SYSTEM
USE NUMBERS AS REQUIRED

— ◁ PRIVATE TELEPHONE SYSTEM

— ◁1 SWITCHBOARD

— ◁2 WALL PHONE

— ◁3 ANY OTHER ITEM ON SAME SYSTEM
USE NUMBERS AS REQUIRED

— ⬠ WATCHMAN SYSTEM DEVICE,
ANY TYPE

— ⬠1 CENTRAL STATION

— ⬠2 KEY STATION

— ⬠3 ANY OTHER ITEM ON SAME SYSTEM
USE NUMBERS AS REQUIRED

— ◁ SOUND SYSTEM, ANY TYPE

— ◁1 AMPLIFIER

— ◁2 MICROPHONE

— ◁3 INTERIOR SPEAKER

— ◁4 EXTERIOR SPEAKER

— ◁5 ANY OTHER ITEM ON SAME SYSTEM
USE NUMBERS AS REQUIRED

FIGURE 7-7   Electrical 4.

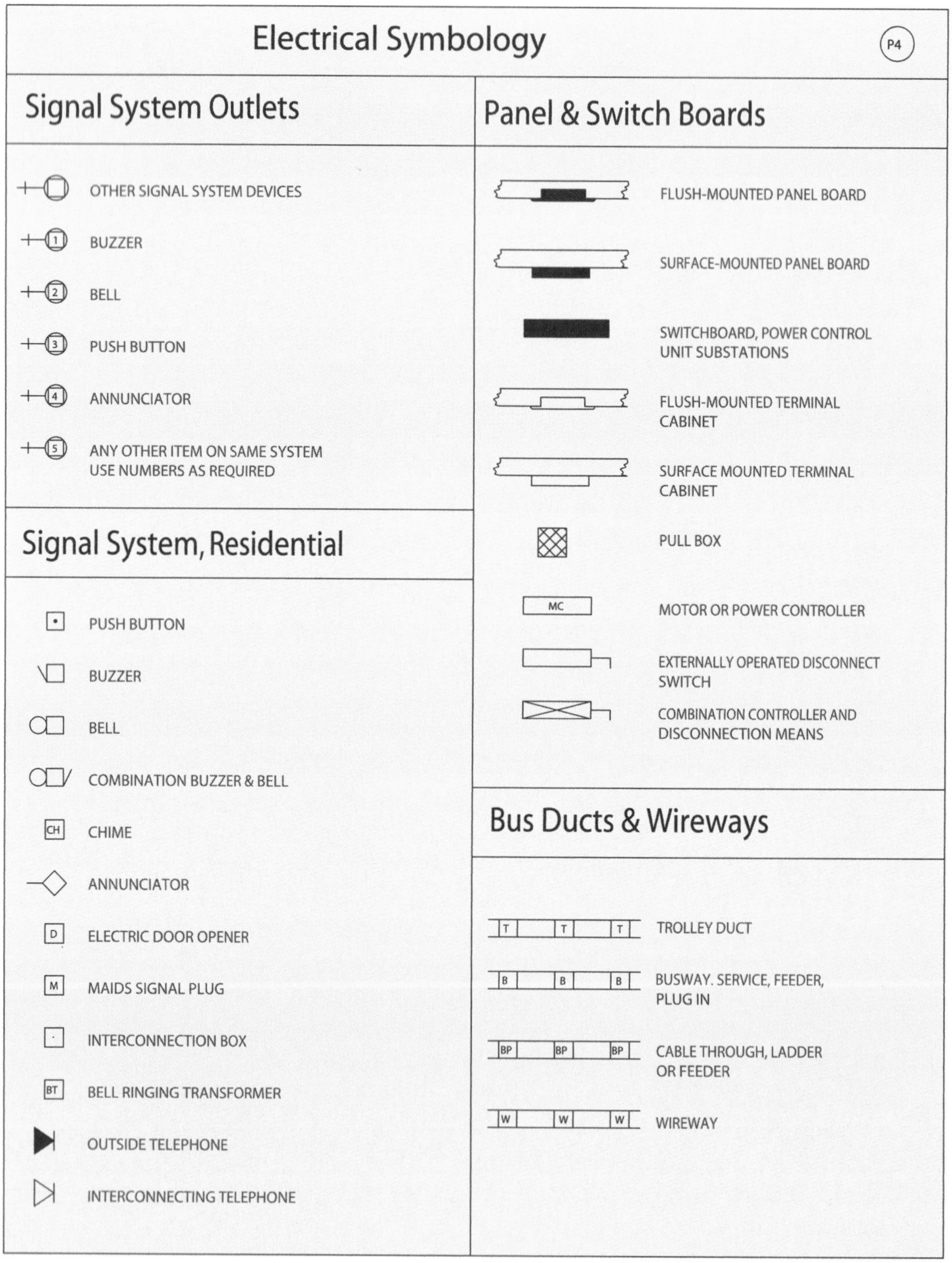

Electrical Symbology   (P4)

Signal System Outlets

⊢◯	OTHER SIGNAL SYSTEM DEVICES
⊢①	BUZZER
⊢②	BELL
⊢③	PUSH BUTTON
⊢④	ANNUNCIATOR
⊢⑤	ANY OTHER ITEM ON SAME SYSTEM USE NUMBERS AS REQUIRED

Signal System, Residential

⊡	PUSH BUTTON
◺	BUZZER
⊂▢	BELL
⊂▢◿	COMBINATION BUZZER & BELL
CH	CHIME
◇	ANNUNCIATOR
D	ELECTRIC DOOR OPENER
M	MAIDS SIGNAL PLUG
⊡	INTERCONNECTION BOX
BT	BELL RINGING TRANSFORMER
▶	OUTSIDE TELEPHONE
▷	INTERCONNECTING TELEPHONE

Panel & Switch Boards

	FLUSH-MOUNTED PANEL BOARD
	SURFACE-MOUNTED PANEL BOARD
	SWITCHBOARD, POWER CONTROL UNIT SUBSTATIONS
	FLUSH-MOUNTED TERMINAL CABINET
	SURFACE MOUNTED TERMINAL CABINET
▨	PULL BOX
MC	MOTOR OR POWER CONTROLLER
	EXTERNALLY OPERATED DISCONNECT SWITCH
⊠	COMBINATION CONTROLLER AND DISCONNECTION MEANS

Bus Ducts & Wireways

T T T	TROLLEY DUCT
B B B	BUSWAY. SERVICE, FEEDER, PLUG IN
BP BP BP	CABLE THROUGH, LADDER OR FEEDER
W W W	WIREWAY

7.3 Panelboards

Panelboards, control boards, and cabinets are shown in Figure 7–7. Commercial and industrial plans will often have letters and or numbers written next to the panels or cabinets to identify them. They are often identified by purpose, voltage, area of the building, or floor and are lettered according to the number of panels. Figure 7–8 shows an example of how a panel might be identified.

FIGURE 7–8   Example of how a panel may be identified.

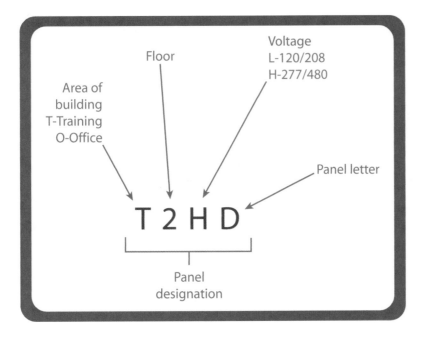

7.4 Wires and Raceways

Wire and raceway symbols are shown in Figure 7–9. A home run symbol will often have a series of letters and numbers written next to it (see Figure 7–10). The first series of letters and numbers will indicate the panelboard from which the circuits are fed. The numbers following the panelboard identifier indicate which circuits within that panel feed the conductors. Multiple arrows drawn on the end of a house run are used to indicate the number of circuits the home run contains. Hash marks are used to identify the number of conductors, as in Figure 7–9.

FIGURE 7-9   Electrical 5.

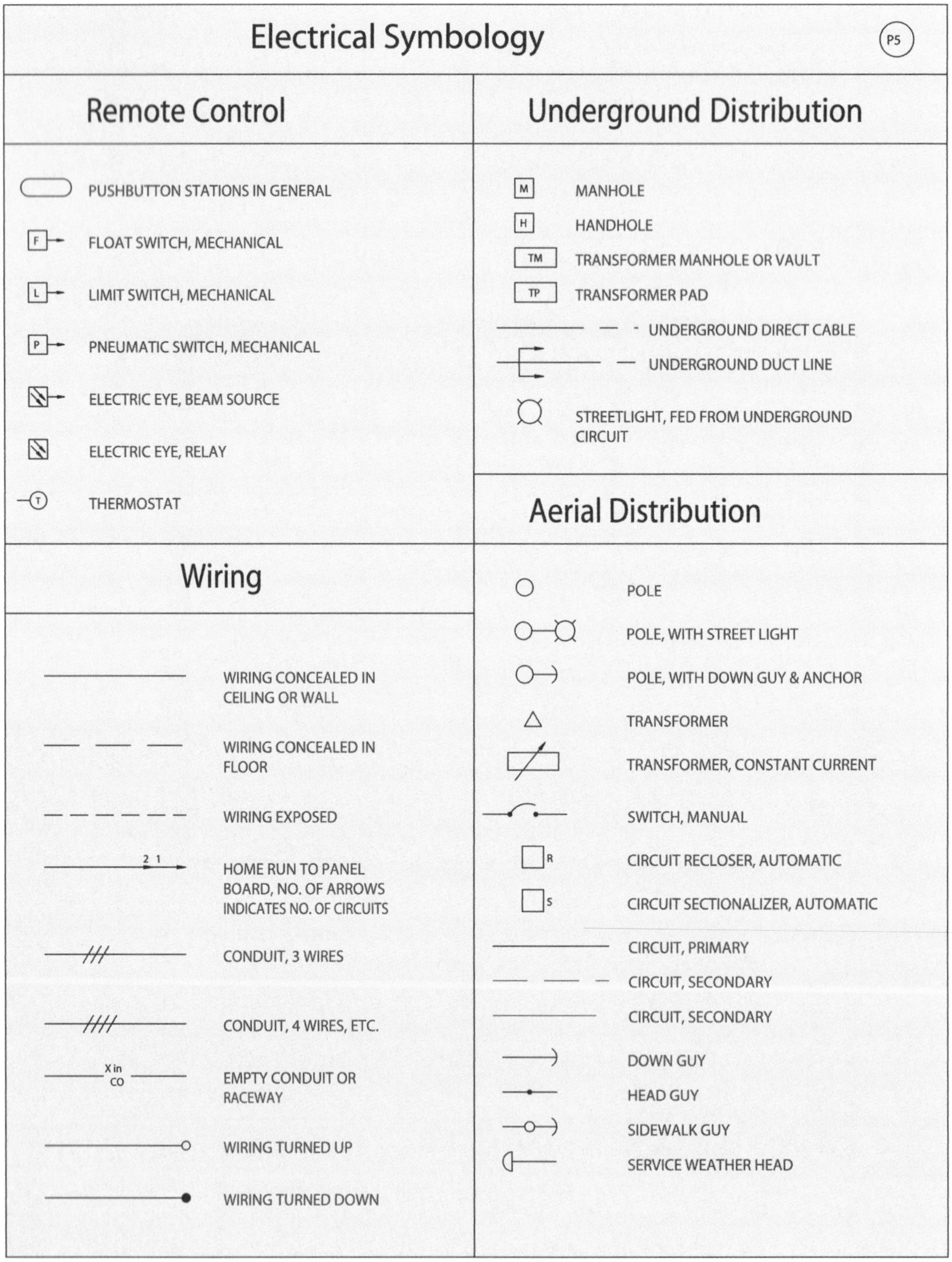

Electrical Symbology
(P5)

Remote Control

⬭	PUSHBUTTON STATIONS IN GENERAL
F	FLOAT SWITCH, MECHANICAL
L	LIMIT SWITCH, MECHANICAL
P	PNEUMATIC SWITCH, MECHANICAL
	ELECTRIC EYE, BEAM SOURCE
	ELECTRIC EYE, RELAY
T	THERMOSTAT

Wiring

————	WIRING CONCEALED IN CEILING OR WALL
— — — —	WIRING CONCEALED IN FLOOR
- - - - -	WIRING EXPOSED
2 1	HOME RUN TO PANEL BOARD, NO. OF ARROWS INDICATES NO. OF CIRCUITS
—///—	CONDUIT, 3 WIRES
—////—	CONDUIT, 4 WIRES, ETC.
X in CO	EMPTY CONDUIT OR RACEWAY
———o	WIRING TURNED UP
———●	WIRING TURNED DOWN

Underground Distribution

M	MANHOLE
H	HANDHOLE
TM	TRANSFORMER MANHOLE OR VAULT
TP	TRANSFORMER PAD
— — -	UNDERGROUND DIRECT CABLE
— —	UNDERGROUND DUCT LINE
	STREETLIGHT, FED FROM UNDERGROUND CIRCUIT

Aerial Distribution

○	POLE
○—○	POLE, WITH STREET LIGHT
○—→	POLE, WITH DOWN GUY & ANCHOR
△	TRANSFORMER
	TRANSFORMER, CONSTANT CURRENT
	SWITCH, MANUAL
R	CIRCUIT RECLOSER, AUTOMATIC
S	CIRCUIT SECTIONALIZER, AUTOMATIC
————	CIRCUIT, PRIMARY
— — —	CIRCUIT, SECONDARY
	CIRCUIT, SECONDARY
—→	DOWN GUY
•——	HEAD GUY
—○—→	SIDEWALK GUY
⊏—	SERVICE WEATHER HEAD

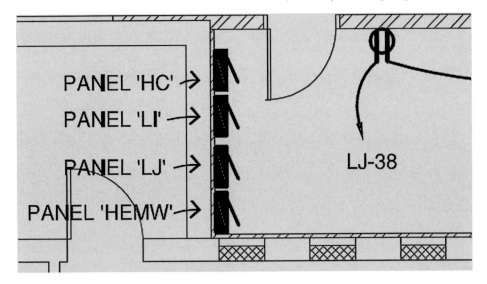

FIGURE 7-10 A home run will typically identify the panel that it is fed from, and which circuits it connects to. This home run is fed from Panel LJ and is connected to circuit 38. (Courtesy of Ulteig Engineers.)

ELECTRONIC SYMBOLOGY

Electronic symbology is graphic representation of electronic components and devices as used in circuit repair and design. There are some shared symbols between electricity and electronics. There is, however, a great deal of difference between the two fields. As an electrician you may need experience in both fields. There are far more symbols than those covered in this publication. The field of electronics is developing new devices and symbols every day. To keep current on electronic symbology, it is necessary to check print symbol libraries or read current electronic publications.

The following pages depict and explain some of the common ANSI graphic standard symbols for electronic devices commonly used by electrical workers. Note that there are hundreds of pages of specialized and standard electronic symbology, and that Figure 7–11 is only a small sample of common symbols.

ABBREVIATIONS

Abbreviations are used on all types of construction drawings to reduce clutter. Although most engineers and designers will try to use abbreviations that are easily understood, there is not particular set of abbreviations that must be used. Most construction drawings will have a listing of the abbreviations used on the drawings. Appendix D of this book outlines some of the commonly used abbreviations.

FIGURE 7-11   Electronic 1.

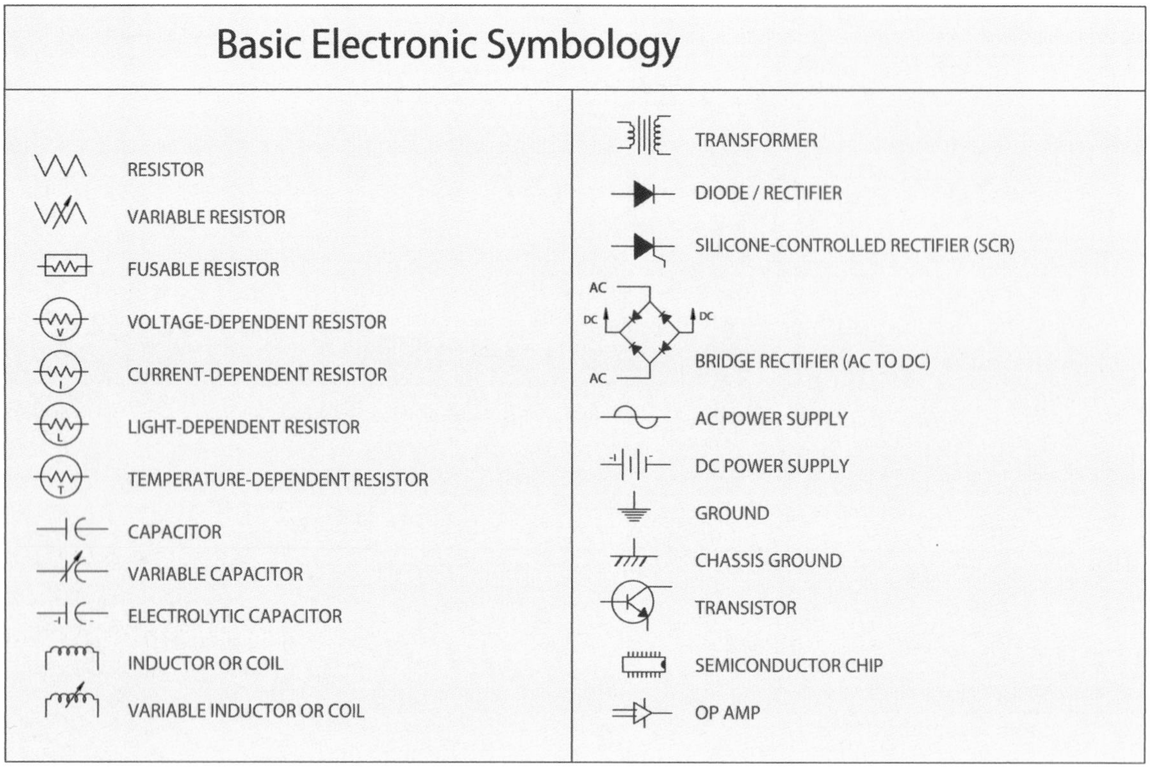

Basic Electronic Symbology

ᐱᐱ RESISTOR	ᗱᕃᕃᕃ TRANSFORMER
Vᗯᐱᐱ VARIABLE RESISTOR	▶▶ DIODE / RECTIFIER
ᕮᗯᗰᔕ FUSABLE RESISTOR	▶▶ SILICONE-CONTROLLED RECTIFIER (SCR)
⊕ VOLTAGE-DEPENDENT RESISTOR	AC
⊕ CURRENT-DEPENDENT RESISTOR	DC ◇ DC
⊕ LIGHT-DEPENDENT RESISTOR	AC BRIDGE RECTIFIER (AC TO DC)
⊕ TEMPERATURE-DEPENDENT RESISTOR	∿ AC POWER SUPPLY
⊣⊢ CAPACITOR	⊣⊦⊩ DC POWER SUPPLY
⊣⊬⊢ VARIABLE CAPACITOR	⏚ GROUND
⊣⊢ ELECTROLYTIC CAPACITOR	⏦⏦⏦ CHASSIS GROUND
ᗰᗰᗰ INDUCTOR OR COIL	⊗ TRANSISTOR
ᗰᗰᗰ VARIABLE INDUCTOR OR COIL	⊟ SEMICONDUCTOR CHIP
	⊳▷ OP AMP

SUMMARY

Symbols are simple graphic or alphanumeric representations of an object or device.

Symbols vary by designer, and therefore a tradesperson should always refer to the legend to verify the correct symbol.

Construction drawings typically have a legend defining the symbols and abbreviations used in the drawings.

REVIEW QUESTIONS

1. What is the American organization responsible for standardizing symbols?
2. What is a symbol?
3. If a symbol is nonstandard, unknown, or new, where should the electrical worker find its meaning?
4. Must the symbol look like the device?
5. Draw the symbol for a duplex receptacle.
6. Draw the symbol for a door switch.
7. Draw the symbol for a surface-mounted panelboard.
8. Draw the symbol for a capacitor.
9. Draw the symbol for a four-way switch.
10. Draw the symbol for a lighting outlet with a pull switch.
11. Draw the symbol for a house run with three circuits.
12. Draw the symbol for a residential chime.
13. What does the abbreviation TS stand for?
14. What does the abbreviation DSC stand for?
15. What does the abbreviation LED stand for?

8

Mechanical Symbology

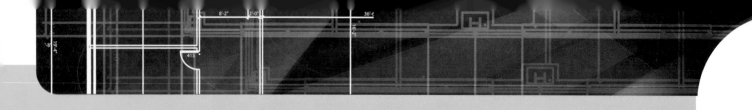

OVERVIEW

This chapter builds upon the information presented in Chapter 7, covering mechanical symbology consisting of heating ventilating and air conditioning (HVAC), plumbing, and piping symbols. It is important for the electrical worker to understand these symbols to avoid making electrical runs where ducts and pipes are to go. On a commercial or industrial job, large pipes and HVAC duct work usually take precedence over electrical runs due to the fact that it is easier to reroute an electrical conduit than it is an HVAC duct or plumbing pipe. It is important not to make print-reading mistakes that would require costly rework.

OBJECTIVES

After completing this chapter, you should be able to:

• Explain why the electrical worker needs to understand mechanical symbology.

• Recognize and interpret various common plumbing, piping, and HVAC symbols.

PLUMBING SYMBOLS

Plumbing symbols are used on residential, commercial, and industrial prints. Plumbing symbols usually represent the device in a somewhat pictorial way. These symbols may appear in both plan and elevation views (Figure 8–1).

FIGURE 8–1 Elevation and plan view of a toilet symbol.

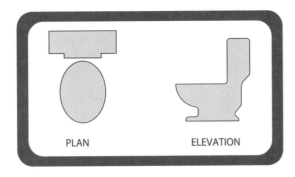

Most architects and engineers use the ANSI standard graphic symbols or minor modifications of these standards for plumbing. As mentioned previously, if the print symbols used are nonstandard a print legend should be supplied. Examples of these standard symbols are shown in Figures 8–2 and 8–3 (see pages 126 and 127).

FIGURE 8–2   Plumbing 1.

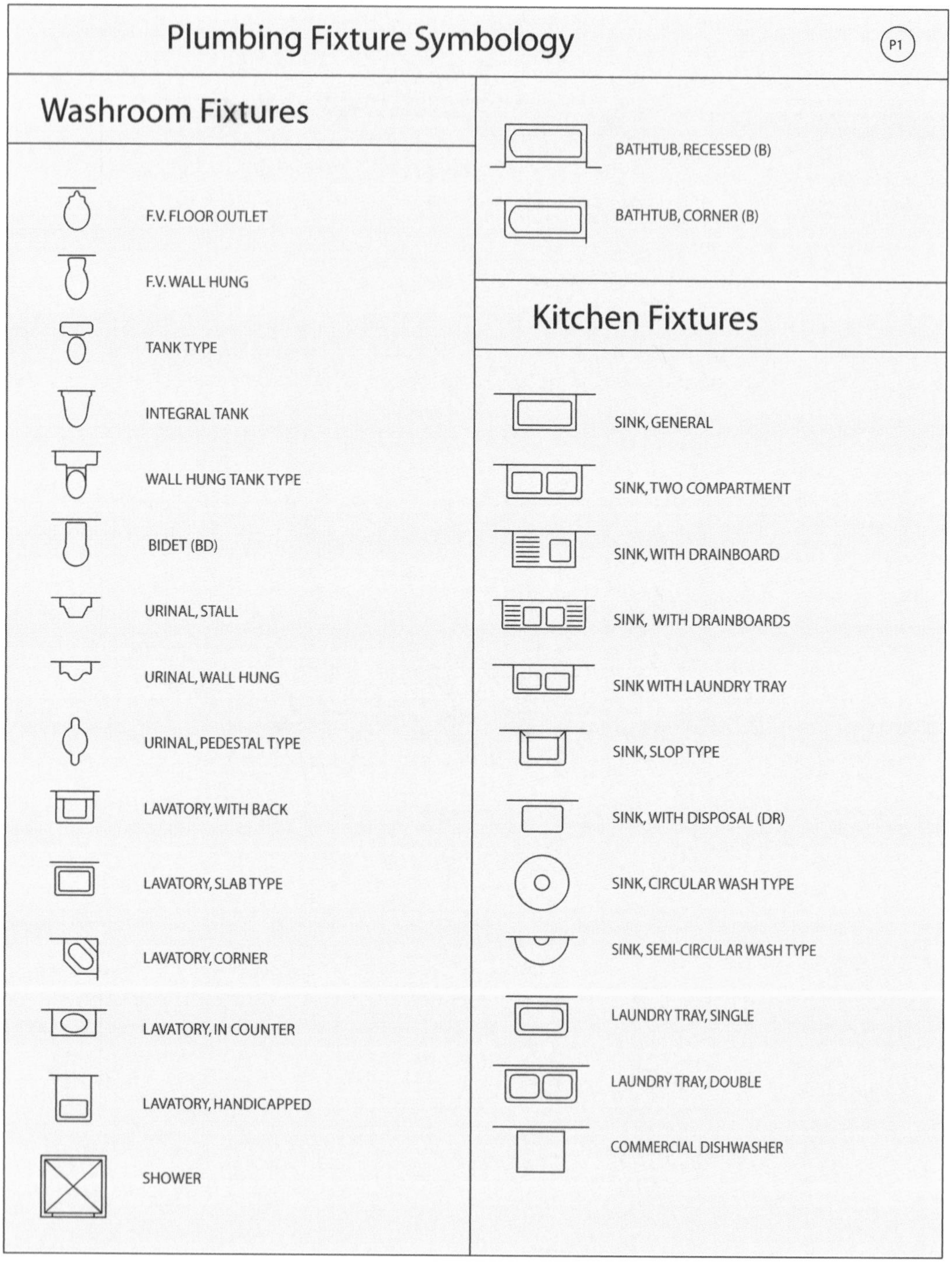

Plumbing Fixture Symbology   (P1)

Washroom Fixtures

F.V. FLOOR OUTLET

F.V. WALL HUNG

TANK TYPE

INTEGRAL TANK

WALL HUNG TANK TYPE

BIDET (BD)

URINAL, STALL

URINAL, WALL HUNG

URINAL, PEDESTAL TYPE

LAVATORY, WITH BACK

LAVATORY, SLAB TYPE

LAVATORY, CORNER

LAVATORY, IN COUNTER

LAVATORY, HANDICAPPED

SHOWER

BATHTUB, RECESSED (B)

BATHTUB, CORNER (B)

Kitchen Fixtures

SINK, GENERAL

SINK, TWO COMPARTMENT

SINK, WITH DRAINBOARD

SINK, WITH DRAINBOARDS

SINK WITH LAUNDRY TRAY

SINK, SLOP TYPE

SINK, WITH DISPOSAL (DR)

SINK, CIRCULAR WASH TYPE

SINK, SEMI-CIRCULAR WASH TYPE

LAUNDRY TRAY, SINGLE

LAUNDRY TRAY, DOUBLE

COMMERCIAL DISHWASHER

FIGURE 8-3   Plumbing 2.

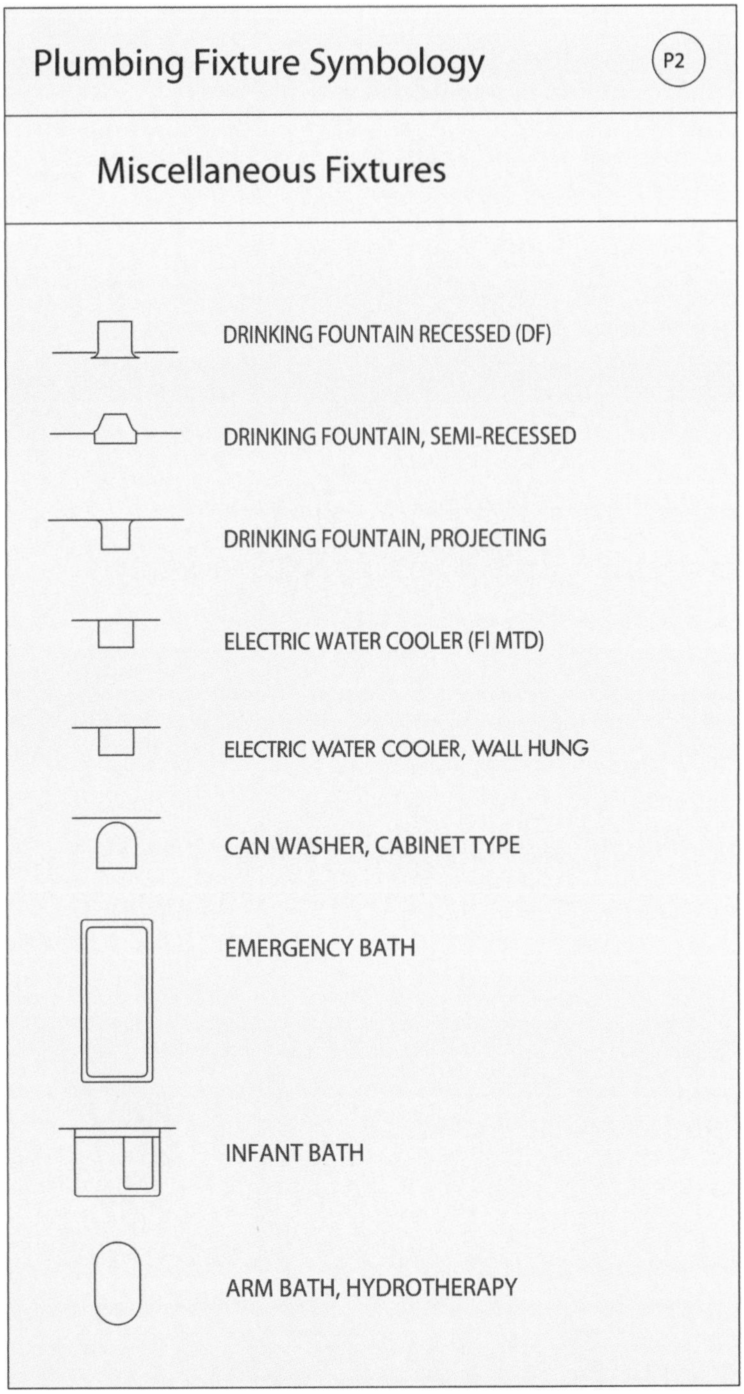

Plumbing Fixture Symbology   (P2)

Miscellaneous Fixtures

DRINKING FOUNTAIN RECESSED (DF)

DRINKING FOUNTAIN, SEMI-RECESSED

DRINKING FOUNTAIN, PROJECTING

ELECTRIC WATER COOLER (FI MTD)

ELECTRIC WATER COOLER, WALL HUNG

CAN WASHER, CABINET TYPE

EMERGENCY BATH

INFANT BATH

ARM BATH, HYDROTHERAPY

PIPE SYMBOLS

Pipe symbols depict fittings, valves, and other devices. Figures 8–4 through 8–6 (see pages 128–130) show examples of these symbols. Note that pipe symbol libraries also have different line types for the different piping runs and materials being piped (Figure 8–6).

Piping runs are often shown in isometric drawing form. These piping isometrics help indicate the relationships of the various fixtures and pipes in a semipictorial fashion (Figure 8–7, page 131).

FIGURE 8–4   Piping 1.

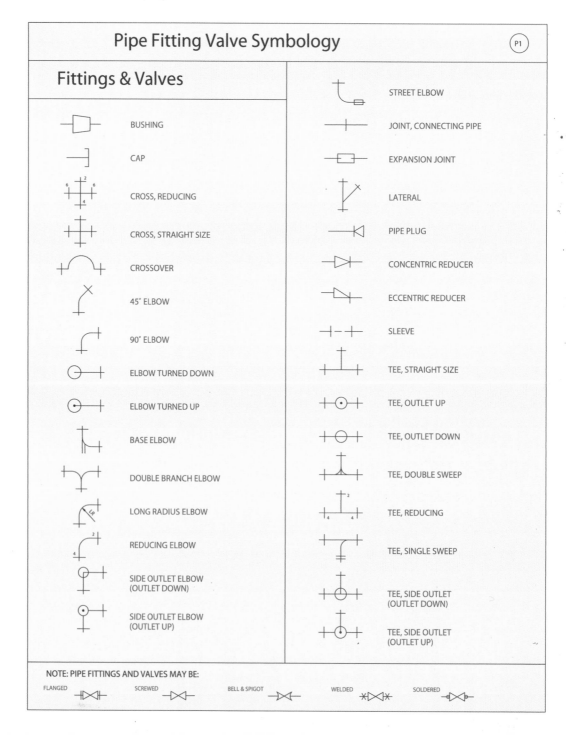

FIGURE 8-5   Piping 2.

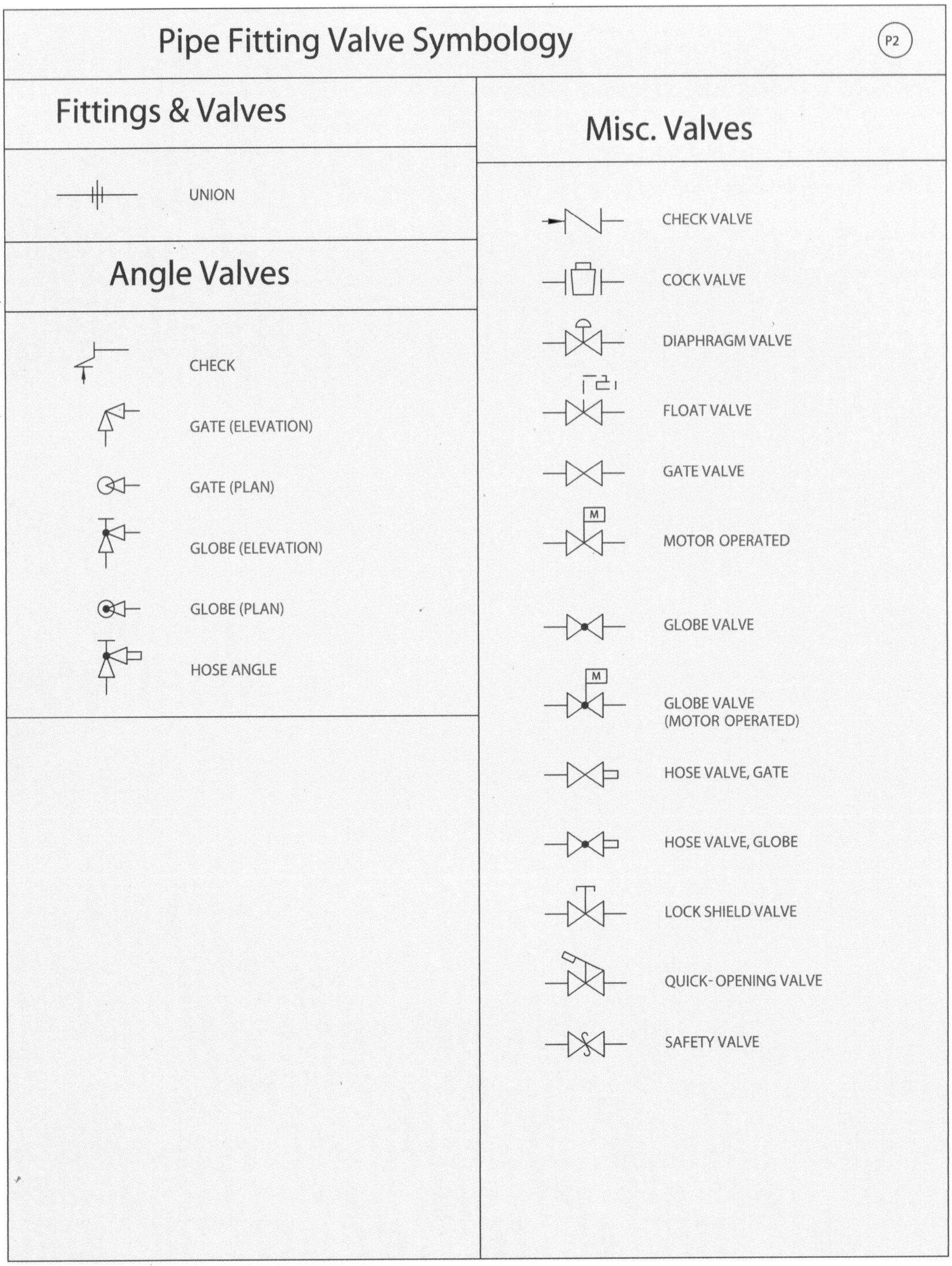

Pipe Fitting Valve Symbology   (P2)

Fittings & Valves	Misc. Valves

Fittings & Valves

UNION

Angle Valves

CHECK

GATE (ELEVATION)

GATE (PLAN)

GLOBE (ELEVATION)

GLOBE (PLAN)

HOSE ANGLE

Misc. Valves

CHECK VALVE

COCK VALVE

DIAPHRAGM VALVE

FLOAT VALVE

GATE VALVE

MOTOR OPERATED

GLOBE VALVE

GLOBE VALVE (MOTOR OPERATED)

HOSE VALVE, GATE

HOSE VALVE, GLOBE

LOCK SHIELD VALVE

QUICK-OPENING VALVE

SAFETY VALVE

FIGURE 8-6   Piping 3.

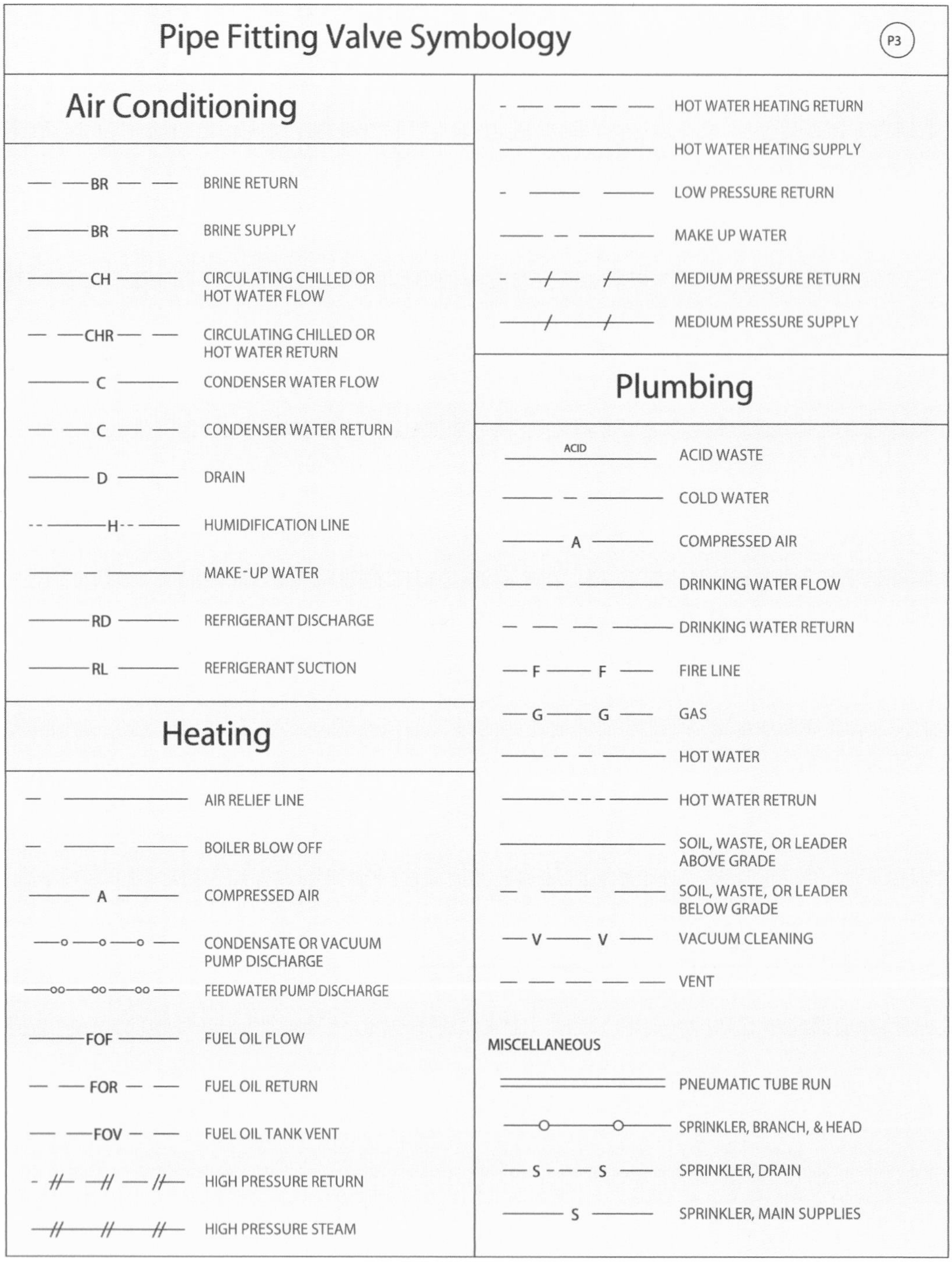

Pipe Fitting Valve Symbology   (P3)

Air Conditioning

— —BR — —	BRINE RETURN
———BR———	BRINE SUPPLY
———CH———	CIRCULATING CHILLED OR HOT WATER FLOW
— —CHR— —	CIRCULATING CHILLED OR HOT WATER RETURN
——— C ———	CONDENSER WATER FLOW
— — C — —	CONDENSER WATER RETURN
——— D ———	DRAIN
-- ———H-- —	HUMIDIFICATION LINE
——— — — ———	MAKE-UP WATER
———RD———	REFRIGERANT DISCHARGE
———RL———	REFRIGERANT SUCTION

Heating

— ———————	AIR RELIEF LINE
— ————— ———	BOILER BLOW OFF
——— A ———	COMPRESSED AIR
—o—o—o—	CONDENSATE OR VACUUM PUMP DISCHARGE
—oo—oo—oo—	FEEDWATER PUMP DISCHARGE
———FOF———	FUEL OIL FLOW
— — FOR — —	FUEL OIL RETURN
———FOV— —	FUEL OIL TANK VENT
- #— —#—#—	HIGH PRESSURE RETURN
—#——#——#—	HIGH PRESSURE STEAM

— · — — · —	HOT WATER HEATING RETURN
———————	HOT WATER HEATING SUPPLY
- ——— ———	LOW PRESSURE RETURN
——— — — ———	MAKE UP WATER
- —/— —/—	MEDIUM PRESSURE RETURN
———/———/—	MEDIUM PRESSURE SUPPLY

Plumbing

———ACID———	ACID WASTE
——— — — ———	COLD WATER
——— A ———	COMPRESSED AIR
——— — ———	DRINKING WATER FLOW
— — — ———	DRINKING WATER RETURN
—F——F—	FIRE LINE
—G——G—	GAS
——— — — ———	HOT WATER
——— - - - ———	HOT WATER RETRUN
———————	SOIL, WASTE, OR LEADER ABOVE GRADE
— ——— ———	SOIL, WASTE, OR LEADER BELOW GRADE
—V——V—	VACUUM CLEANING
— — — — ———	VENT

MISCELLANEOUS

═══════	PNEUMATIC TUBE RUN
———o———o———	SPRINKLER, BRANCH, & HEAD
—S - - - S———	SPRINKLER, DRAIN
——— S ———	SPRINKLER, MAIN SUPPLIES

FIGURE 8–7   Piping isometric.

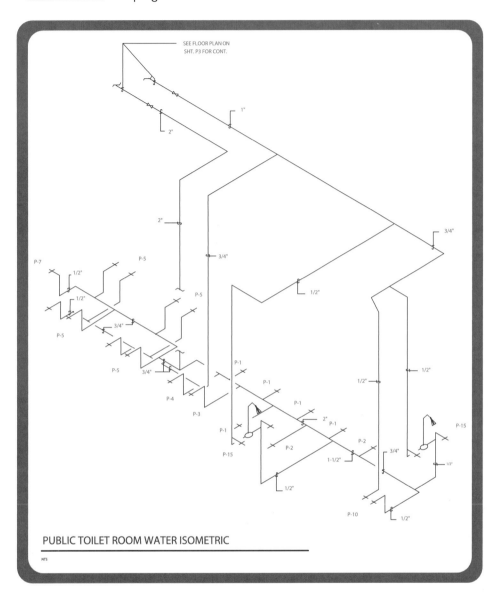

PUBLIC TOILET ROOM WATER ISOMETRIC

NTS

HVAC SYMBOLS

The ANSI symbols for HVAC are divided into three areas. These areas are heating, ventilating, and air conditioning—as indicated in Figures 8–8 through 8–10, respectively (see pages 132–134). An electrician will often have to make an electrical connection to HVAC equipment and devices. To properly plan the electrical installation, the electrician must be able to recognize the mechanical symbols.

FIGURE 8–8 HVAC 1.

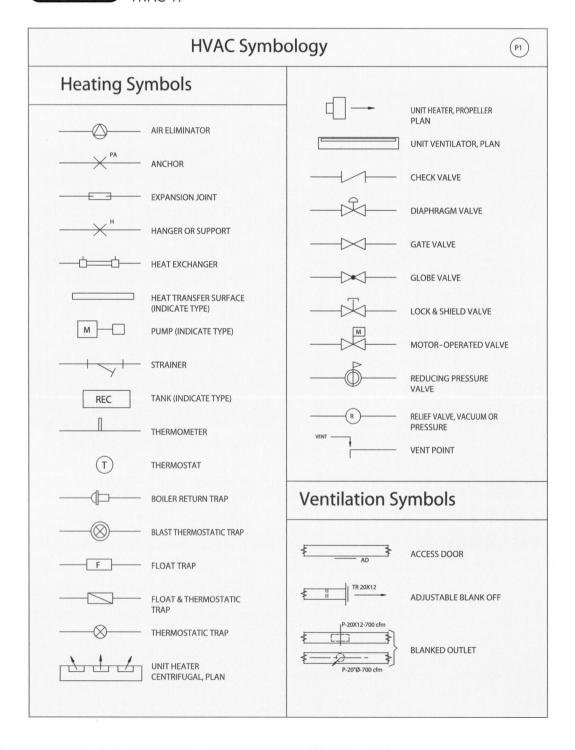

FIGURE 8-9   HVAC 2.

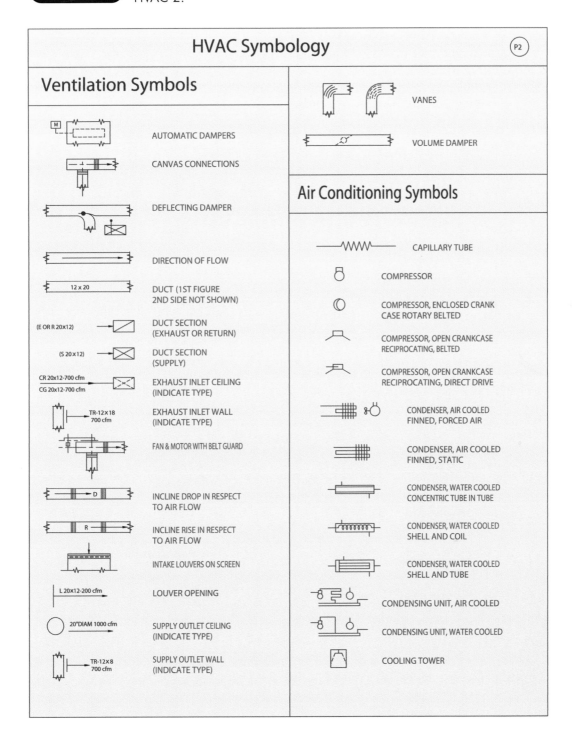

HVAC Symbology   (P2)

Ventilation Symbols

AUTOMATIC DAMPERS

CANVAS CONNECTIONS

DEFLECTING DAMPER

DIRECTION OF FLOW

12 x 20   DUCT (1ST FIGURE 2ND SIDE NOT SHOWN)

(E OR R 20x12)   DUCT SECTION (EXHAUST OR RETURN)

(S 20x12)   DUCT SECTION (SUPPLY)

CR 20x12-700 cfm
CG 20x12-700 cfm   EXHAUST INLET CEILING (INDICATE TYPE)

TR-12x18 700 cfm   EXHAUST INLET WALL (INDICATE TYPE)

FAN & MOTOR WITH BELT GUARD

D   INCLINE DROP IN RESPECT TO AIR FLOW

R   INCLINE RISE IN RESPECT TO AIR FLOW

INTAKE LOUVERS ON SCREEN

L 20x12-200 cfm   LOUVER OPENING

20"DIAM 1000 cfm   SUPPLY OUTLET CEILING (INDICATE TYPE)

TR-12x8 700 cfm   SUPPLY OUTLET WALL (INDICATE TYPE)

VANES

VOLUME DAMPER

Air Conditioning Symbols

CAPILLARY TUBE

COMPRESSOR

COMPRESSOR, ENCLOSED CRANK CASE ROTARY BELTED

COMPRESSOR, OPEN CRANKCASE RECIPROCATING, BELTED

COMPRESSOR, OPEN CRANKCASE RECIPROCATING, DIRECT DRIVE

CONDENSER, AIR COOLED FINNED, FORCED AIR

CONDENSER, AIR COOLED FINNED, STATIC

CONDENSER, WATER COOLED CONCENTRIC TUBE IN TUBE

CONDENSER, WATER COOLED SHELL AND COIL

CONDENSER, WATER COOLED SHELL AND TUBE

CONDENSING UNIT, AIR COOLED

CONDENSING UNIT, WATER COOLED

COOLING TOWER

FIGURE 8–10   HVAC 3.

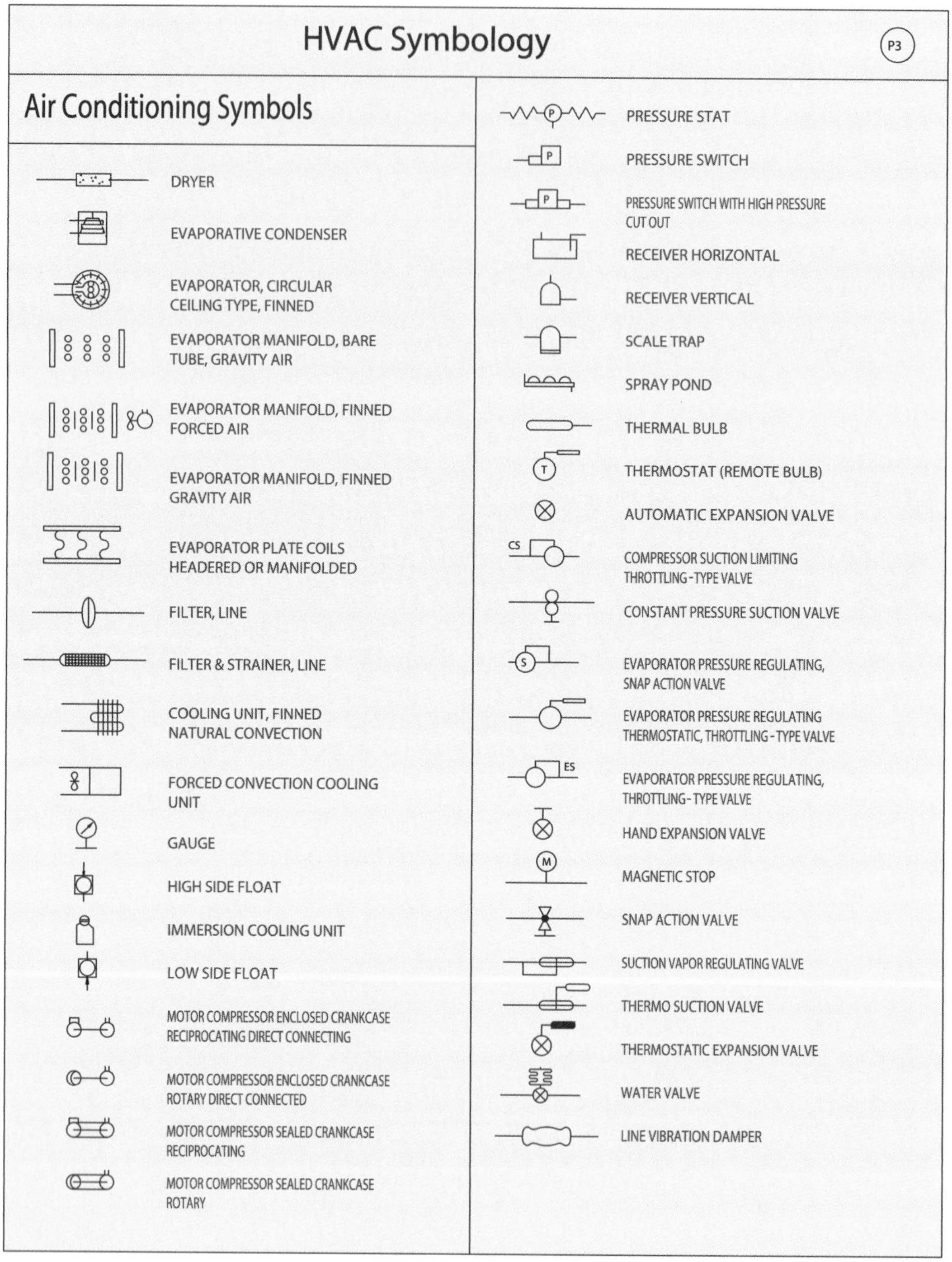

HVAC Symbology   (P3)

Air Conditioning Symbols

	DRYER
	EVAPORATIVE CONDENSER
	EVAPORATOR, CIRCULAR CEILING TYPE, FINNED
	EVAPORATOR MANIFOLD, BARE TUBE, GRAVITY AIR
	EVAPORATOR MANIFOLD, FINNED FORCED AIR
	EVAPORATOR MANIFOLD, FINNED GRAVITY AIR
	EVAPORATOR PLATE COILS HEADERED OR MANIFOLDED
	FILTER, LINE
	FILTER & STRAINER, LINE
	COOLING UNIT, FINNED NATURAL CONVECTION
	FORCED CONVECTION COOLING UNIT
	GAUGE
	HIGH SIDE FLOAT
	IMMERSION COOLING UNIT
	LOW SIDE FLOAT
	MOTOR COMPRESSOR ENCLOSED CRANKCASE RECIPROCATING DIRECT CONNECTING
	MOTOR COMPRESSOR ENCLOSED CRANKCASE ROTARY DIRECT CONNECTED
	MOTOR COMPRESSOR SEALED CRANKCASE RECIPROCATING
	MOTOR COMPRESSOR SEALED CRANKCASE ROTARY

	PRESSURE STAT
	PRESSURE SWITCH
	PRESSURE SWITCH WITH HIGH PRESSURE CUT OUT
	RECEIVER HORIZONTAL
	RECEIVER VERTICAL
	SCALE TRAP
	SPRAY POND
	THERMAL BULB
	THERMOSTAT (REMOTE BULB)
	AUTOMATIC EXPANSION VALVE
	COMPRESSOR SUCTION LIMITING THROTTLING-TYPE VALVE
	CONSTANT PRESSURE SUCTION VALVE
	EVAPORATOR PRESSURE REGULATING, SNAP ACTION VALVE
	EVAPORATOR PRESSURE REGULATING THERMOSTATIC, THROTTLING-TYPE VALVE
	EVAPORATOR PRESSURE REGULATING, THROTTLING-TYPE VALVE
	HAND EXPANSION VALVE
	MAGNETIC STOP
	SNAP ACTION VALVE
	SUCTION VAPOR REGULATING VALVE
	THERMO SUCTION VALVE
	THERMOSTATIC EXPANSION VALVE
	WATER VALVE
	LINE VIBRATION DAMPER

The graphic symbols for HVAC are used individually or in groups, as shown on the transformer pictured in Figure 8–11. The purpose for using and being able to read these mechanical symbols is to help simplify the process of reading blueprints by identifying mechanical devices in a clear and simplified manner. Reading and understanding these mechanical prints is invaluable in construction work for the electrical worker. Proper reading and interpreting of mechanical and electrical prints will save time and rework expense on the job.

FIGURE 8–11 Transformer.

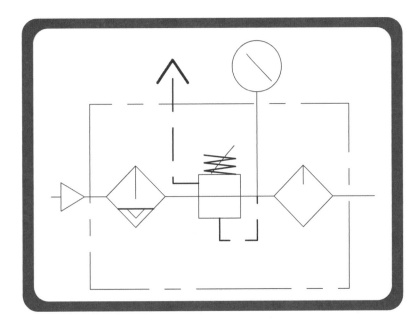

SUMMARY

To plan an electrical installation an electrician must be able to identify the mechanical equipment that requires an electrical connection.

It is important for an electrician to understand the mechanical symbols so as to not interfere with HVAC and plumbing equipment.

REVIEW QUESTIONS

1. Why is it important for the electrical worker to understand mechanical prints?

2. What are the three common types of mechanical prints and symbols?

3. Of the three mechanical symbols in Question 2, which has symbols that look most pictorially like the object they represent?

4. What is a piping isometric?

5. By reading the schematic on a symbolic drawing for pipe or HVAC prints, how can you tell the size of the duct?

NOTE: Study the symbols in this chapter and on the CD print sets to become more familiar with their uses.

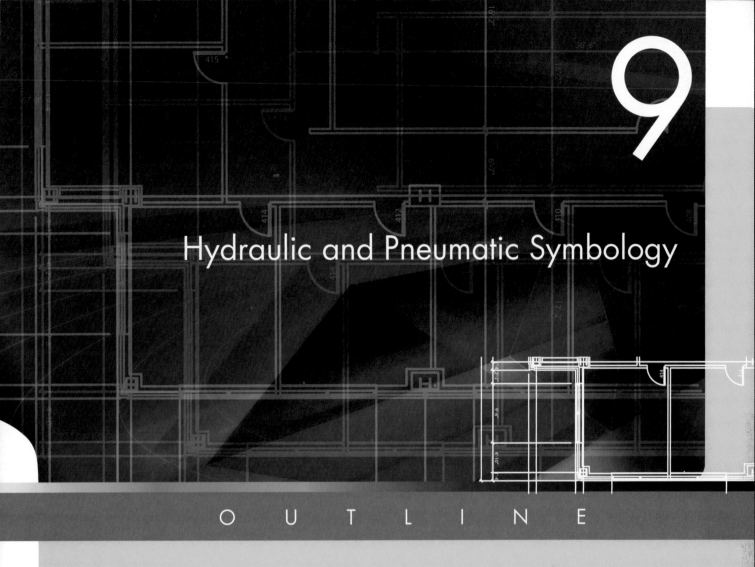

9

Hydraulic and Pneumatic Symbology

OUTLINE

OVERVIEW

This chapter is a continuation of the information presented in Chapters 7 and 8 and covers hydraulic and pneumatic drawings and symbols. In the area of electrical construction or maintenance, the electrical worker often comes into contact with hydraulic and pneumatic devices that are controlled by electronics. It is important for the electrical worker to be familiar with the appropriate symbolic representations of these devices.

OBJECTIVES

After completing this chapter, you should be able to:

• Recognize and identify common hydraulic and pneumatic symbols.

• Differentiate between hydraulic and pneumatic symbols.

HYDRAULIC AND PNEUMATIC SYMBOLOGY

The ANSI graphic standard symbols for hydraulics and pneumatics fall in the category of fluid power symbols. The basic difference between a hydraulic device and a pneumatic device is in the drawing of the direction arrow. If the device direction arrow is darkened, it is a hydraulic device. If the device arrow is outlined, it is a pneumatic device. It helps to remember that pneumatic devices operate with air.

Air is colorless. Therefore, the direction arrow is not colored (Figure 9–1). In fluid-power symbology there are device and line symbols. The line symbols are similar to those in electronics (Figure 9–2). Fluid-power devices can be divided into several areas: motors, pumps, controls, valves, and conditioners. Figures 9–3 through 9–6 depict these various devices.

FIGURE 9–1 Air versus hydraulic.

PNEUMATIC

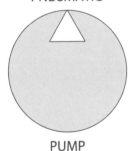

PUMP

HYDRAULIC

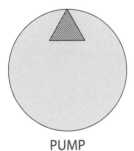

PUMP

FIGURE 9–2 Line symbols.

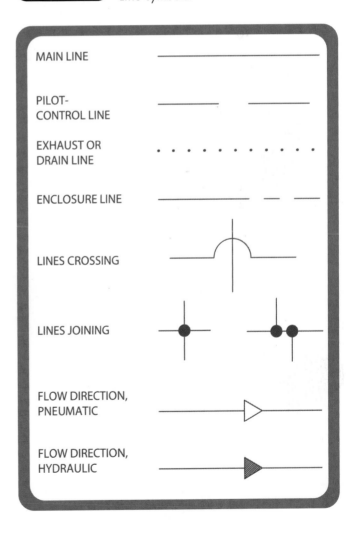

MAIN LINE

PILOT-CONTROL LINE

EXHAUST OR DRAIN LINE

ENCLOSURE LINE

LINES CROSSING

LINES JOINING

FLOW DIRECTION, PNEUMATIC

FLOW DIRECTION, HYDRAULIC

FIGURE 9-3   Fluid power 1.

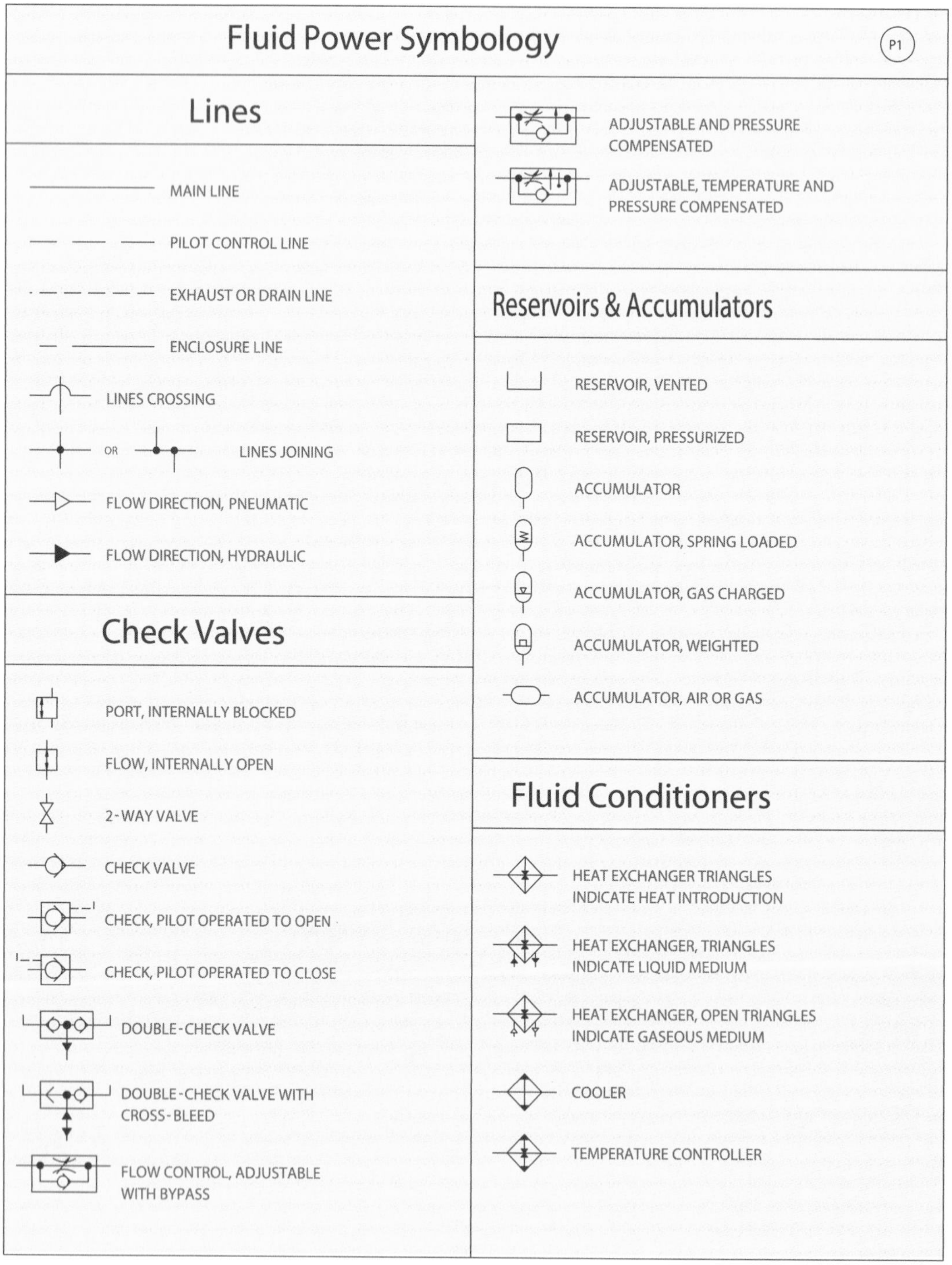

Fluid Power Symbology   (P1)

Lines

MAIN LINE

PILOT CONTROL LINE

EXHAUST OR DRAIN LINE

ENCLOSURE LINE

LINES CROSSING

OR   LINES JOINING

FLOW DIRECTION, PNEUMATIC

FLOW DIRECTION, HYDRAULIC

Check Valves

PORT, INTERNALLY BLOCKED

FLOW, INTERNALLY OPEN

2-WAY VALVE

CHECK VALVE

CHECK, PILOT OPERATED TO OPEN

CHECK, PILOT OPERATED TO CLOSE

DOUBLE-CHECK VALVE

DOUBLE-CHECK VALVE WITH CROSS-BLEED

FLOW CONTROL, ADJUSTABLE WITH BYPASS

ADJUSTABLE AND PRESSURE COMPENSATED

ADJUSTABLE, TEMPERATURE AND PRESSURE COMPENSATED

Reservoirs & Accumulators

RESERVOIR, VENTED

RESERVOIR, PRESSURIZED

ACCUMULATOR

ACCUMULATOR, SPRING LOADED

ACCUMULATOR, GAS CHARGED

ACCUMULATOR, WEIGHTED

ACCUMULATOR, AIR OR GAS

Fluid Conditioners

HEAT EXCHANGER TRIANGLES INDICATE HEAT INTRODUCTION

HEAT EXCHANGER, TRIANGLES INDICATE LIQUID MEDIUM

HEAT EXCHANGER, OPEN TRIANGLES INDICATE GASEOUS MEDIUM

COOLER

TEMPERATURE CONTROLLER

FIGURE 9-4   Fluid power 2.

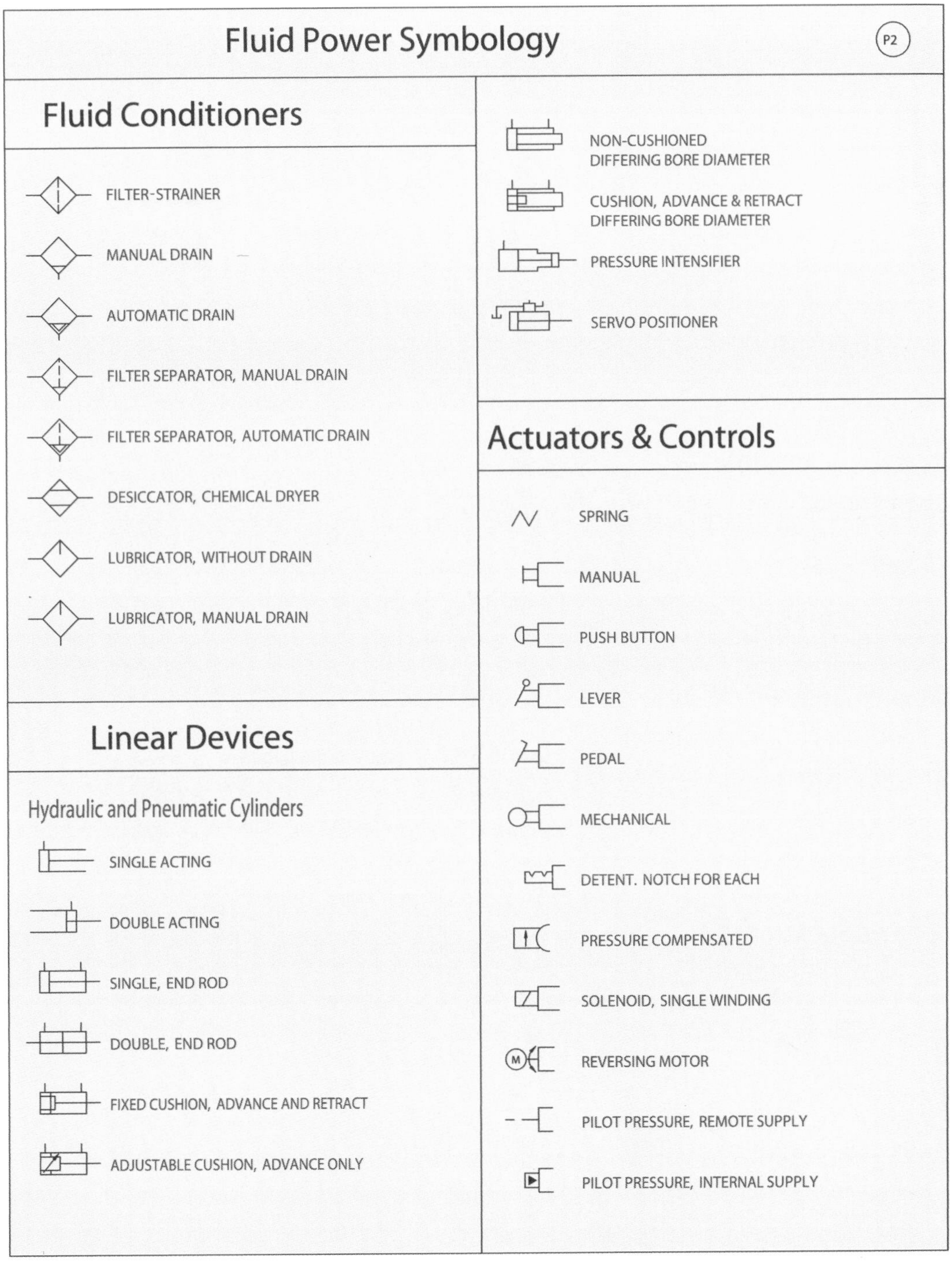

Fluid Power Symbology   (P2)

Fluid Conditioners

FILTER-STRAINER

MANUAL DRAIN

AUTOMATIC DRAIN

FILTER SEPARATOR, MANUAL DRAIN

FILTER SEPARATOR, AUTOMATIC DRAIN

DESICCATOR, CHEMICAL DRYER

LUBRICATOR, WITHOUT DRAIN

LUBRICATOR, MANUAL DRAIN

Linear Devices

Hydraulic and Pneumatic Cylinders

SINGLE ACTING

DOUBLE ACTING

SINGLE, END ROD

DOUBLE, END ROD

FIXED CUSHION, ADVANCE AND RETRACT

ADJUSTABLE CUSHION, ADVANCE ONLY

NON-CUSHIONED
DIFFERING BORE DIAMETER

CUSHION, ADVANCE & RETRACT
DIFFERING BORE DIAMETER

PRESSURE INTENSIFIER

SERVO POSITIONER

Actuators & Controls

SPRING

MANUAL

PUSH BUTTON

LEVER

PEDAL

MECHANICAL

DETENT. NOTCH FOR EACH

PRESSURE COMPENSATED

SOLENOID, SINGLE WINDING

REVERSING MOTOR

PILOT PRESSURE, REMOTE SUPPLY

PILOT PRESSURE, INTERNAL SUPPLY

FIGURE 9–5   Fluid power 3.

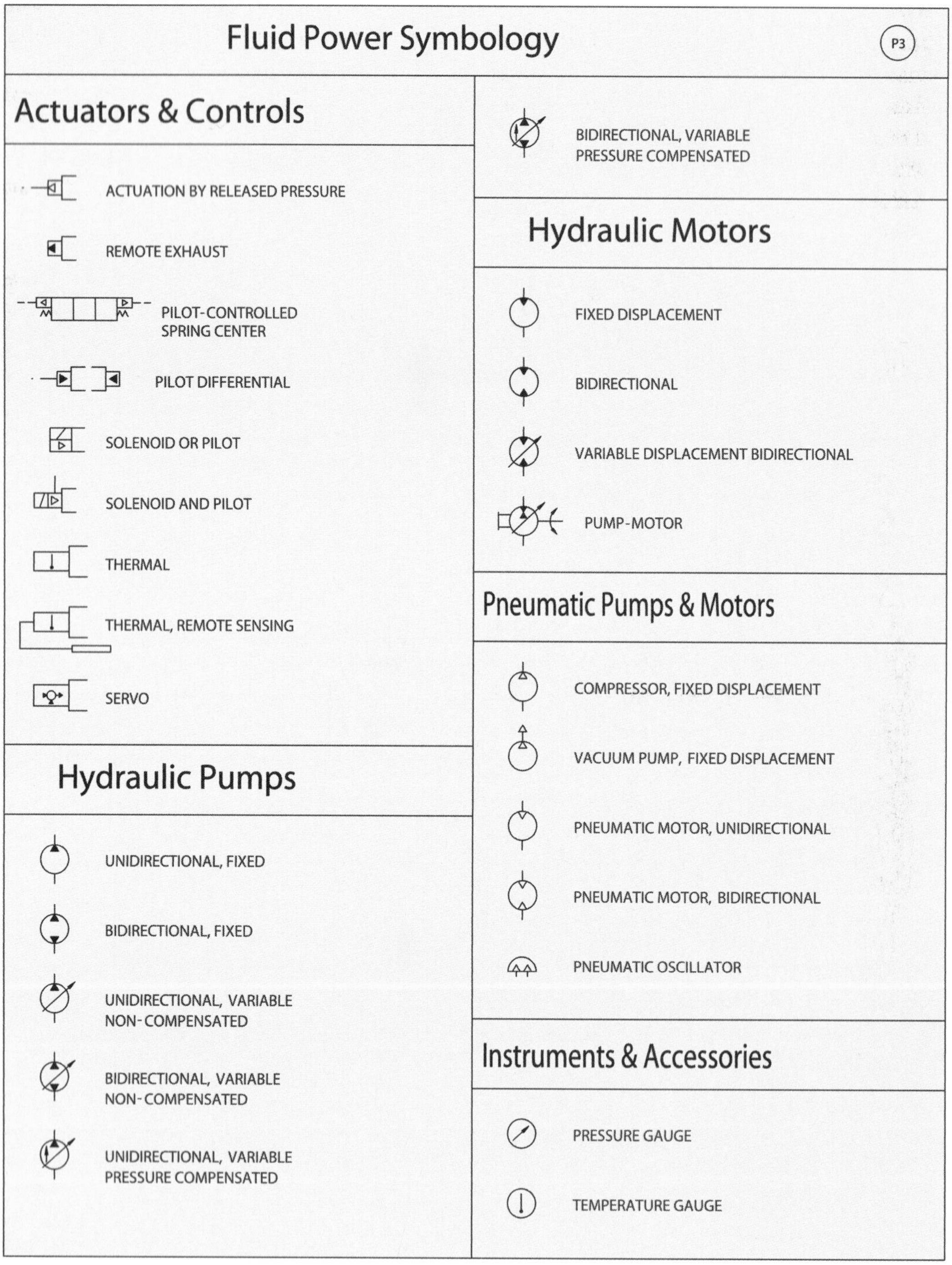

Fluid Power Symbology   (P3)

Actuators & Controls

ACTUATION BY RELEASED PRESSURE

REMOTE EXHAUST

PILOT-CONTROLLED SPRING CENTER

PILOT DIFFERENTIAL

SOLENOID OR PILOT

SOLENOID AND PILOT

THERMAL

THERMAL, REMOTE SENSING

SERVO

Hydraulic Pumps

UNIDIRECTIONAL, FIXED

BIDIRECTIONAL, FIXED

UNIDIRECTIONAL, VARIABLE NON-COMPENSATED

BIDIRECTIONAL, VARIABLE NON-COMPENSATED

UNIDIRECTIONAL, VARIABLE PRESSURE COMPENSATED

BIDIRECTIONAL, VARIABLE PRESSURE COMPENSATED

Hydraulic Motors

FIXED DISPLACEMENT

BIDIRECTIONAL

VARIABLE DISPLACEMENT BIDIRECTIONAL

PUMP-MOTOR

Pneumatic Pumps & Motors

COMPRESSOR, FIXED DISPLACEMENT

VACUUM PUMP, FIXED DISPLACEMENT

PNEUMATIC MOTOR, UNIDIRECTIONAL

PNEUMATIC MOTOR, BIDIRECTIONAL

PNEUMATIC OSCILLATOR

Instruments & Accessories

PRESSURE GAUGE

TEMPERATURE GAUGE

FIGURE 9-6   Fluid power 4.

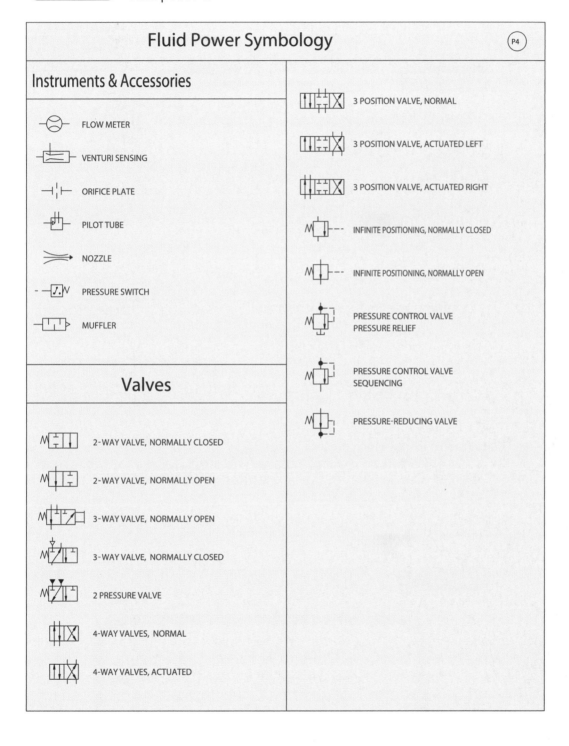

Fluid Power Symbology   (P4)

Instruments & Accessories

FLOW METER

VENTURI SENSING

ORIFICE PLATE

PILOT TUBE

NOZZLE

PRESSURE SWITCH

MUFFLER

Valves

2-WAY VALVE, NORMALLY CLOSED

2-WAY VALVE, NORMALLY OPEN

3-WAY VALVE, NORMALLY OPEN

3-WAY VALVE, NORMALLY CLOSED

2 PRESSURE VALVE

4-WAY VALVES, NORMAL

4-WAY VALVES, ACTUATED

3 POSITION VALVE, NORMAL

3 POSITION VALVE, ACTUATED LEFT

3 POSITION VALVE, ACTUATED RIGHT

INFINITE POSITIONING, NORMALLY CLOSED

INFINITE POSITIONING, NORMALLY OPEN

PRESSURE CONTROL VALVE
PRESSURE RELIEF

PRESSURE CONTROL VALVE
SEQUENCING

PRESSURE-REDUCING VALVE

Valves may be used individually or may be packaged into two or three envelope packages. These packages are actually separate sections of the valve (Figure 9–7). Each envelope may also contain from one to four ports, which are the actual openings in or out of the valve (Figure 9–8).

FIGURE 9–7 Valve envelopes.

ONE

TWO

THREE

FIGURE 9–8 Valve ports.

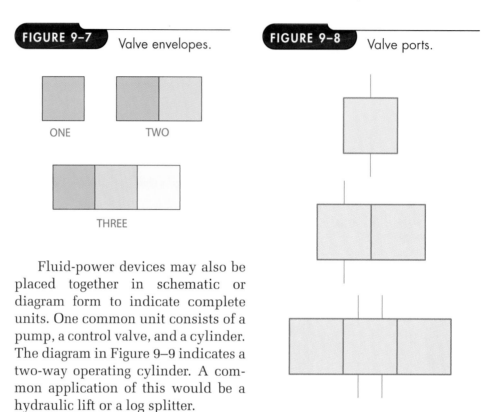

Fluid-power devices may also be placed together in schematic or diagram form to indicate complete units. One common unit consists of a pump, a control valve, and a cylinder. The diagram in Figure 9–9 indicates a two-way operating cylinder. A common application of this would be a hydraulic lift or a log splitter.

These fluid-power symbols are used with electrical and electronic control symbols in flow process diagrams. Electrical control of pneumatic devices is very common in industrial construction. In commercial construction, HVAC systems use both electrical and pneumatic controls. This is a major reason for understanding symbology for all of the other crafts.

FIGURE 9–9 Group devices.

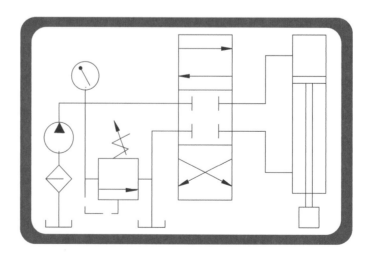

SUMMARY

• Electrical control of pneumatic devices is common in industrial construction, and it is therefore important for an electrician to understand the respective symbols.

REVIEW QUESTIONS

1. Why is it important for the electrical worker to understand hydraulic and pneumatic prints?

2. What is the basic difference between hydraulics and pneumatics as related to symbolic representation?

3. In fluid-power schematics and symbolic diagrams, what does a solid black line indicate?

4. In fluid-power schematics and symbolic diagrams, what does a dashed line indicate?

5. When are fluid-power and electrical symbols used on the same drawing?

NOTE: Study the symbols in this chapter and on the CD print sets to become more familiar with their uses.

10

Specialized Symbology

OUTLINE

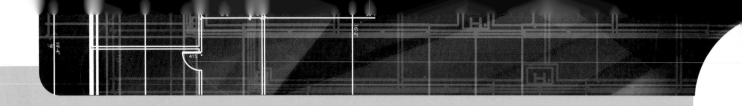

OVERVIEW

This is the final chapter dealing with symbology. Specialized symbology is covered in the areas of programmable logic control (PLC), ladder logic, motor control, process control, instrumentation, structured wiring, and telecommunications distribution. As the electrical worker moves into the areas of process control, structural wiring, and telecommunications, it is necessary to recognize and understand the ever-changing symbology that accompanies these existing and emerging fields.

OBJECTIVES

After completing this chapter, you should be able to:

• Recognize and identify common PLC and motor control symbols.

• Recognize and identify common process control and instrumentation symbols.

• Recognize and identify common structured wiring and telecommunications symbols, block diagrams, and pin diagrams.

INTRODUCTION

There are various other types of unusual or specialized symbology used in industry. Specialized symbols are those related to a small, specific area of drafting. They show devices used in that area depicted by symbols. Specialized symbology is used in the areas of programmable logic control (PLC) and motor control. Flow process symbols are also specialized symbols that incorporate some additional standard symbols. Burglar alarms use some standard electrical symbols and some specialized symbols. Another area with specialized symbology is closed-loop and programmed power distribution.

Ladder and power diagrams are more abstract than real. The diagrams are schematic in form. Schematic form shows the hookup relation of the connectors but not necessarily the size or actual location of the devices (Figure 10–1).

FIGURE 10–1   Logic and power.

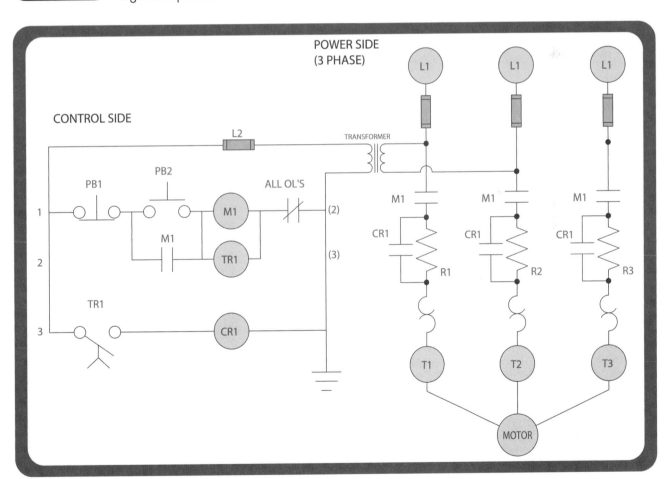

PROGRAMMABLE LOGIC CONTROL AND MOTOR CONTROL SYMBOLS

There are a variety of graphic symbols used in PLC and motor connection diagrams. As in other tasks, the engineer may use symbols that come from the ANSI standards. If other symbols are used, a symbol library or key should be included on the print. It is important to have a working knowledge of these symbols, which are shown in Figures 10–2 and 10–3.

When the term PLC comes up, it is also common to hear the terms *ladder logic* and *ladder logic diagram.* Previously, industry had used a pictorial representation for motor control circuits. This type of representation is appropriate for simple hookups such as the simple motor starter shown in Figure 10–4. The problem arises as circuits and controls become more and more complex and pictorial representation no longer suffices. The ladder logic diagram is used to simplify and lay out complex control circuits in a logical order. Logic diagrams are also used to program PLCs.

FIGURE 10–2   PLC 1.

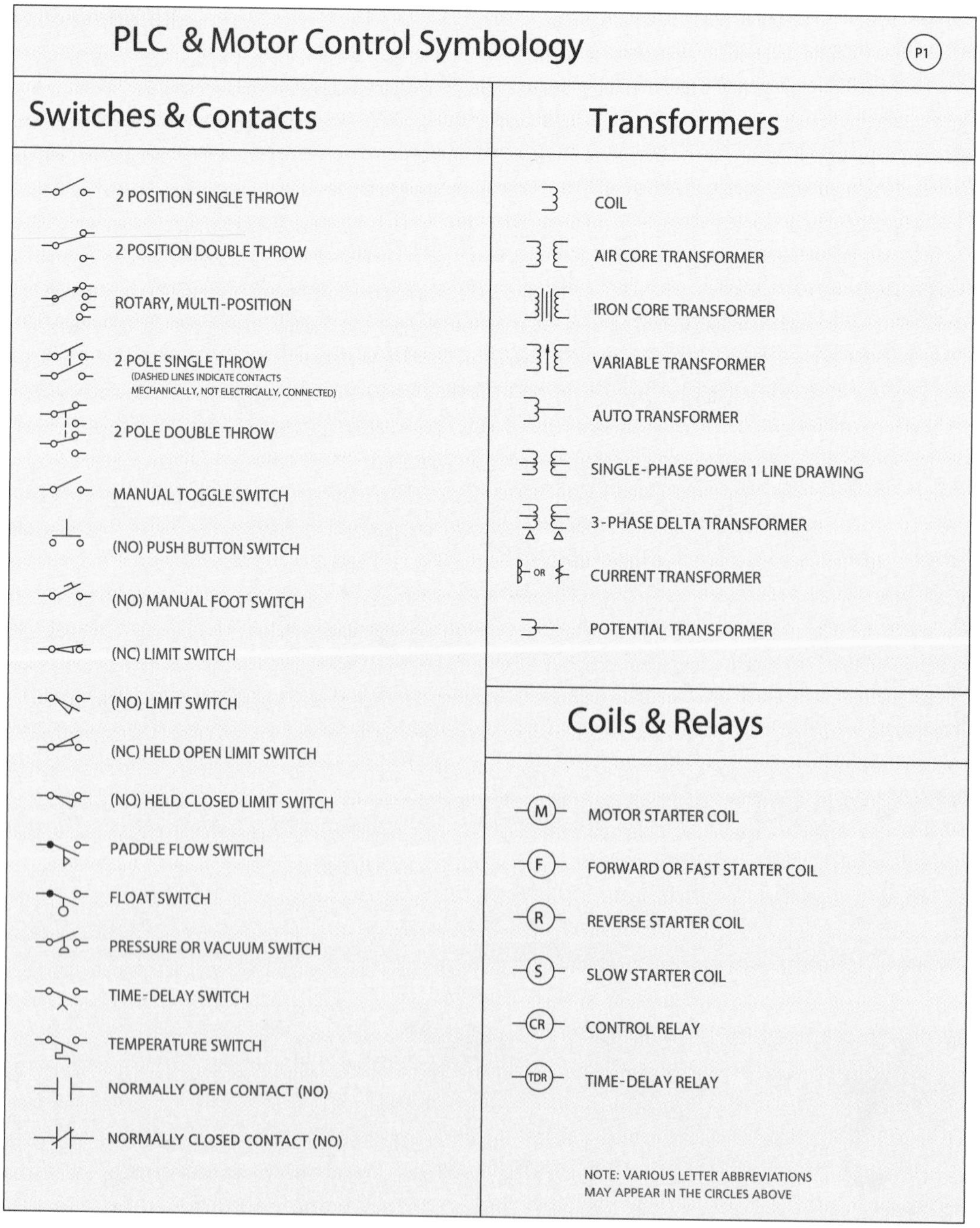

PLC & Motor Control Symbology

Switches & Contacts

- 2 POSITION SINGLE THROW
- 2 POSITION DOUBLE THROW
- ROTARY, MULTI-POSITION
- 2 POLE SINGLE THROW (DASHED LINES INDICATE CONTACTS MECHANICALLY, NOT ELECTRICALLY, CONNECTED)
- 2 POLE DOUBLE THROW
- MANUAL TOGGLE SWITCH
- (NO) PUSH BUTTON SWITCH
- (NO) MANUAL FOOT SWITCH
- (NC) LIMIT SWITCH
- (NO) LIMIT SWITCH
- (NC) HELD OPEN LIMIT SWITCH
- (NO) HELD CLOSED LIMIT SWITCH
- PADDLE FLOW SWITCH
- FLOAT SWITCH
- PRESSURE OR VACUUM SWITCH
- TIME-DELAY SWITCH
- TEMPERATURE SWITCH
- NORMALLY OPEN CONTACT (NO)
- NORMALLY CLOSED CONTACT (NO)

Transformers

- COIL
- AIR CORE TRANSFORMER
- IRON CORE TRANSFORMER
- VARIABLE TRANSFORMER
- AUTO TRANSFORMER
- SINGLE-PHASE POWER 1 LINE DRAWING
- 3-PHASE DELTA TRANSFORMER
- CURRENT TRANSFORMER
- POTENTIAL TRANSFORMER

Coils & Relays

- M MOTOR STARTER COIL
- F FORWARD OR FAST STARTER COIL
- R REVERSE STARTER COIL
- S SLOW STARTER COIL
- CR CONTROL RELAY
- TDR TIME-DELAY RELAY

NOTE: VARIOUS LETTER ABBREVIATIONS MAY APPEAR IN THE CIRCLES ABOVE

FIGURE 10-3   PLC 2.

PLC & Motor Control Symbology   (P2)

Protection Devices	Output Devices

Protection Devices

▭▭▭▭ FUSE

○∿∿○ FUSE

⊸)(⊸ OVERLOAD

⊸)(∿∿ OVERLOAD THERMAL

⊸)(⌐⌐ OVERLOAD MAGNETIC

OL'S
⊸//⊢ LADDER LOGIC OVERLOAD

Output Devices

Ⓛ LIGHT INDICATOR

Ⓜ MOTOR

▪Ⓐ DC MOTOR ARMATURE

Ⓜ 3-PHASE MOTOR

⊣HTR⊢ HEATER

▭▭▭ LED INDICATOR

○∿○ SOLENOID

FIGURE 10-4   Motor starter.

Blue = Control
Red = Power

WIRING DIAGRAM

Motor Power Circuit Fuses

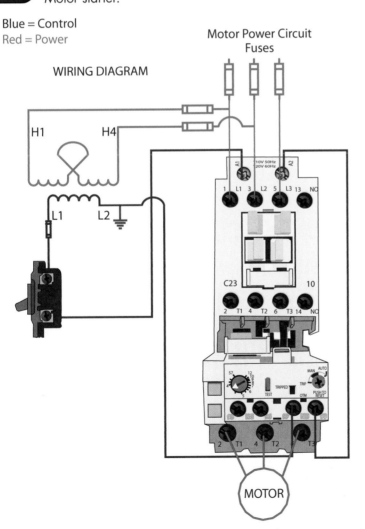

Another term common with PLCs is the term function block programming or FBP. Function block programming uses *function block diagrams*, instead of *ladder diagrams* to show motor control and other control circuitry. Function block diagrams allow more complex motor control circuits to be illustrated. Symbols used in FBP include input tags, output tags, and the function block themselves. In Figure 10–5, input tags include the start tag and the stop tag. The only output tag is the motor starter tag. There are two function blocks shown, BAND_01 is a boolean *and* function block and BOR_01 is a boolean *or* function block. These function blocks are tied to the input tags, output tags, and each other with logical connectors, represented by dotted lines. These lines represent digital signals. Other signals, such as analog signals, are represented by different types of lines.

FIGURE 10–5   Function block programming.

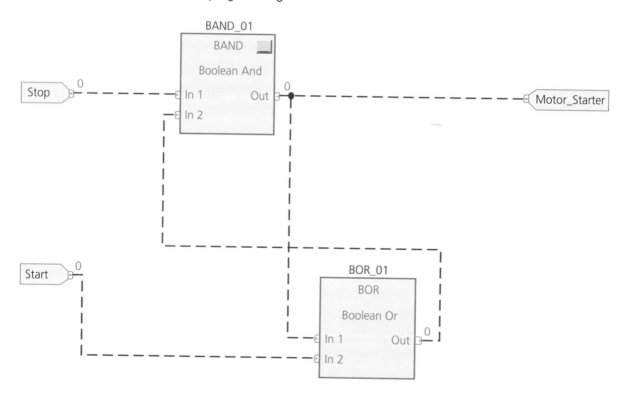

LADDER LOGIC DIAGRAMS

Ladder logic diagrams are a representation of control voltage and elements used to simplify the understanding of control circuits. The diagrams are drawn in the shape of a ladder with the rails depicting supply voltage, as in L1 and L2. The rungs of the ladder contain the control logic path. Those rungs are sequentially numbered outside the left rail from the top down. The numbers outside the right rail are reference numbers. They refer to the rung number on which the control element output or outputs are located.

If a line appears under the reference number, the contact is normally closed. If no line appears, the contact is normally open. The numbers that appear on lines inside the rungs of the logic diagram are actual wire numbers as hooked to terminal blocks. The only other part of the ladder logic diagram is the various contacts and coils that are part of the control circuit. These contacts and coils show on the rungs as circles and contacts (Figure 10–6). Ladder logic diagrams contain three basic component types. They are input devices, logic or control devices, and output devices. Some of these devices are shown in Figure 10–7.

PROCESS CONTROL SYMBOLS

Process control is monitoring and controlling a process or series of processes. Relays, sensors, or PLCs may be used to accomplish the task. Commonly controlled processes are flow, level, speed, temperature, light-

FIGURE 10-6 Ladder logic.

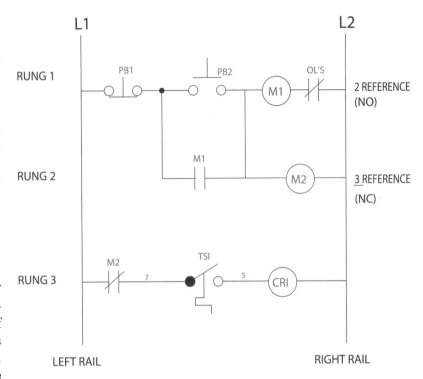

FIGURE 10-7 Logic devices.

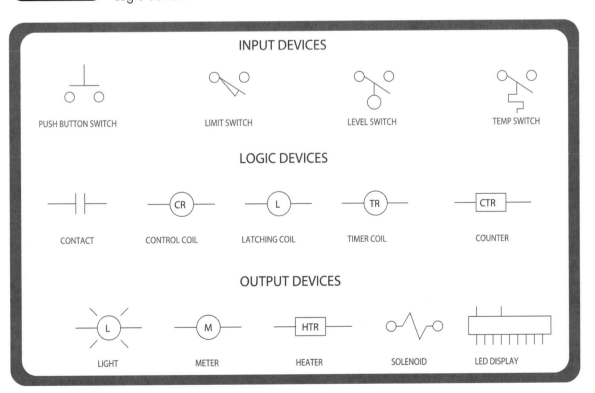

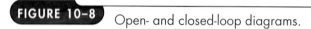

FIGURE 10-8 Open- and closed-loop diagrams.

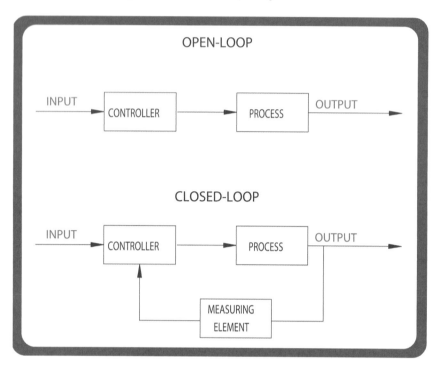

ing, and a variety of manufacturing processes. Process control takes the shape of two basic systems. They are open- and closed-loop diagrams (Figure 10–8).

A process control or flow process diagram may appear in a totally block diagram or in a semi-pictorial format. The block diagram lists all elements in blocks and uses arrows to show the flow directional path between each element, as shown in Figure 10–9.

The semipictorial process control diagram is more common. It depicts some devices in pictorial form and others, such as valves, using standard graphic symbology. A variety of specialized instrument or function symbols may appear as part of this combined flow process control diagram, as shown in Figure 10–10.

FIGURE 10-9 Process loops.

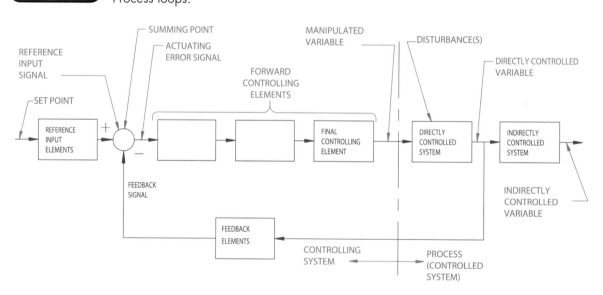

INSTRUMENT SYMBOLOGY

There is a variety of schematic symbol shapes that have various letters and abbreviations to specify their function. These symbols may be custom made or go by a national standard such as that used by the Instrument Society of America (ISA). A set of basic symbol shapes used by the ISA is listed in Figure 10–11.

FIGURE 10-10 Flow diagram.

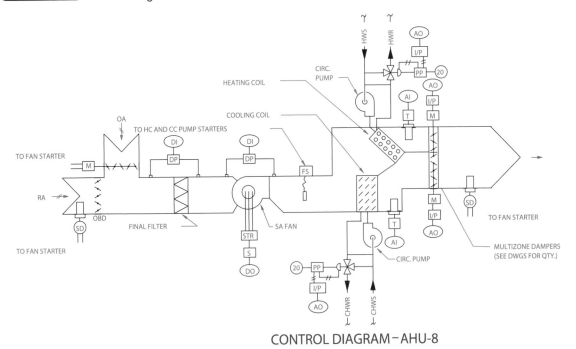

CONTROL DIAGRAM–AHU-8

FIGURE 10-11 ISA symbols.

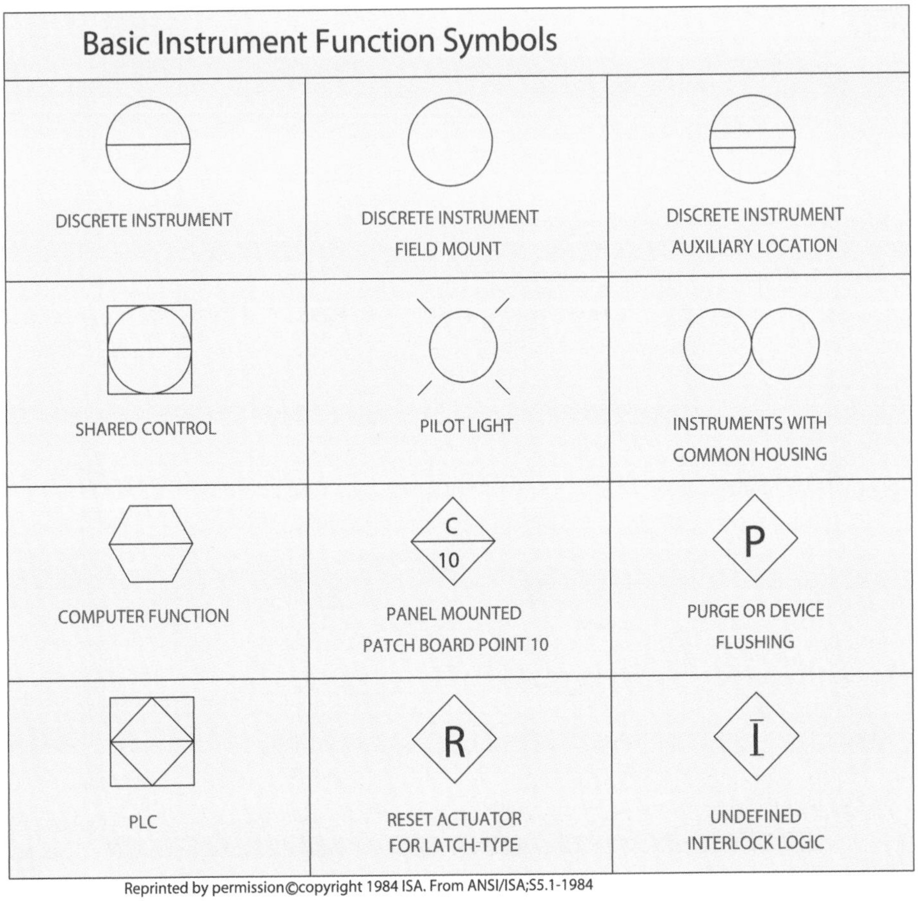

Basic Instrument Function Symbols

DISCRETE INSTRUMENT	DISCRETE INSTRUMENT FIELD MOUNT	DISCRETE INSTRUMENT AUXILIARY LOCATION
SHARED CONTROL	PILOT LIGHT	INSTRUMENTS WITH COMMON HOUSING
COMPUTER FUNCTION	PANEL MOUNTED PATCH BOARD POINT 10	PURGE OR DEVICE FLUSHING
PLC	RESET ACTUATOR FOR LATCH-TYPE	UNDEFINED INTERLOCK LOGIC

STRUCTURED WIRING SYMBOLS

Structured cabling, programmed power distribution systems, fiber optics, and telecommunications distribution systems are appearing more often in commercial, industrial, and residential construction. Each of these systems has specific terminology and print representations.

Structured cabling, twisted pairs, and fiber optics are rapidly growing and changing. Partially as a result of this, and due to the detailed installation training and procedures, there are no currently available standard symbols for these systems. Some prints use standard electrical symbols, such as the telephone triangle shown in Figure 7–6 and specific notes for connection information. The industry installer may have pictorial drawings, block diagrams, or pin connection diagrams (Figures 10–12 through 10–14).

Pictorials show a simplified picture-like representation of the device or panel. They may also show an isometric representation of the building or room, with the devices and conduits shown in simplified picture form. Block diagrams use various shapes of blocks with letters or notes to indicate what each block represents. They are usually accompanied by a legend to explain notation. Pin connection diagrams may show blocks of connections or individually numbered pin diagrams for cable ends or connections.

FIGURE 10–12   Outlet pictorial.

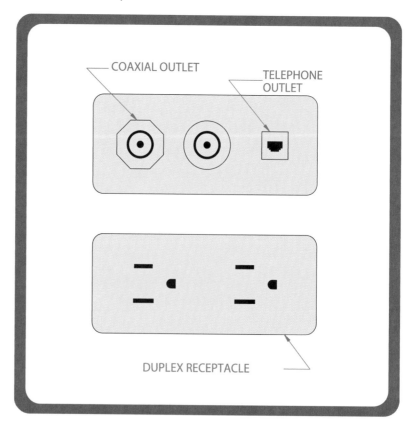

FIGURE 10-13 Block diagrams.

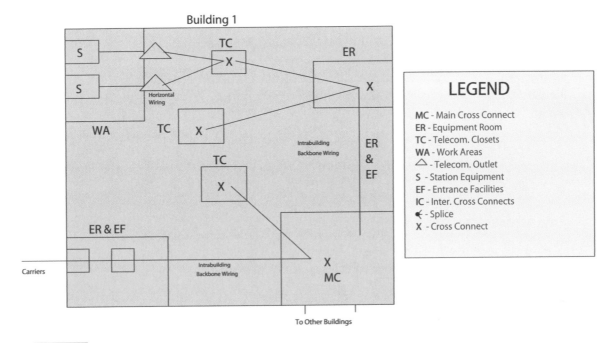

FIGURE 10-14 Pin connection diagrams.

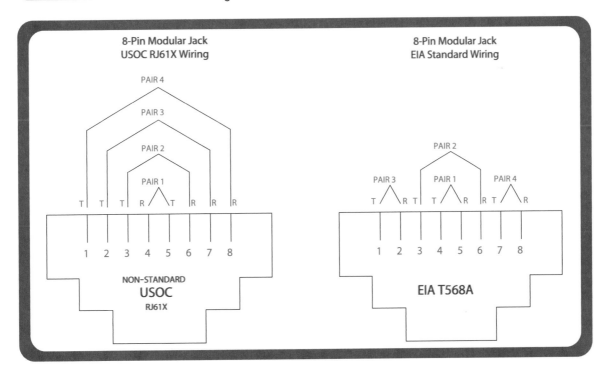

TELECOMMUNICATIONS DISTRIBUTION

Telecommunications distribution systems are being used more widely in the industry. These systems are used to control a variety of voltages, inputs, and outputs. The current *Telecommunications Distribution Methods* manual lists symbols for a variety of devices, conduits, and controls. These symbols are listed in Figure 10–15. These symbols may become more widely used, or other symbols may be created. As always, if you find new unknown symbols, consult the drawing legend, manufacturers' information sheets, electrical manuals, or industrial standards.

FIGURE 10–15 Telecommunications distribution.

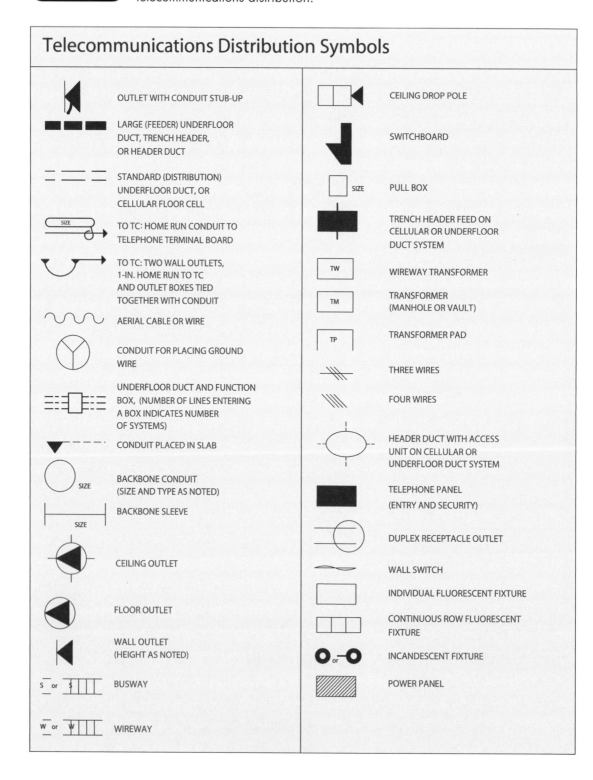

Telecommunications Distribution Symbols

OUTLET WITH CONDUIT STUB-UP	CEILING DROP POLE
LARGE (FEEDER) UNDERFLOOR DUCT, TRENCH HEADER, OR HEADER DUCT	SWITCHBOARD
STANDARD (DISTRIBUTION) UNDERFLOOR DUCT, OR CELLULAR FLOOR CELL	PULL BOX
TO TC: HOME RUN CONDUIT TO TELEPHONE TERMINAL BOARD	TRENCH HEADER FEED ON CELLULAR OR UNDERFLOOR DUCT SYSTEM
TO TC: TWO WALL OUTLETS, 1-IN. HOME RUN TO TC AND OUTLET BOXES TIED TOGETHER WITH CONDUIT	WIREWAY TRANSFORMER
AERIAL CABLE OR WIRE	TRANSFORMER (MANHOLE OR VAULT)
CONDUIT FOR PLACING GROUND WIRE	TRANSFORMER PAD
UNDERFLOOR DUCT AND FUNCTION BOX, (NUMBER OF LINES ENTERING A BOX INDICATES NUMBER OF SYSTEMS)	THREE WIRES
CONDUIT PLACED IN SLAB	FOUR WIRES
BACKBONE CONDUIT (SIZE AND TYPE AS NOTED)	HEADER DUCT WITH ACCESS UNIT ON CELLULAR OR UNDERFLOOR DUCT SYSTEM
BACKBONE SLEEVE	TELEPHONE PANEL (ENTRY AND SECURITY)
CEILING OUTLET	DUPLEX RECEPTACLE OUTLET
FLOOR OUTLET	WALL SWITCH
WALL OUTLET (HEIGHT AS NOTED)	INDIVIDUAL FLUORESCENT FIXTURE
BUSWAY	CONTINUOUS ROW FLUORESCENT FIXTURE
WIREWAY	INCANDESCENT FIXTURE
	POWER PANEL

SUMMARY

As an electrician starts to work with process control, instrumentation, communication systems, and many other types of systems, knowledge of the various symbols and diagrams will be essential.

REVIEW QUESTIONS

1. What is the function of the PLC?

2. Where do you find the meaning of nonstandard or unknown symbols?

3. In the motor starter diagram, Figure 10–4, what does the L in a circle represent?

4. On the same drawing, what type of limit switch is used?

5. Describe a ladder logic diagram.

6. What are the three types of devices contained in ladder logic diagrams?

7. What are the two types of process control loops?

8. What is the function of block diagrams and pin connection diagrams?

9. Where are structured wiring and telecommunications distribution systems used?

10. What does a solid line with three hash marks indicate?

NOTE: Study the symbols and diagrams in this chapter and on the CD print sets to become more familiar with their uses.

11

Site Plans

OVERVIEW

Site plans, sometimes referred to as plot plans, are scaled drawings that show the natural features of a property as well as any man-made features. The man-made features would include the building, sidewalks, parking lots, lights, and so on. This chapter covers the information typically found on a site plan and relates that information to the electrical industry.

OBJECTIVES

After completing this chapter, you should be able to:

- Understand how contour lines affect the slope of a property.
- Differentiate the scales used on site plans versus other drawings.
- Identify easements and public utilities on site plans.
- Identify electrical information on site plans.

SYMBOLS AND ABBREVIATIONS

As with all types of construction drawings, site plans have many line types, symbols, and abbreviations used to save clutter and make the drawings easier to read. A very important symbol common to all site plans is the north arrow. This indicates how the property and building sit with respect to the due or magnetic north direction (see Figure 11–1).

FIGURE 11–1   Residential site plan.

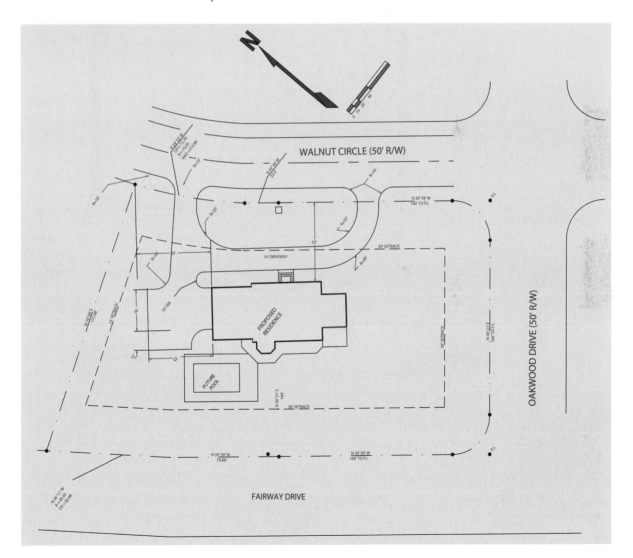

Unfortunately, all designers do not use the same symbols on their drawings. Most construction drawings will have a legend to identify the symbols and abbreviations used. If a symbol or abbreviation is unfamiliar and not identified in the legend, a person could check reference materials (textbooks) or ask the designer of the drawings to define the symbol.

ELEVATION

Elevation is not only a type of construction drawing but a vertical dimension. A change in elevation will be shown on the site plan with contour lines. The elevation of the contour lines may be written on the high side of the contour line, in a break in the contour line, or to the side of the contour line (see Figure 11–2). Contour lines that are far apart represent a slight slope, whereas lines close together represent a steeper slope.

FIGURE 11-2   Contour lines showing a gentle slope will be farther apart than contour lines showing a steeper slope.

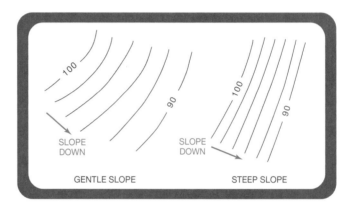

Natural grade is the elevation or slope of the earth before construction begins. Finished grade is the elevation or slope of the earth when the project is finished. The finished grade of a property is often not the same as the natural grade was. Contour lines representing natural grade are represented as a dashed line. Contour lines representing finished grade are represented as solid lines (see Figure 11–3). The elevation of contour

FIGURE 11-3   Contour lines showing natural grade are a dashed line, while contour lines showing finished grade are a solid line.

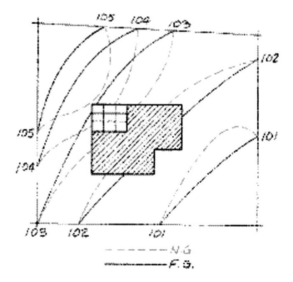

lines could be referenced to sea level, a reference point for the community, or the point of beginning on the drawing. The point of beginning, if shown on the drawing, is the starting point for all horizontal and vertical measurements. It is a benchmark of where the location and elevation are known. It may be a steak or pin buried in the ground, or a point on a street or curb.

PROPERTY LINES

Property lines, which represent the edges of a property, are often found on site plans. The property lines will typically be marked with the length of the line, as well as with the bearing angle—given in degrees, minutes, and seconds from north, south, east, or west. The location of the building and other features on the site plan will be indicated with dimensions from the property lines or the point of beginning (see Figure 11–4).

Professional
Practice!

Having an understanding of contour lines can save a lot of hassle, time, and money. An example would be a new house for which an electrician has to trench the service conductors from the utility pole over to the house. Finished grade will be at a lower elevation than natural grade because of landscaping the homeowner wants.

Imagine that the electrician trenched in the cable before the natural grade was excavated. The trencher depth was set for 24 inches to meet burial require-ments for the National Electrical Code. After the electrician trenched the cable in, the excavator removed dirt to meet final grade. Now the cables will not be 24 inches underground. In fact, they may even get damaged or cut in the process. If the electrician had known the details of the finished grade, he could have trenched the cable in deeper to allow for the earth that was to be removed. Alternatively, he could have waited to trench until the property was at finished grade.

FIGURE 11–4   Bearing angles of the property lines may be shown on the site plan.

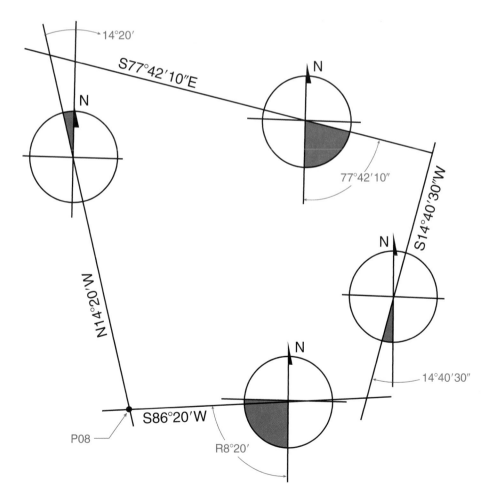

SCALE

The scale used on a site plan is not only smaller than the other drawings in a set of prints but is a different type of scale. Site plans are typically drawn by a civil engineer and will have a scale that uses tenths of a foot rather than inches. Remember that 10.5 feet is not the same as 10 feet 5 inches. There is not much of a difference, but there is a difference. Imagine making 25 measurements that were an inch off. By the time you were done you would be more than 2 feet off from the intended location. Having a tape measure with tenths of a foot rather than inches will save having to convert all dimensions to inches (see Figure 11–5).

FIGURE 11–5 The top tape measure has feet divided into tenths, which is the scale typically used on site plans. The bottom tape measure has feet divided into inches.

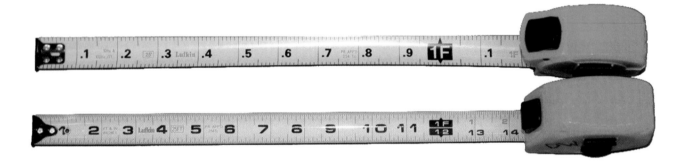

EASEMENTS AND UTILITIES

Site plans will often include the locations of easements and utilities. An easement is a portion of land dedicated to public utilities (or access to utilities and the like) such as water, gas, sewer, and electricity (see Figure 11–6). It is important to know where all buried cables, pipes, and so on are so as to not trench through them. Before digging or trenching, the public utilities must be located. Some areas of the country have a centralized agency that will locate all public utilities on the property. Keep in mind that only the cables and pipes owned by the public utilities will be marked.

Some properties do not have public utilities for sewer and water, but will instead have a well and septic system. A well is typically drilled in the yard near the building, with a plastic pipe run underground to the building. A septic system will have a septic tank for solid waste and a drain field for liquid waste (see Figure 11–7). The drain field allows the liquids to seep into the ground, reducing the frequency of getting a septic tank cleaned out. It is very important to know where the septic tank and drain field will be located so as to not get any electrical cables or raceways buried in the way. Once a drain field is installed, one should never drive across that part of the yard (to prevent crushing the pipes).

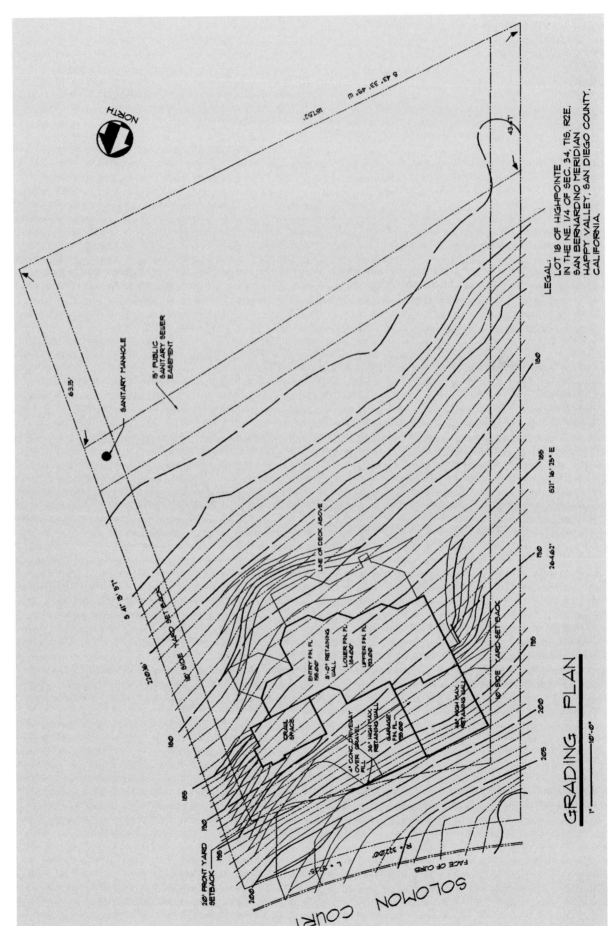

An easement is a portion of the land that is set aside for public utilities.

GRADING PLAN

1" = 10'-0"

LEGAL:
LOT 18 OF HIGHPOINTE
IN THE NE. 1/4 OF SEC. 34, T15, R2E.
SAN BERNARDINO MERIDIAN
HAPPY VALLEY, SAN DIEGO COUNTY,
CALIFORNIA.

SOLOMON COURT

15' PUBLIC
SANITARY SEWER
EASEMENT

SANITARY MANHOLE

NORTH

Professional
Practice!

The use of site plans varies greatly, depending on the type and complexity of a job. Completely understanding the electrical information found on a site plan may require the use of several sheets from the construction drawings.

Most residential jobs only have one site plan, and it may not have any electrical on it (see Figure 11–1). The reason there is no electrical on a residential site plan is that typically there is not much going on outside the building from an electrical standpoint. Most garages are attached, and very few people have yard lights. If a person is planning to have a post light in the yard, a river pump, detached garage with power, or something similar, there will be electrical on a residential site plan.

Commercial jobs, however, will have a lot of electrical work to be done outside the structure (see Figure 11–8). Simply looking at the electrical site plan will not be enough. An electrician will have to look at all of the site plans in a set of construction drawings. The electrical site plan will give electrical circuitry and locations. However, it will not have details of all other work done on site. To have everything on one sheet would simply be too cluttered. If a person does not refer to the other prints, there would be no way to understand all of the other work done on site. That will lead to costly mistakes and delays.

FIGURE 11–7   A septic system separates the solid waste from the liquid waste. The solid waste will be evacuated when the tank fills up while the liquid waste will seep into the ground.

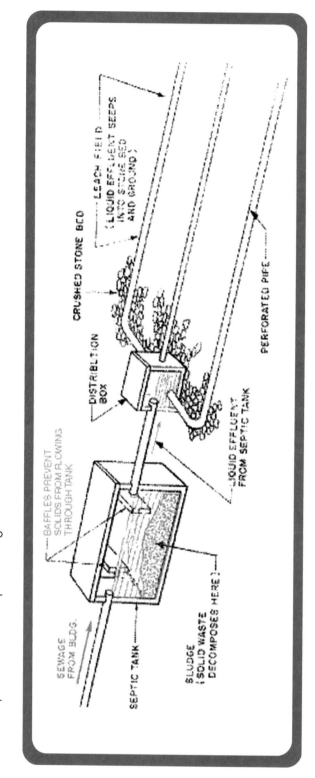

FIGURE 11-8 Commercial site plan.

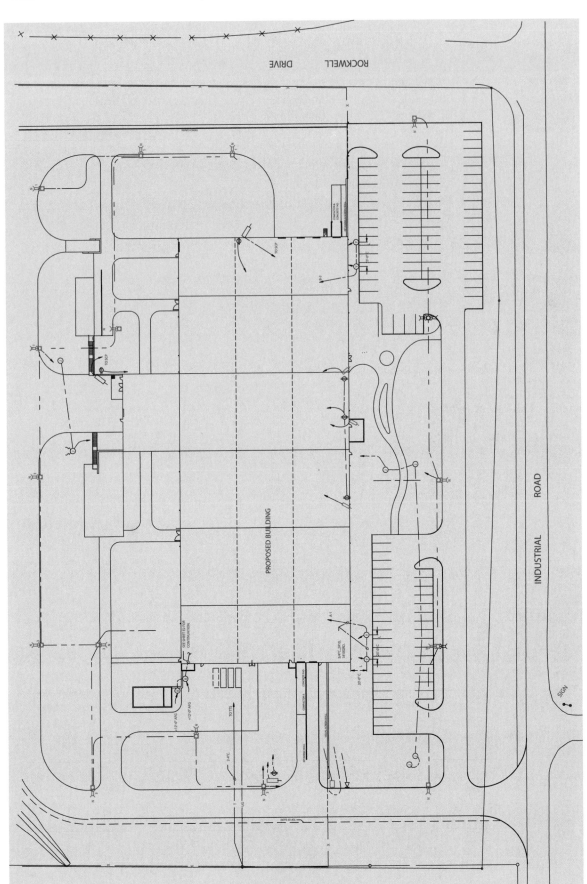

ELECTRICAL INFORMATION

The electrical information found on a site plan is typically anything electrical outside the structure or building. Some jobs (such as small houses) may not have any electrical information except for the location of the utility cables that will feed the house. Many commercial and industrial jobs will have one or more sheets dedicated to the electrical that exists outside the structure or building. The following are examples of electrical information found on site plans:

- Parking lot light and receptacle locations
- Circuitry used for exterior lighting and receptacles
- Sign location and circuitry
- Raceway locations (duct banks) to connect to utility power
- Raceways locations (duct banks) to connect communication circuits
- Utility transformer location
- Exterior well location and circuitry
- River pump location and circuitry
- Exterior security and cameras
- Underground sensors
- Landscape lighting

SUMMARY

Symbols and abbreviations are used to save clutter and make drawings easier to read.

The slope of the earth is indicated on a site plan with contour lines. The contour lines will indicate the elevation of the line.

Property lines show the edges of the property, and are used to locate the building or other features on the property.

Easements are a portion of land set aside for public utilities and will be shown on the site plan.

It is important for an electrician to understand not only the electrical information on a site plan but all information to avoid costly mistakes and delays.

REVIEW QUESTIONS

1. List five man-made features found on a site plan.
2. What symbol found on site plans gives the orientation of the property?
3. If a symbol is unfamiliar, what should be done?
4. What is elevation with respect to site plans?
5. Draw the symbols used to represent natural and finished grade on a site plan.
6. What is different about the scale used on site plans in relation to the scale used on floor plans?
7. What is an easement?
8. List six pieces of electrical information that may be found on a site plan.

SITE PLAN EXERCISE

Use Problem 11–1 to answer the following questions.

1. What is the scale used on the drawing?
2. What is to the south of the property?
3. What is the slope of the drain field?
4. How far is the building from the west property line?
5. What is the length of the north property line?
6. Describe the septic tank.
7. What material is used for the driveway?
8. Where is the well located?
9. How far is the drain field from the north property line?
10. What is the elevation of the southernmost point of the house?
11. How far is the house from the north property line?
12. What size of pipe runs from the house to the septic tank?
13. What material is the sidewalk that surrounds the house?
14. Where is the electrical connection to the utility transformer?
15. When trenching the electrical to the house, what should be avoided?

Problem 11–1

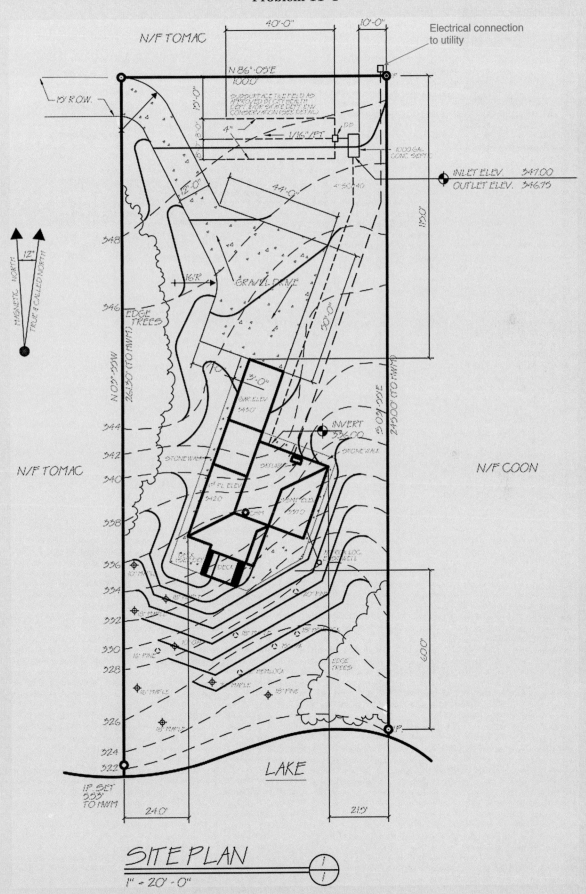

Electrical connection
to utility

SITE PLAN

1" = 20'-0"

Use Problem 11–2 in the text or industrial plan E1 from the CD to answer the following questions.

1. What is the scale of the site plan?

2. On which side of the building are the transformers located?

3. How many and what size of conduits are to be installed for the incoming telephone line?

4. There are two ground-mounted flood lights (fixtures Y) shining toward the southwest corner of the building. Describe the spacing of these lights from the building, as well as the spacing between the two fixtures.

5. What is the quantity of (N) parking lot lights?

6. What size of raceway is feeding the parking lot lights at the north side of the building.

7. What is the height of the camera facing the doorway on the northwest side of the building?

8. How many cameras are shown on the site plan?

9. What size of conduit feeds the parking lot lights on the south side of the building?

10. What is the mounting height of fixture K?

11. How many weatherproof outlets are shown on the site plan?

12. Are the existing power and telephone lines the building is to be fed from overhead or underground?

13. Who is to install the underground primary that feeds from the utility power to the transformers?

14. At what height is the pole-mounted camera on the southwest side of the building to be mounted?

15. What is the pole-mounted camera on the southwest side of the building facing?

Problem 11–2

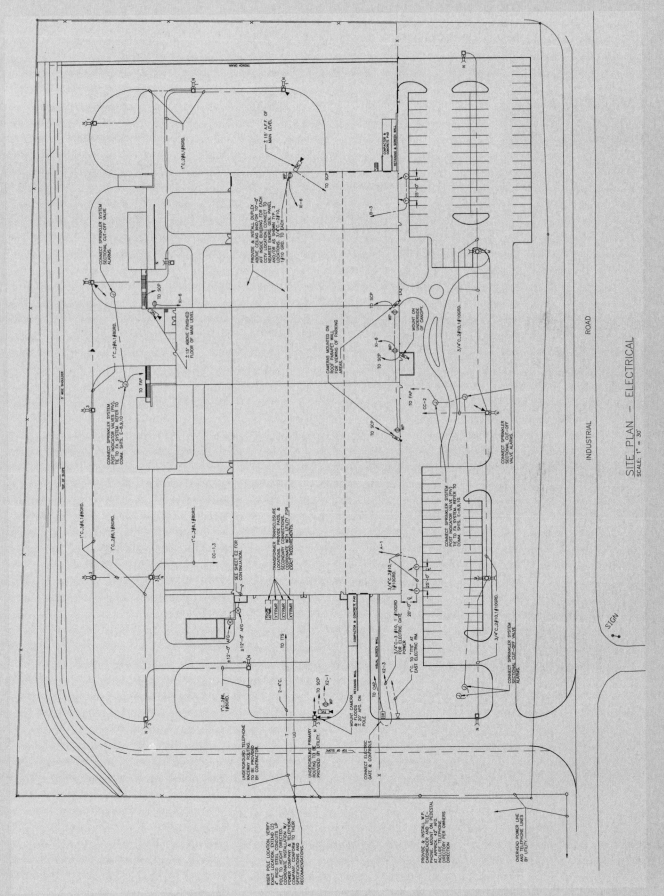

SITE PLAN — ELECTRICAL
SCALE: 1" = 30'

12

Floor Plans

OUTLINE

OVERVIEW

Floor plans are the type of plan most often used by electricians and those in the majority of the other trades. It is important for an electrician to be able to understand electrical floor plans and floor plans of the other disciplines. The information found on the entire set of prints ties together to make the job happen. This chapter discusses the information found on the various floor plans and how the information relates to the electrical industry.

OBJECTIVES

After completing this chapter, you should be able to:

• Identify the features shown on a floor plan.

• Describe the information found on various types of floor plans.

• Explain the importance of understanding the floor plans of the other trades as well as the electrical plans.

• Recognize the commonly used symbols for doors and windows.

• Differentiate between a foundation plan and a lower-level or basement plan.

• Describe the different types of electrical floor plans.

FLOOR PLAN

A floor plan is a scaled view looking downward at one level or floor of a building. It is as if the building were cut at the height of the cutting plane and everything above lifted off (see Figure 12–1). The cutting plane

FIGURE 12-1 A floor plan is a view looking downward as if the roof and everything above an imaginary cut has been lifted off.

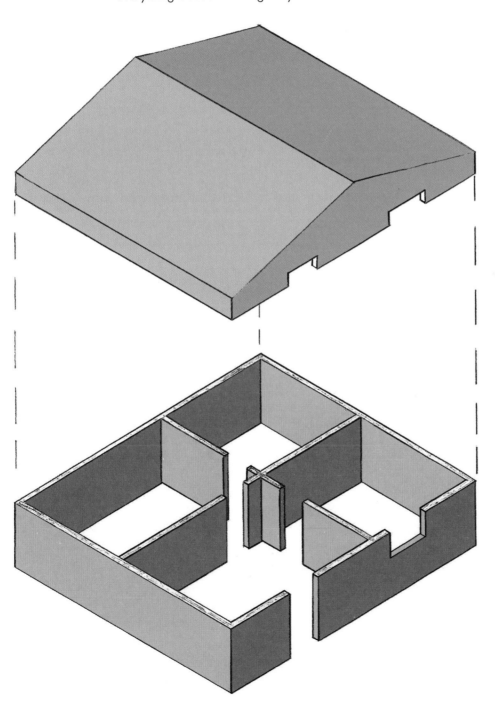

is typically at a height of about 5 feet and runs through the doors and windows. Any object or wall that is cut by the cutting-plane line or is below the cutting-plane line will be shown with a solid line. For example, partition walls, base cabinet units, doors, and windows will be represented with a solid line. Any object above the cutting-plane line will be shown with a hidden line. For example, upper cabinets in a kitchen, cased openings, roof lines, and vaulted ceilings will be represented with a hidden line.

The floor plan shows the layout and relationships of rooms and walls. By looking at a floor plan, a tradesperson should be able to visualize the layout of the building and imagine what the finished project will look like. Floor plans give the relationships of doors, windows, plumbing, electrical, and any other feature permanently installed. The floor plan will also give detailed information about the sizes and placement of these features within a building. Figures 12–2 and 12–3 show the first- and second-floor electrical floor plans for a residence.

There will be at least one floor plan for each level in a building. Most commercial and industrial prints will have several drawings for each floor of a building. The basic layout of walls, doors, and windows for each of the drawings will be the same. However, each drawing will detail different aspects of the job. For example, the first-floor framing plan will have the information necessary to construct and place the walls of the

FIGURE 12–2   Residential first-floor plan.

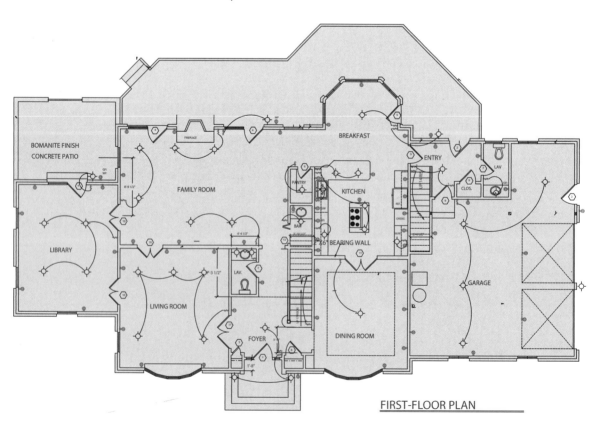

FIRST-FLOOR PLAN

FIGURE 12–3   Residential second-floor plan.

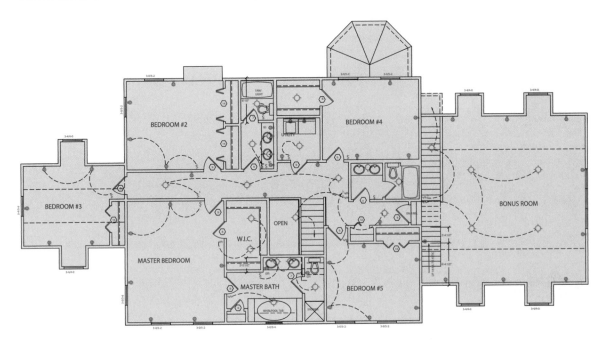

SECOND–FLOOR PLAN

first floor. The first-floor framing plan will not have any information about the lighting, receptacles, plumbing fixtures, and so on. That information would be found on the first-floor lighting plan, power plan, or plumbing plan.

Each of the disciplines will have drawings with the information pertaining to its aspect of the job. There may be several electrical drawings for the first floor alone, including a lighting plan, a power plan, a systems plan, and so on. It is very important that a person completely understand the prints relating to their field and those of the other trades. If an electrician does not pay attention to the plumbing and mechanical plans it is very likely that some of the electrical raceways and equipment will conflict with the plumbing and HVAC equipment and runs. By coordinating with other trades, not only will the job run more smoothly but time and money will be saved.

DIMENSIONING

The floor plans are where most of the dimensioning is done. The dimensions will indicate both size and location. Not every plan will contain every dimension. Only the necessary dimensions for that particular discipline or type of drawing will be included on that print. Having only the necessary dimensions on each type of plan will prevent the print from becoming too cluttered.

FieldNote!

Often a raceway will be run below a concrete floor. The raceway will go down into the concrete below the electrical panel and be stubbed up within a wall or in front of a wall some other place in the building. The electrician will have to install the raceway and have it in the proper place before the concrete is poured. This requires the electrician to locate where the walls are to be before most of the building has been constructed. By using dimensions on the floor plan, an electrician can accurately locate the walls and get the raceways stubbed up in the proper locations (see Figure 12–4).

FIGURE 12–4 Many times raceways are run under the concrete and stubbed up in the appropriate location.

Although most electrical devices do not have location dimensions, occasionally a device must be installed in a very specific location. If a device does require a specific location, it may have horizontal and vertical dimensions. Horizontal dimensions to electrical devices are typically to center. Most vertical dimensions to devices are to the edge. Many times the vertical dimension for receptacles and switches will be identified by a note on the drawing that gives the height for a typical receptacle or switch. Any switch or receptacle height that deviates from the typical dimension will have a specific note giving the height of that particular device.

If a person wants to be able to lay out and draw notes on plan, it is convenient that they receive a print without any notes or dimensions. This is not necessarily helpful with determining locations, but it gives a person a clean print for jotting down notes and information about a job. Computer-aided design makes it easy to have a drawing printed with as little or as much information as you want.

SYMBOLS

There are many symbols shown on the floor plan. The symbols will give the appropriate information without cluttering the print. Although most designers use the standardized symbols put forth by the American National Standards Institute, this is not required. Designers might use a different symbol to represent the same object. Chapter 7 covered the commonly used symbols in the electrical industry. Again, it is important for the electrician to understand electrical symbols as well as the symbols used by other trades.

Most construction drawings will have a legend to identify the symbols used on the sheets (see Figure 12–5). If a symbol is unfamiliar and not identified in the legend, a person could check reference materials (textbooks) or ask the designer of the drawings to define the symbol.

FIGURE 12-5 Commercial and industrial prints have a legend to identify the symbols and abbreviations used in the construction drawings.

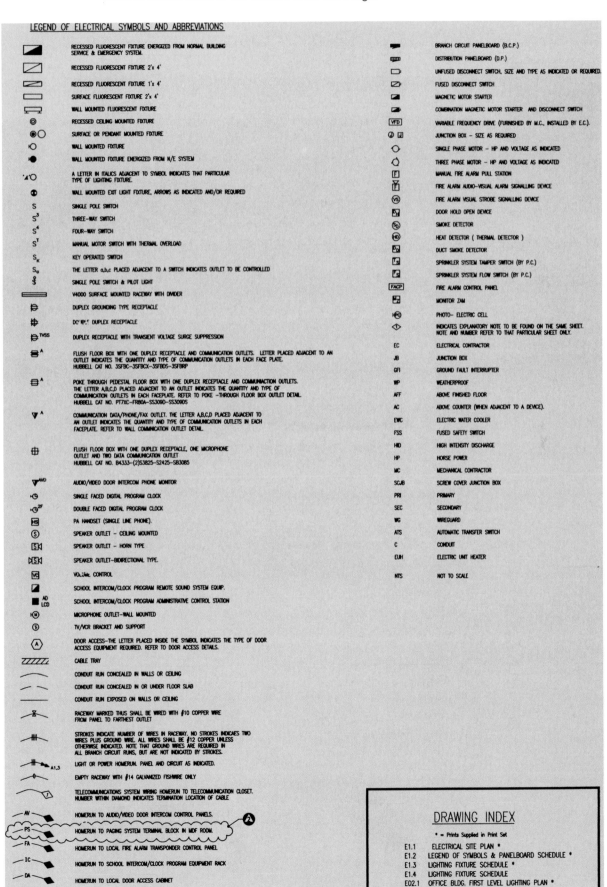

LEGEND OF ELECTRICAL SYMBOLS AND ABBREVIATIONS

RECESSED FLUORESCENT FIXTURE ENERGIZED FROM NORMAL BUILDING SERVICE & EMERGENCY SYSTEM.

RECESSED FLUORESCENT FIXTURE 2'x 4'

RECESSED FLUORESCENT FIXTURE 1'x 4'

SURFACE FLUORESCENT FIXTURE 2'x 4'

WALL MOUNTED FLUORESCENT FIXTURE

RECESSED CEILING MOUNTED FIXTURE

SURFACE OR PENDANT MOUNTED FIXTURE

WALL MOUNTED FIXTURE

WALL MOUNTED FIXTURE ENERGIZED FROM N/E SYSTEM

A LETTER IN ITALICS ADJACENT TO SYMBOL INDICATES THAT PARTICULAR TYPE OF LIGHTING FIXTURE.

WALL MOUNTED EXIT LIGHT FIXTURE, ARROWS AS INDICATED AND/OR REQUIRED

S   SINGLE POLE SWITCH

S³   THREE-WAY SWITCH

S⁴   FOUR-WAY SWITCH

Sᵀ   MANUAL MOTOR SWITCH WITH THERMAL OVERLOAD

Sₖ   KEY OPERATED SWITCH

Sₐ   THE LETTER a,b,c PLACED ADJACENT TO A SWITCH INDICATES OUTLET TO BE CONTROLLED

§   SINGLE POLE SWITCH & PILOT LIGHT

V4000 SURFACE MOUNTED RACEWAY WITH DIVIDER

DUPLEX GROUNDING TYPE RECEPTACLE

DG,WP,C DUPLEX RECEPTACLE

DUPLEX RECEPTACLE WITH TRANSIENT VOLTAGE SURGE SUPPRESSION

FLUSH FLOOR BOX WITH ONE DUPLEX RECEPTACLE AND COMMUNICATION OUTLETS. LETTER PLACED ADJACENT TO AN OUTLET INDICATES THE QUANTITY AND TYPE OF COMMUNICATION OUTLETS IN EACH FACE PLATE. HUBBELL CAT NO. 3SFBC-3SFBCX-3SFBDS-3SFBRP

POKE THROUGH PEDESTAL FLOOR BOX WITH ONE DUPLEX RECEPTACLE AND COMMUNIACTION OUTLETS. THE LETTER A,B,C,D PLACED ADJACENT TO AN OUTLET INDICATES THE QUANTITY AND TYPE OF COMMUNICATION OUTLETS IN EACH FACEPLATE. REFER TO POKE -THROUGH FLOOR BOX OUTLET DETAIL. HUBBELL CAT NO. PT7XC-FR80A-SS3090-SS3090S

COMMUNICATION DATA/PHONE/FAX OUTLET. THE LETTER A,B,C,D PLACED ADJACENT TO AN OUTLET INDICATES THE QUANTITY AND TYPE OF COMMUNICATION OUTLETS IN EACH FACEPLATE. REFER TO WALL COMMUNICATION OUTLET DETAIL.

FLUSH FLOOR BOX WITH ONE DUPLEX RECEPTACLE, ONE MICROPHONE OUTLET AND TWO DATA COMMUNICATION OUTLET HUBBELL CAT NO. B4333-(2)S3825-S2425-SB3085

AUDIO/VIDEO DOOR INTERCOM PHONE MONITOR

SINGLE FACED DIGITAL PROGRAM CLOCK

DOUBLE FACED DIGITAL PROGRAM CLOCK

PA HANDSET (SINGLE LINE PHONE).

SPEAKER OUTLET - CEILING MOUNTED

SPEAKER OUTLET - HORN TYPE

SPEAKER OUTLET-BIDIRECTIONAL TYPE.

VOLUME CONTROL

SCHOOL INTERCOM/CLOCK PROGRAM REMOTE SOUND SYSTEM EQUIP.

SCHOOL INTERCOM/CLOCK PROGRAM ADMINISTRATIVE CONTROL STATION

MICROPHONE OUTLET-WALL MOUNTED

TV/VCR BRACKET AND SUPPORT

DOOR ACCESS-THE LETTER PLACED INSIDE THE SYMBOL INDICATES THE TYPE OF DOOR ACCESS EQUIPMENT REQUIRED. REFER TO DOOR ACCESS DETAILS.

CABLE TRAY

CONDUIT RUN CONCEALED IN WALLS OR CEILING

CONDUIT RUN CONCEALED IN OR UNDER FLOOR SLAB

CONDUIT RUN EXPOSED ON WALLS OR CEILING

RACEWAY MARKED THUS SHALL BE WIRED WITH #10 COPPER WIRE FROM PANEL TO FARTHEST OUTLET

STROKES INDICATE NUMBER OF WIRES IN RACEWAY. NO STROKES INDICAES TWO WIRES PLUS GROUND WIRE. ALL WIRES SHALL BE #12 COPPER UNLESS OTHERWISE INDICATED. NOTE THAT GROUND WIRES ARE REQUIRED IN ALL BRANCH CIRCUIT RUNS, BUT ARE NOT INDICATED BY STROKES.

LIGHT OR POWER HOMERUN. PANEL AND CIRCUIT AS INDICATED.

EMPTY RACEWAY WITH #14 GALVANIZED FISHWIRE ONLY

TELECOMMUNICATIONS SYSTEM WIRING HOMERUN TO TELECOMMUNICATION CLOSET. NUMBER WITHIN DIAMOND INDICATES TERMINATION LOCATION OF CABLE

HOMERUN TO AUDIO/VIDEO DOOR INTERCOM CONTROL PANELS.

HOMERUN TO PAGING SYSTEM TERMINAL BLOCK IN MDF ROOM.

HOMERUN TO LOCAL FIRE ALARM TRANSPONDER CONTROL PANEL

HOMERUN TO SCHOOL INTERCOM/CLOCK PROGRAM EQUIPMENT RACK

HOMERUN TO LOCAL DOOR ACCESS CABINET

BRANCH CIRCUIT PANELBOARD (B.C.P.)

DISTRIBUTION PANELBOARD (D.P.)

UNFUSED DISCONNECT SWITCH, SIZE AND TYPE AS INDICATED OR REQUIRED.

FUSED DISCONNECT SWITCH

MAGNETIC MOTOR STARTER

COMBINATION MAGNETIC MOTOR STARTER AND DISCONNECT SWITCH

VFD   VARIABLE FREQUENCY DRIVE (FURNISHED BY M.C., INSTALLED BY E.C.)

JUNCTION BOX - SIZE AS REQUIRED

SINGLE PHASE MOTOR - HP AND VOLTAGE AS INDICATED

THREE PHASE MOTOR - HP AND VOLTAGE AS INDICATED

MANUAL FIRE ALARM PULL STATION

FIRE ALARM AUDIO-VISUAL ALARM SIGNALLING DEVICE

FIRE ALARM VISUAL STROBE SIGNALLING DEVICE

DOOR HOLD OPEN DEVICE

SMOKE DETECTOR

HEAT DETECTOR (THERMAL DETECTOR)

DUCT SMOKE DETECTOR

SPRINKLER SYSTEM TAMPER SWITCH (BY P.C.)

SPRINKLER SYSTEM FLOW SWITCH (BY P.C.)

FACP   FIRE ALARM CONTROL PANEL

MONITOR ZAM

PHOTO- ELECTRIC CELL

INDICATES EXPLANATORY NOTE TO BE FOUND ON THE SAME SHEET. NOTE AND NUMBER REFER TO THAT PARTICULAR SHEET ONLY.

EC   ELECTRICAL CONTRACTOR

JB   JUNCTION BOX

GFI   GROUND FAULT INTERRUPTER

WP   WEATHERPROOF

AFF   ABOVE FINISHED FLOOR

AC   ABOVE COUNTER (WHEN ADJACENT TO A DEVICE).

EWC   ELECTRIC WATER COOLER

FSS   FUSED SAFETY SWITCH

HID   HIGH INTENSITY DISCHARGE

HP   HORSE POWER

MC   MECHANICAL CONTRACTOR

SCJB   SCREW COVER JUNCTION BOX

PRI   PRIMARY

SEC   SECONDARY

WG   WIREGUARD

ATS   AUTOMATIC TRANSFER SWITCH

C   CONDUIT

EUH   ELECTRIC UNIT HEATER

NTS   NOT TO SCALE

DRAWING INDEX

* = Prints Supplied in Print Set

FieldNote!

Notations
A few words or sentences used to convey additional information.
Foundation Plan
A drawing that shows a plan view of foundation walls, footings, and load-bearing posts or columns.

ABBREVIATIONS

Abbreviations are also an important part of the floor plan. They minimize clutter on a print by not having to write the entire word. Most construction drawings will have a place on one of the sheets that will define the abbreviations (see Figure 12–5). If an abbreviation is unfamiliar and not identified in the legend, a person could check reference materials (textbooks) or ask the designer of the drawings to define the abbreviation. Appendix D outlines abbreviations commonly used in construction drawings.

NOTATIONS

Notations (notes) are often found on floor plans. A notation might consist of only a few words or of several sentences. The notes are used to convey additional information. The notations can be specific and refer only to one specific point on a drawing. Specific notations will typically be written on the drawing where the note applies or will have a leader pointing to the location where the note applies (see Figure 12–6). The notations can also be general in nature and refer to all of the drawings (see Figure 12–7).

12.1 Door and Window Symbols

Being able to fully understand and visualize a print requires a person to understand what types of doors and windows are being installed by simply looking at the drawing. Figure 12–8 (see page 187) shows a plan and pictorial view of some of the commonly used doors. Figure 12–9 (see page 188) shows plan, elevation, and pictorial views of some of the commonly used windows.

12.2 Foundation Plan

A **foundation plan** looks similar to a floor plan in that it is a view looking down as if everything above the cutting plane line were removed. It does not, however, give the same information as a floor plan. The foundation plan is a view of the foundation walls, footings, and any load-bearing posts or columns. It does not show partition walls, plumbing fixtures, cabinetry, and the like. This is the drawing that will be used by concrete workers to build the footings and foundation walls.

Any wall or post cut by the cutting-plane line will be a solid line. The footings, which are underground or under the concrete, will be shown with a hidden line (see Figure 12–10, page 189). Residential drawings will sometimes have a plan that serves as both the foundation plan and the basement floor plan. Commercial and industrial drawings will have a foundation plan separate from any lower-level or basement plans.

FIGURE 12-6 A specific note gives detail about a specific point on a construction drawing. Specific notes often have a leader pointing to the location the note applies.

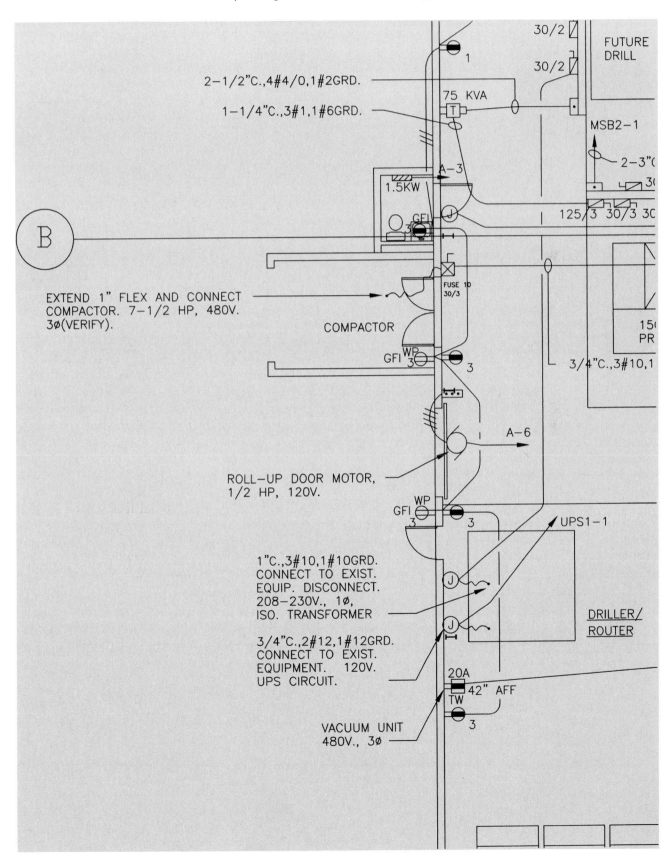

2-1/2"C.,4#4/0,1#2GRD.

1-1/4"C.,3#1,1#6GRD.

FUTURE DRILL

75 KVA

MSB2-1

2-3"(

A-3

1.5KW

125/3 30/3 30

GFI

B

FUSE 10
30/3

EXTEND 1" FLEX AND CONNECT
COMPACTOR. 7-1/2 HP, 480V.
3ø(VERIFY).

COMPACTOR

15
PR

WP
GFI 3   3

3/4"C.,3#10,1

A-6

ROLL-UP DOOR MOTOR,
1/2 HP, 120V.

WP
GFI 3   3

UPS1-1

1"C.,3#10,1#10GRD.
CONNECT TO EXIST.
EQUIP. DISCONNECT.
208-230V., 1ø,
ISO. TRANSFORMER

DRILLER/
ROUTER

3/4"C.,2#12,1#12GRD.
CONNECT TO EXIST.
EQUIPMENT. 120V.
UPS CIRCUIT.

20A

42" AFF
TW

VACUUM UNIT
480V., 3ø

3

FIGURE 12-7   General notes give general information that applies to more than one location or plan.

LIGHTING FIXTURE NOTES:

1. INCLUDE ALL MOUNTING & INSTALLATION HARDWARE FOR EACH FIXTURE FOR A COMPLETE OPERATING FACILITY.

2. EACH FIXTURE SHALL BE "UL" LABELED & INSTALLED FOR USE INTENDED.

3. ALL INCANDESCENT LAMPS SHALL BE 130 VOLTS. INCLUDE EXTENDED SERVICE LAMPS WHERE APPLICABLE.

4. ALL FLUORESCENT UNITS SHALL BE 277 VOLTS, WITH PHILIPS LAMPS & BALLASTS; ALL 48" LENGTH WHITE LIGHTING TO BE 40 WATT, T–10 RAPID START LAMPS; ALL 48" LENGTH GOLD OR RED LIGHTING TO BE 40 WATT T–12 RAPID START LAMPS; ALL 96" WHITE LIGHTING TO BE 75 WATT T–12 SLIMLINE LAMPS; ALL 96" GOLD LIGHTING TO BE 75 WATT, T–12 SLIMLINE LAMPS. REFER TO LIGHTING FIXTURE SCHEDULE.

5. ALL HIGH INTENSITY DISCHARGES (HID) UNITS SHALL BE 277 VOLTS WHERE INSTALLED ON BUILDING. EXTERIOR POLE-MOUNTED UNITS SHALL BE 277 VOLTS. ALL GROUND MOUNTED UNITS SHALL BE 208 VOLTS.

6. VERIFY ALL CEILING/WALL CONSTRUCTION PRIOR TO ORDERING/INSTALLING UNIT. INCLUDE NECESSARY ALLOWANCES AND ADJUSTMENTS.

7. ALL FIXTURES SHALL BE AS SPECIFIED OR APPROVED EQUALS. WHEN SUBMITTING EQUALS, INCLUDE ALL PERTINENT INFORMATION (PHOTOMETRIC DATA, INSTALLATION REQUIREMENTS, ETC.).

FIGURE 12-8 Pictorial and plan views of commonly used doors.

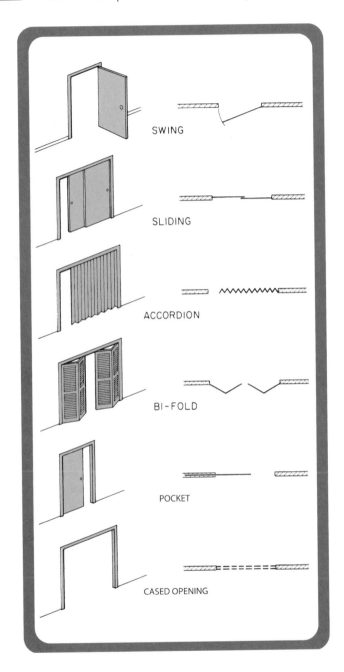

SWING

SLIDING

ACCORDION

BI-FOLD

POCKET

CASED OPENING

FIGURE 12-9 Pictorial, plan, and elevation views of commonly used windows.

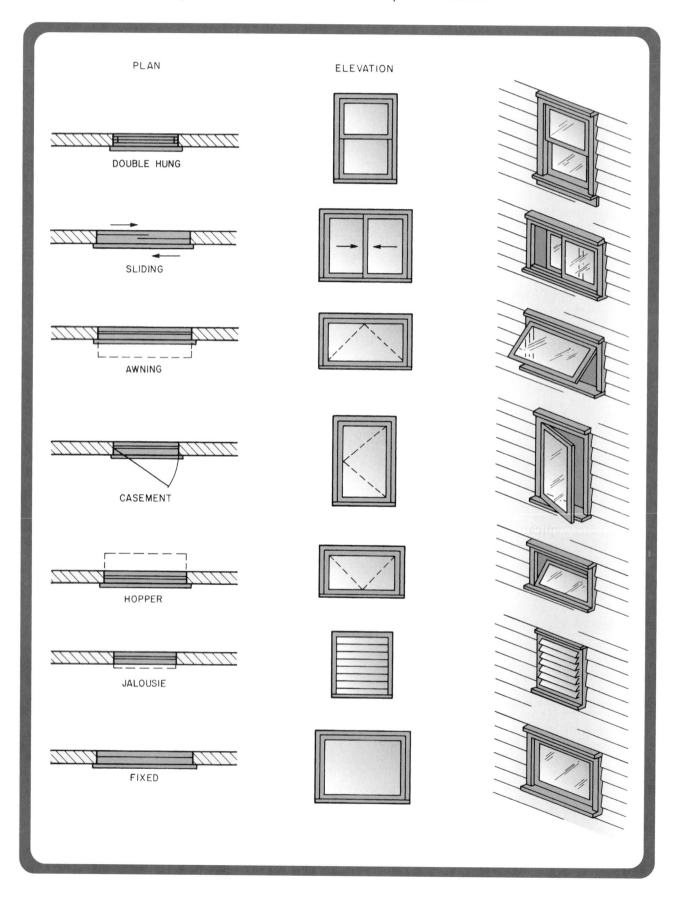

PLAN

ELEVATION

DOUBLE HUNG

SLIDING

AWNING

CASEMENT

HOPPER

JALOUSIE

FIXED

FIGURE 12-10 Residential foundation plan.

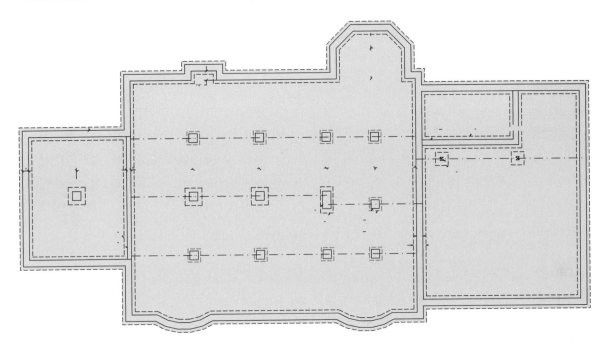

DEMOLITION PLAN

On a renovation or remodel, existing features of a building will often have to be removed. A **demolition plan** will indicate which items are to be demolished or moved. Each discipline may have a demolition plan indicating the demolition that pertains to that particular discipline.

Most renovations will involve electrical that has to be removed and therefore will have an electrical demolition plan. The items to be removed or demolished typically have hash marks through them (see Figure 12–11). Be sure to check the legend on the prints to verify which symbols that designer used for demolition. Removing something that was not meant to be removed can be a costly error.

ELECTRICAL FLOOR PLANS

12.3 Residential

Residential blueprints are not typically drawn up by an architect or engineer. Instead, they are drawn by the contractor or by a draftsperson at a lumberyard. These prints are typically only a few pages long and will not be divided into disciplines in the way that commercial and industrial prints are. The electrical contractor will usually be the one to draw in the symbols and lay out all of the electrical on the drawing according to the National Electrical Code and what the customer wants. Because the electrical contractor pencils in most of the symbols, there will typically not be a legend, the symbols will not be perfect, and the circuitry will usually not be laid out (see Figure 12–12).

Demolition Plan
A drawing that indicates existing features of a building that will have to be demolished or moved.

FieldNote!

Occasionally on a renovation, items such as lighting will be taken down during construction and saved or reinstalled after the remodel. If this is the case, care must be taken to not lose any parts or destroy them. If the customer intends to reinstall or save any fixtures, these will be identified in the specifications or by a note.

FIGURE 12–11 A demolition plan often uses hash marks to indicate which parts of the existing building are to be removed or demolished. (Courtesy of Ulteig Engineers.)

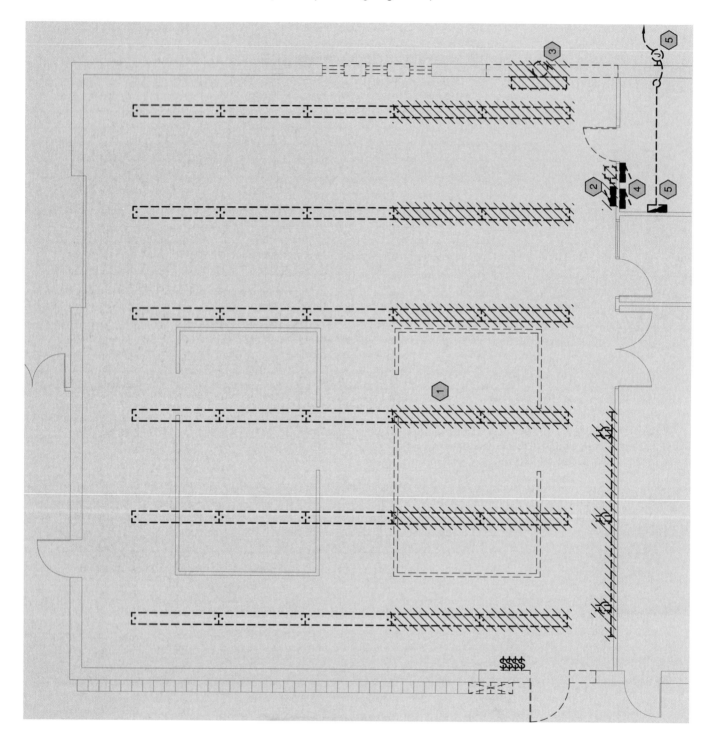

The set of drawings printed for an electrician on a residential job will typically have very few dimensions, and no information from the other disciplines (see Figure 12–13). Having a clean plan without most of the dimensions and other trade information makes it easier to draw in the electrical and make it legible. The downfall about not having info on the other trades is that you will not know where ductwork, plumbing, and the like will be. It is important for an electrician to get together with the other trades on the job to coordinate locations before the installation begins. This will save a lot of headaches, time, and money.

FIGURE 12–12 Residential plans often have the electrical sketched in by the electrical contractor.

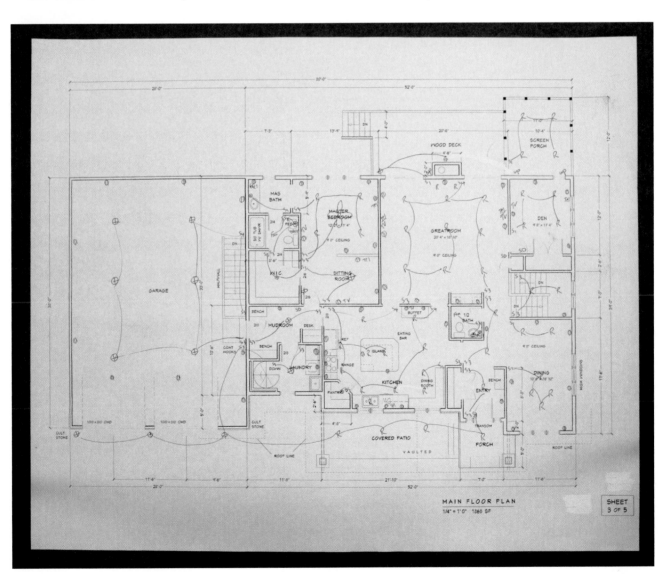

FIGURE 12-13 CAD makes it very easy to have a plan printed with only the layout of the walls. This eliminates the clutter when drawing in your own symbols and circuitry.

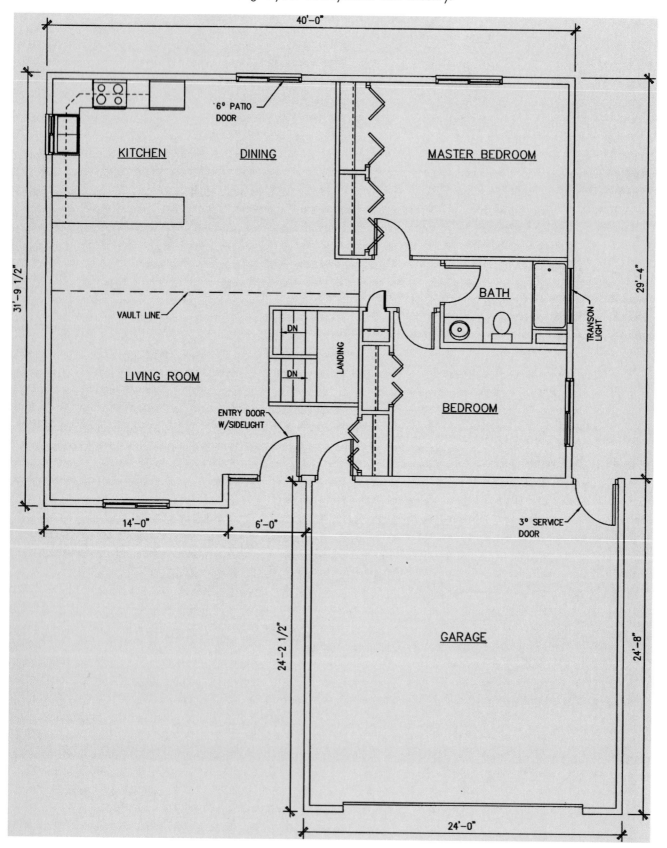

12.4 Commercial and Industrial

Commercial and industrial blueprints are usually drawn and designed by an engineer or architect. The full set of construction drawings may be several hundred pages long. The drawings will be separated by discipline, such as architectural, electrical, plumbing, and so on. The electrical prints will be divided into the various types of drawings (e.g., site, plan, elevation) and electrical categories (e.g., lighting, power, systems).

Power Plan

The **power plan** is the drawing that will convey the information about where devices and equipment are to go. The power plan will also indicate how the circuitry is divided and how the raceways are to be run.

The electrical panels will be identified with a panelboard symbol and a series of letters, which are the panelboard identifiers. Panelboard identifiers are used in buildings with multiple panels to distinguish which panel is which. The panelboard identifier will typically indicate the voltage of the panel where there are multiple voltages on the premises. For example, a panel may have *LK* written next to it (see Figure 12–15). The *L* in this case represents lower voltage (typically $^{120}/_{208}$), and the *K* indicates panel K. Another panel might have *HA* written next to it. The *H* in this case represents higher voltage (typically $^{277}/_{480}$), and the *A* indicates panel A.

Power Plan
A drawing in plan view that conveys where electrical devices and equipment are to be installed.

FieldNote!

Many new homes are being constructed with recessed lights. Recessed lights take up a considerable amount of space above the finished ceiling (see Figure 12–14). In addition, recessed lights require that a room that would otherwise have only one surface-mounted light fixture have several recessed light fixtures. This will often create a conflict with ductwork and plumbing. It is very important that the electrician communicate with the plumbing and HVAC installer to see where their runs will be and what accommodations can be made.

FIGURE 12–15 A power plan shows the location of devices and equipment as well as the circuitry. (Courtesy of Ulteig Engineers.)

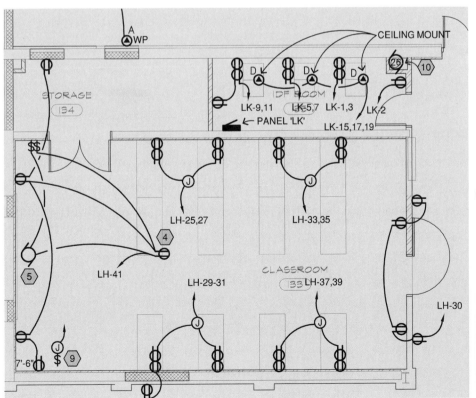

FIGURE 12–14

Recessed lights have an enclosure (can) that is above the sheetrock.

Panelboards will be typically separated by voltage and will be lettered starting with *A*. If there are four lower-voltage ($^{120}/_{208}$) panels they will be identified by the letters *LA* through *LD*. The *L* and *H* (which, as stated, stand for "lower" and "higher" voltage) represent a comparison of the voltages available at the premises. The *L* represents the lower voltage available on the premises, and the *H* represents the higher voltage available on the premises. Larger buildings many times have additional distinguishing letters, numbers, and abbreviations used to identify their panelboards by floor, area of the building, or purpose.

The power plan will show where the receptacles and equipment are to be located. The location of receptacles is typically approximate, meaning that the electrician can place the box on the nearest stud. Occasionally, a receptacle will have a very specific location and dimensions or a note will give details on its exact location.

A home run symbol with a panel identifier will identify which panel the home run goes to and which circuit in that panel it connects to. For example, in Figure 12–15 the home run with LK-2 indicates that the receptacle is to be fed from circuit 2 in panel LK. Curved lines are used to represent which equipment and devices are tied together by conduit runs or by circuitry, or by both. The home run symbol and the curved lines connecting devices will sometimes have hash marks. The hash marks represent how many circuit conductors are installed in the raceway. Multiple arrows at the end of the home run depiction are also used to identify the number of circuits (see Figure 12–16).

FIGURE 12–16 Arrows are often used to indicate the number of circuits in a home run. Hash marks identify the number of conductors.

A1.3

Lighting Plan

Lighting Plan
A drawing in plan view that gives information on lighting.

The **lighting plan** will have the information necessary to install the lights. The plan will have a lighting symbol with a capital letter, a number, or a combination of the two near it to identify which lighting fixture is to be installed (see Figure 12–17). The letter will correspond to a letter on the lighting schedule, which will give specific information about the light (see Figure 12–18).

FIGURE 12-17 A lighting plan shows the location of light fixtures, how the switching is accomplished, and the circuitry. (Courtesy of Ulteig Engineers.)

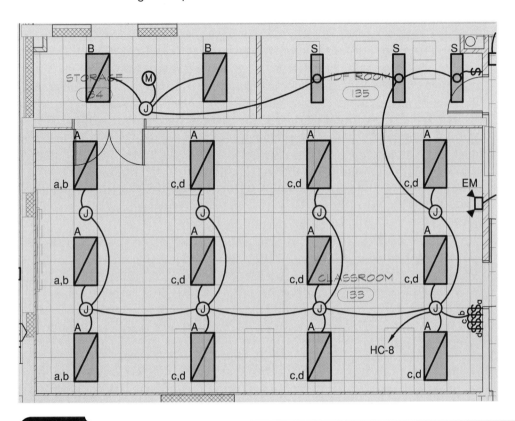

FIGURE 12-18 Light fixture schedule.

LIGHT FIXTURE SCHEDULE

TYPE	MANUFACTURER	CATALOG NUMBER	LAMPS	MOUNTING	REMARKS
A	LITHONIA	WA440A277	4-40W	CEILING	WRAPAROUND ACRYLIC LENS
			FLUOR.	SURFACE	
B	KEYSTONE	2A440EXA277GPWS	4-40W	CEILING	2X4 GRID TROFFER
			FLUORESCENT	SURFACE	ACRYLIC LENS. AIR RETURN
C	KEYSTONE	2A440PWSGPW277S	4-40W	CEILING	2X4 GRID TROFFER
			FLUORESCENT	SURFACE	1/2"X1/2"X1/2" SILVER
					PARABOLIC LOUVER
					AIR RETURN
D	PHOENIX	DL-300D60-3	1-300W	WALL	DUAL ARM DOCK
			R40	7'-0" AFF	LIGHT, 62" LENGTH W/ VERT
					ADJUSTMENT,WIREGUARD,&
					BUILT-IN SWITCH

The lighting symbols drawn on the plan will sometimes have a lower-case letter written next to them (see Figure 12–17). This letter indicates which switch is controlling the lights. If there are two lowercase letters next to a lighting symbol, it would mean that part of the light is controlled by one switch and the other part by another switch. This is common with fluorescent lighting fixtures.

A curved line will often tie together lighting fixtures fed from the same switch or on the same circuit. A home run symbol with the panel identifier written next to it will indicate which panel the lights are fed from, and which circuit the lights connect to.

The lighting plan will also indicate how the lighting is switched. It will indicate which lights are being controlled, and where the switch is to be located. The switches will often have a lowercase letter by them to indicate which lights they switch (see Figure 12–16). The height of the switches will typically be a standard height that is indicated by a note or in the specifications. If the height of a switch varies from the standard height, it will be indicated on the plan near that particular switch.

The **reflected ceiling plan** is also used for the placement of lighting fixtures. The plan is viewed as if a person were looking at a mirrored floor and could see the reflection of the ceiling (see Figure 12–19). This is the plan used by the ceiling tile installers. It will tell them how to space the ceiling tiles and where lights are to be located.

Systems Plan

There may be several types of systems that will be on a **systems plan**. There may also be a separate plan for each system. Examples of items found on a systems plan are fire alarms, security, speakers, microphones, television, computer, and telephone.

The systems plan will also show where a raceway will be required for pulling in cables. An example would be raceways that need to be installed in the concrete for a data jack in the floor. Another example would be boxes installed in the wall with a pipe that extends out above the ceiling. Often a note will indicate that a ¾-inch pipe must be run up from each communication box and stubbed out above the ceiling tile or along the ceiling. That pipe will then be used to pull the computer or telephone cable down to the box in the wall.

Cable trays are often used to support the various communications cable used on a project. The path of the cable tray throughout a project will be indicated on the systems plan (see Figure 12–20). The symbols used to represent the various systems are not consistent from one designer to another. It is very important to check the legend to know what each of the symbols represents.

Reflective Ceiling Plan
A drawing that indicates how the finished ceiling is to be laid out. It has information on the placement of ceiling tiles, lights, and any other device located in or on the ceiling.

Systems Plan
A drawing in plan view that contains information on the various electrical systems of a premises. Examples would be fire alarm, communications, and security.

FIGURE 12-19 A reflected ceiling plan shows a view of the ceiling as if it were reflected off a mirrored floor.

CEILING OF TWO ROOMS

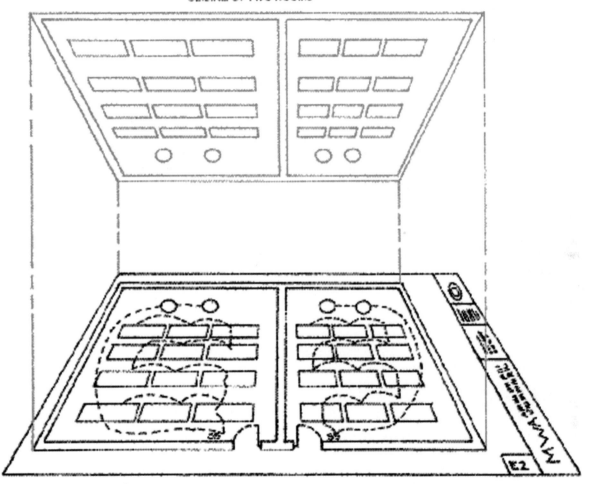

REFLECTED CEILING PLAN

FIGURE 12-20 A systems plan gives the locations of the various system devices and equipment on the premises. This plan has a cable tray installed for all the system cables. (Courtesy of Ulteig Engineers.)

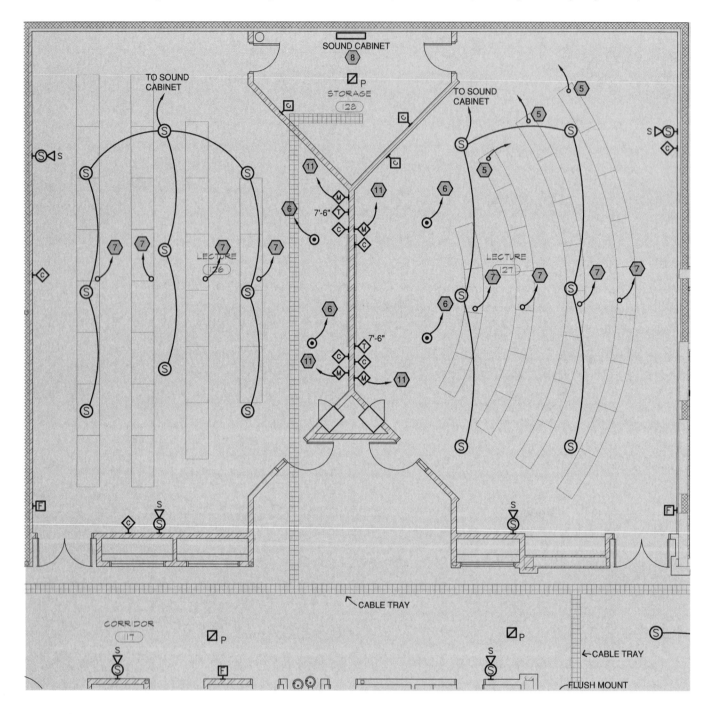

SUMMARY

- The floor plan is a scaled view looking downward at one floor or level of a building.
- An electrician should be able to understand the electrical floor plan and the floor plans pertaining to other disciplines.
- The floor plan shows relationships of rooms, walls, doors, windows, plumbing, electrical, and many other features of a building.
- A floor plan will only have the dimensions necessary to convey the information on that particular print.
- A set of construction drawings will typically have a legend to identify the symbols and abbreviations used in the drawings.
- A foundation plan is a drawing giving information about the foundation walls, footings, and support columns or posts.
- Demolition plans are used in a remodel to give detailed information about the areas of a building that will be moved or demolished.
- Residential electrical plans are typically drawn up by the electrical contractor, not an electrical engineer. The circuitry is laid out by the electrician on the job.

- Commercial electrical drawings are typically drawn by an engineer and will have the placement of lighting, devices, equipment, and circuitry laid out on the floor plan. There are three main types of electrical floor plans.

1. *Power plan.* Information on circuitry and the placement of devices and equipment.
2. *Lighting plan.* Information of circuitry and the placement of lighting fixtures and lighting control.
3. *Systems plan.* Information on the placement of various electrical systems, such as the following:

   - Fire alarm system
   - Security system
   - Speakers
   - Microphones
   - Television
   - Computer
   - Telephone

REVIEW QUESTIONS

1. List three items on a floor plan that would be shown with a hidden line.
2. Why would there be more than one drawing for one level of a building?
3. Why is it important to understand the floor plans of the other disciplines?
4. Where would a person look to find out the meaning of a symbol or abbreviation?
5. Write an example of a specific note found on an electrical floor plan.
6. Write an example of a general note found on an electrical floor plan.
7. What is the difference between a foundation plan and a basement plan?
8. What type of information would be found on a demolition plan?
9. Who typically draws in the electrical on a residential floor plan?
10. List three pieces of information that would be found in an electrical power plan.
11. List three pieces of information that would be found in an electrical lighting plan.
12. List three pieces of information that could be found on an electrical systems plan.

FLOOR PLAN EXERCISE

Refer to Problem 12–1 (page 201) to answer the following questions:

1. Which panel and circuit are feeding the electric watercooler in corridor 117?

2. How many duplex receptacles are in classroom 118?

3. How many double duplex receptacles are in room 122 (faculty offices)?

4. Which panel and circuit are the double duplex receptacles in room 122 (faculty offices) fed from?

5. Which panel and circuit are feeding the above-counter receptacles in room 132 (community room)?

6. Describe which receptacles are on the same circuit as the watercooler.

7. How many receptacles are installed at 7 feet 6 inches in classroom 118?

8. How many circuits feed the receptacles in room 122 (faculty offices)?

9. At what height is the receptacle over the counter in room 119 (conference room) mounted?

10. Which panel and circuit are feeding the receptacles in room 121 (storage)?

Refer to Problem 12–2 (pages 202-203) to answer the following questions:

1. Where is the water heater located?

2. What surrounds the patio?

3. Which room has a fireplace?

4. What does the hidden line in the dining room represent?

5. How many receptacle outlets are shown in the kitchen?

6. How many walk-out exterior doors are there in this residence?

7. Where is the cooktop located?

8. How many exterior receptacles are there?

9. Is there a basement in this house?

10. Describe the window in the living room.

11. How many switches control the light in the rear entryway?

12. How many receptacles are in the garage?

13. Describe the wall between the family room and living room.

14. What do the dashed lines in the kitchen represent?

15. How many exterior lights are there?

Problem 12–1

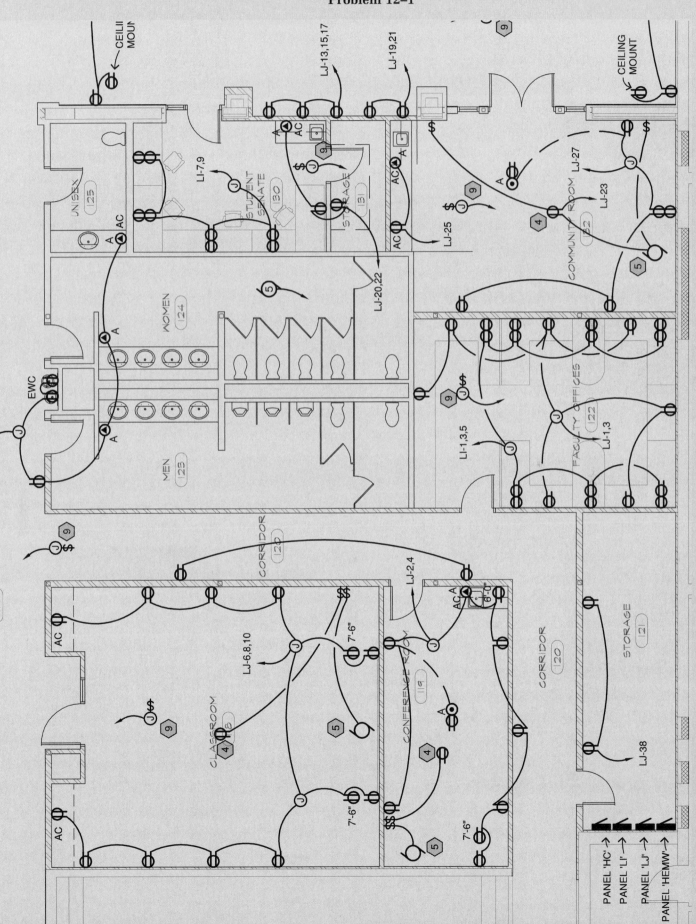

Problem 12–2

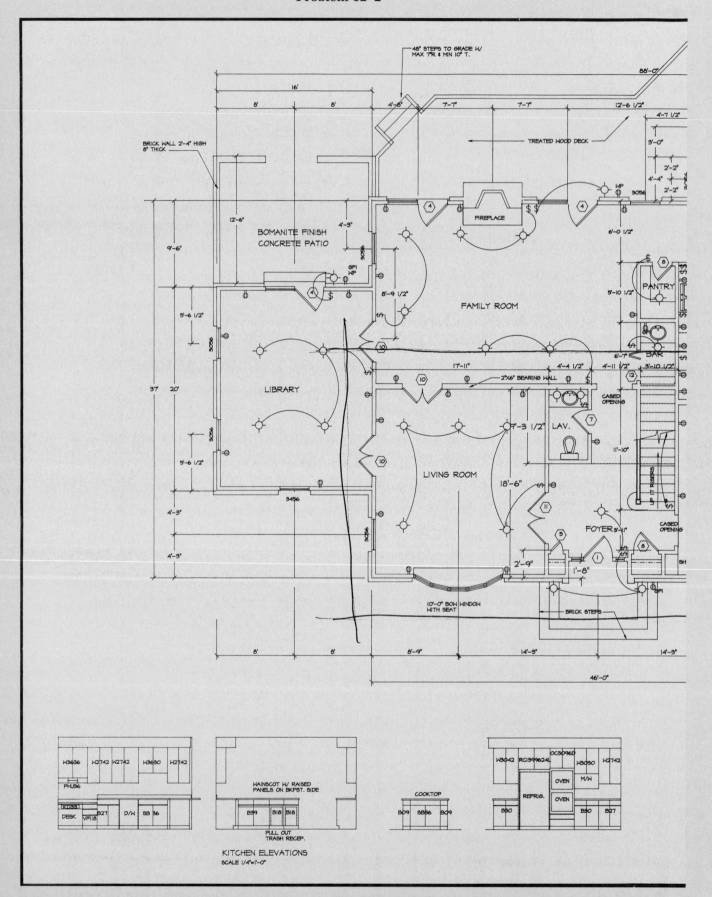

KITCHEN ELEVATIONS
SCALE 1/4"=1'-0"

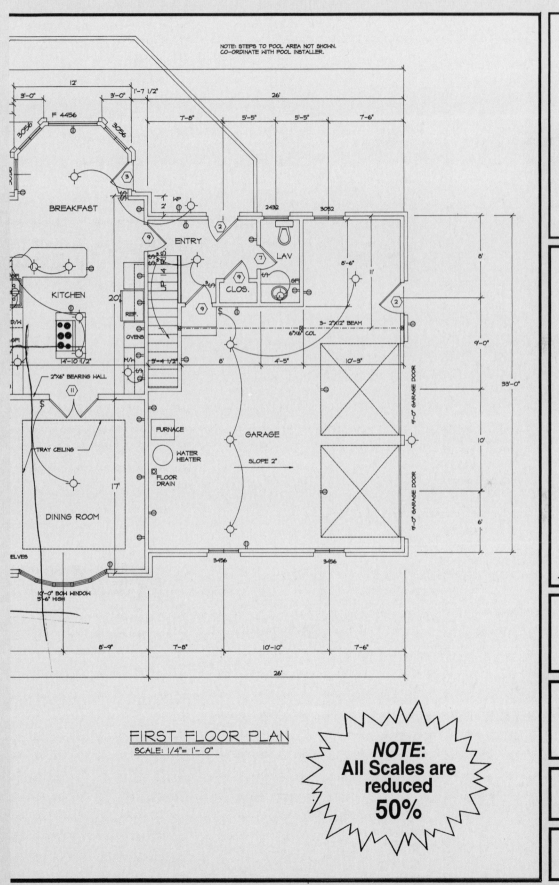

NOTE: STEPS TO POOL AREA NOT SHOWN.
CO-ORDINATE WITH POOL INSTALLER.

BREAKFAST

F 4456

KITCHEN

2"X6" BEARING WALL

TRAY CEILING

DINING ROOM

SHELVES

10'-0" BOW WINDOW
5'-6" HIGH

ENTRY

CLOS.

LAV

REF.

OVENS

M/W

FURNACE

WATER
HEATER

FLOOR
DRAIN

GARAGE

SLOPE 2"

3- 2"X12" BEAM
6"X6" COL.

9'-0" GARAGE DOOR

9'-0" GARAGE DOOR

3456   3456

FIRST FLOOR PLAN
SCALE: 1/4"= 1'- 0"

NOTE:
All Scales are
reduced
50%

ANTHONY KAMPWERTH
ARCHITECT

FRANKLIN RESIDENCE
FAIRWAY SUBDIVISION

PROBLEM
12-1 is
this full plan

FIRST FLOOR PLAN

REVISIONS:
4/21/92

PROJECT:
FILE NO: FRANKLINA1
DATE: 4-13-92

SHEET NO.

A1

13

Elevations

O U T L I N E

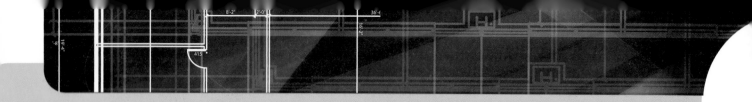

OVERVIEW

There are two main types of elevations: exterior and interior. Exterior elevations give a view of what the building looks like from the outside. Interior elevations give a view of an internal part of the building. Elevations not only give important construction specifics on how a project is built but are extremely helpful in visualizing a project.

OBJECTIVES

After completing this chapter, you should be able to:

• Identify the various elevation views and relate them to the construction site.

• Recognize the type of wall finish used on a drawing and its relevance to the electrical field.

• Understand the dimensions given on an elevation drawing.

• Identify the type of roof and pitch by looking at elevation drawings.

• Identify the types of doors and windows in an elevation drawing.

• Describe the information found on an interior elevation.

FIGURE 13-1 Elevations may be named by compass direction.

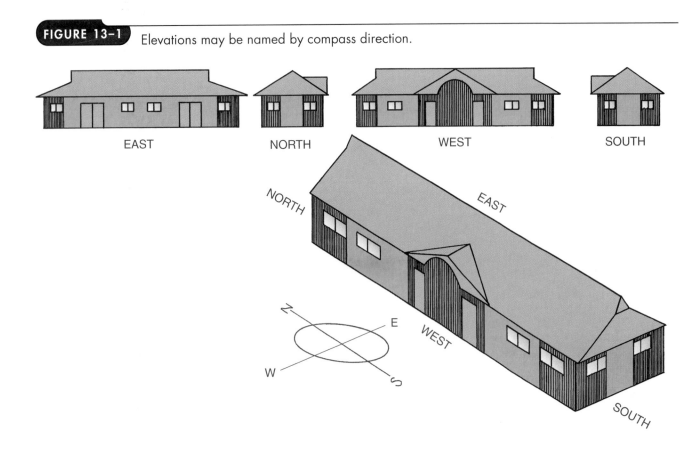

EAST   NORTH   WEST   SOUTH

EXTERIOR ELEVATIONS

Exterior elevations are used to give a view of what a building looks like from the exterior. There are typically four main elevations in a set of construction drawings. The four main elevations are north, south, east, and west. This is based on the side of the house that faces a particular direction. For example, the side of the house that faces south would be the south elevation (see Figure 13–1).

Some plans are drawn without knowing where the building will be located, making it impossible to give the view a direction. In this case, they will be labeled as "front," "right side" (or "right"), "left side" (or "left"), and "rear" elevations (see Figure 13–2). This is based on the main entrance of the building. The view of the building you are looking at while facing the main entrance is the *front* elevation. The side of the building to the right while facing the main entrance is the *right* elevation. The side of the building to the left while facing the main entrance is the *left* elevation. The side opposite the main entrance is the *rear* elevation.

The front view is typically drawn to the same scale as the floor plan. The other three views are sometimes drawn to a smaller scale than the floor plan and all placed on the same sheet. The scale of the drawings will typically be printed below the drawing.

FIGURE 13-2 Elevations may be named by their relation to the front of the building.

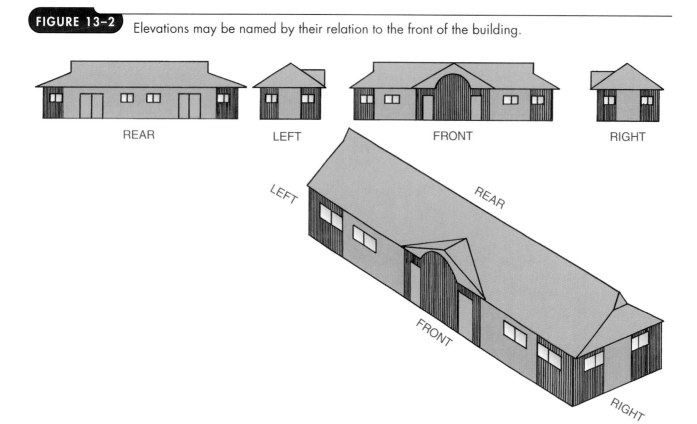

REAR   LEFT   FRONT   RIGHT

13.1 Exterior Wall Finish

Exterior elevations will indicate the wall finish to be used on the building. This does not mean that the entire building will be drawn showing the finished wall material. Often only parts of the building will show the wall finish. The wall finishes will either just fade off or there will be a break line where it ends. Some elevations will not show the exterior finish at all. They will be an outline of the building and openings only (see Figure 13–3). Not showing all details of the exterior finish minimizes clutter on the prints and makes it easier to differentiate the openings, jogs, and other distinguishing features on the building.

Notes are often used to indicate the specific types of wall finish to be used, and to give other information that cannot be easily communicated by just looking at a drawing. Figure 13–4 contains leaders that point to the wall to indicate the color of brick to be used in that section.

13.2 Dimensions

Vertical dimensions are sometimes given on elevations. The string of dimensions will typically include the following information:

- The distance from grade to the first floor, second floor, third floor, and so on.
- The distance from grade to the tops of windows and doors.
- The distance from grade to footings, and the thickness of footings.

FIGURE 13-3 Right elevation of a residence.

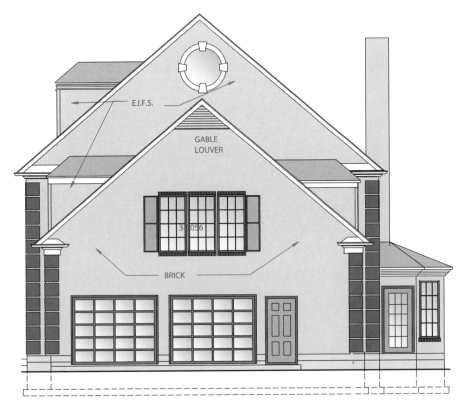

E.I.F.S.

GABLE
LOUVER

3.5056

BRICK

RIGHT ELEVATION

FieldNote!

An electrician must know the exterior finish of the building to make the proper provisions for mounting exterior lights. If the finish is to be brick, a masonry box will be installed. If it is lap siding, a siding block will be installed. If the exterior is Drivit, a box will be mounted on the sheathing that is the thickness of the Drivit. The elevation drawings are the best means of determining what type of wall finish will be used—and if there are multiple wall finishes, where each ends (see Figure 13–4).

FIGURE 13-4 Front elevation of a residence.

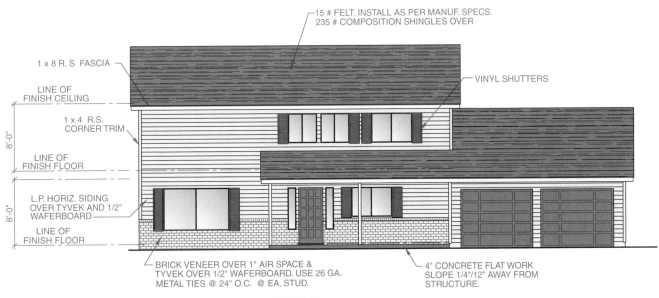

15 # FELT. INSTALL AS PER MANUF. SPECS.
235 # COMPOSITION SHINGLES OVER

1 x 8 R. S FASCIA

LINE OF
FINISH CEILING

VINYL SHUTTERS

1 x 4 R.S.
CORNER TRIM

8'-0"

LINE OF
FINISH FLOOR

L.P. HORIZ. SIDING
OVER TYVEK AND 1/2"
WAFERBOARD

8'-0"

LINE OF
FINISH FLOOR

BRICK VENEER OVER 1" AIR SPACE &
TYVEK OVER 1/2" WAFERBOARD. USE 26 GA.
METAL TIES @ 24" O.C. @ EA. STUD.

4" CONCRETE FLAT WORK
SLOPE 1/4"/12" AWAY FROM
STRUCTURE.

FRONT ELEVATION
1/4" = 1'-0"

FIGURE 13-5   Elevations will sometimes give vertical dimensions of the building.

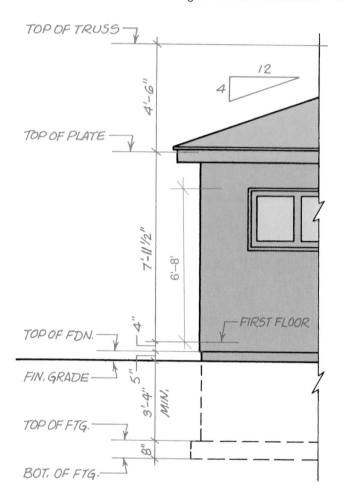

Roof Pitch

A representation of the slope or angle of a roof, typically written as a fraction of the rise over the run [such as $4/12$, meaning a rise of 4 (inches) for every run of 12 (inches)].

Rise

The amount a roof increases in height in a predetermined horizontal distance.

Run

A horizontal distance used to determine the pitch of a roof.

Dormer

A projection out of a sloped roof, typically containing a window that provides additional space in a room with a sloped ceiling.

The foundation walls and the footings of the building will be shown with a hidden line. The depths of the foundation and any elevation changes will be indicated (see Figure 13–5).

13.3 Roof Information

The **roof pitch** is often found on elevation drawings. This is shown with a right triangle drawn above the roof line (see Figure 13–5). Two numbers will be shown beside the triangle. The number beside the vertical line of the triangle is the **rise,** and the number beside the horizontal line of the triangle is the **run.** This indicates the pitch or the rise over run of the roof. If a roof has a rise of 4 and a run of 12, this means that the roof rises 4 inches for every 12 inches traveled horizontally. This is commonly referred to as a $4/12$ pitch.

The layout of the roof and the style of the roof are easily understood by looking at the elevations. Is the roof a gable, hip, or shed roof? Are there any **dormers?** All of this information is clearly identified on the elevations. The type of material used on the roof will be identified either by a symbol or a note.

13.4 Windows and Doors

The elevations will also contain information about windows and doors. The types and sizes of windows used might be identified by a dimension or a notation. The windows shown in Figure 13–6 have letters written in each window that will correspond to the window schedule. If the window swings open, the window will have an arrow with the point of the arrow

FIGURE 13–6   Front side and rear building elevations.

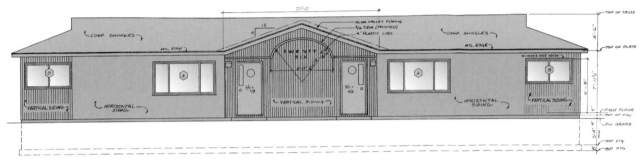

FRONT ELEVATION

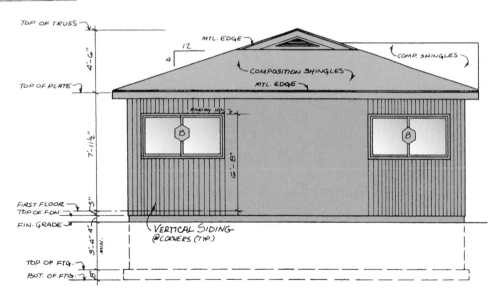

SIDE ELEVATION
SCALE = 1/4" = 1'-0"

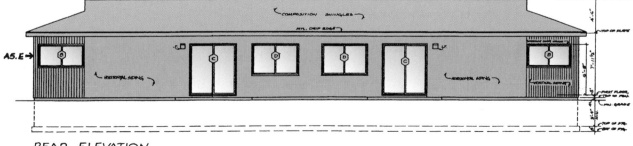

REAR ELEVATION
SCALE = 3/16" = 1'-0"

pointing toward the hinge side of the window. Figure 13–7 shows plan, elevation, and pictorial views of commonly used windows.

Information on doors can also be found on elevation drawings. Floor plans and schedules will convey the same information or more, but the elevation is giving the information from a different perspective that may be easier to visualize.

INTERIOR ELEVATIONS

Interior elevations are a type of detail drawing. They are typically drawn to a larger scale than the floor plan. Interior elevations give important information about how an interior feature is to be built. Interior elevations are dimensioned to indicate size and location. Notations and symbols are used to indicate construction materials and finish. The drawing gives a pictorial view of how the feature is to be constructed and how it is to look when finished.

A common interior elevation is a cabinet drawing. These drawings typically include the height and width of all sections of cabinets. It will show the location of appliances, sinks, and so on (see Figure 13–8). The point of an arrow on a cabinet door represents the hinge side of the door.

The kitchen cabinet details are an interior elevation that is often referred to by electricians. The following are a few examples of the information an electrician will use from a set of cabinet elevations:

- *Height to the bottom of the upper cabinets.* This is used when under-cabinet lights will be mounted below the upper cabinets. The electrician will need to know at what height the cable will need to be sticking out of the wall.

- *Height to the top of the upper cabinets.* This is used when outlets will be mounted above the upper cabinets for plugging in lights and the like.

- *Location of the cabinet above the microwave.* The outlet the microwave plugs into will be mounted in the cabinet above the microwave.

- *Location of the oven or cooktop.* The kitchen elevations will indicate at what height a cable should be left out to tie into the appliances.

FIGURE 13-7 Pictorial, plan, and elevation views of commonly used windows.

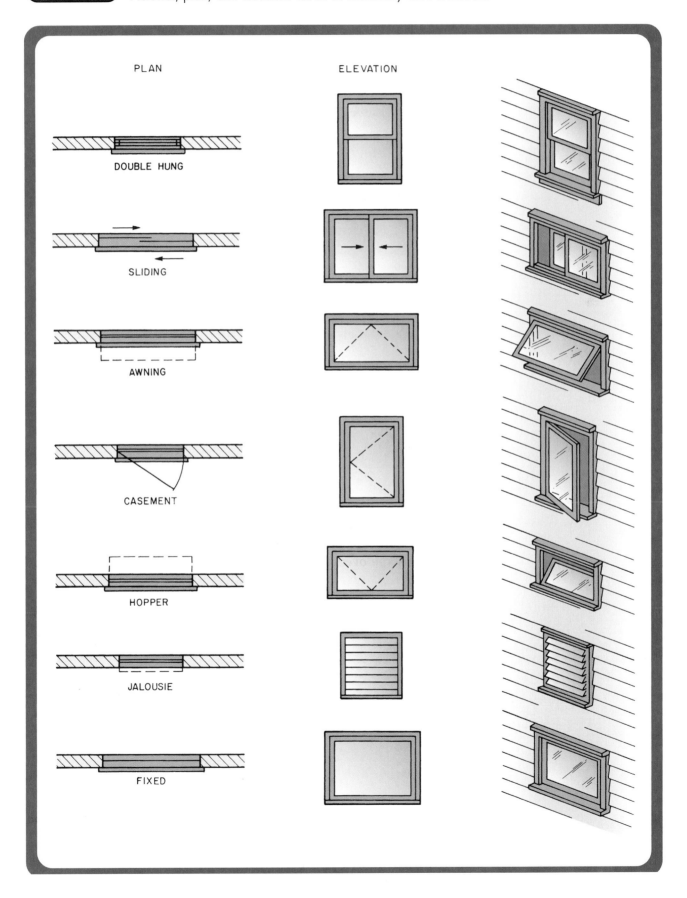

PLAN

ELEVATION

DOUBLE HUNG

SLIDING

AWNING

CASEMENT

HOPPER

JALOUSIE

FIXED

FIGURE 13-8 Interior elevations such as cabinet elevations often have size and location dimensions.

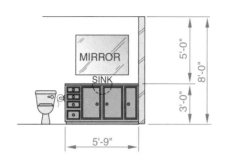

E — BATH ELEVATION
SCALE: 1/4" = 1'-0"

D — BATH ELEVATION
SCALE: 1/4" = 1'-0"

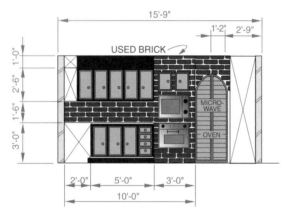

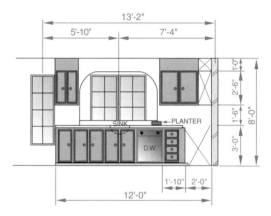

B — KITCHEN ELEVATION
SCALE: 1/4" = 1'-0"

A — KITCHEN ELEVATION
SCALE: 1/4" = 1'-0"

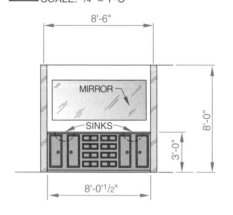

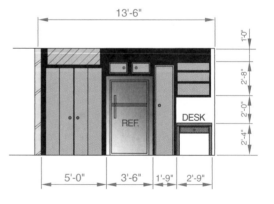

F — MASTER BATH ELEVATION
SCALE: 1/4" = 1'-0"

C — KITCHEN ELEVATION
SCALE: 1/4" = 1'-0"

SUMMARY

Interior and exterior are the two types of elevations found in a set of construction drawings.

- Understanding the various types of exterior elevation views will help to visualize a project.
- Exterior elevations will be identified by their relationship to north, south, east, and west, or will be identified by their relationship to the main entrance of the building (front, right side, left side, or rear).
- The exterior elevations will identify the finished wall material, as well as where it begins and ends.

- The type of roof and roof pitch are identified in the exterior elevations.
- A pictorial view of windows and doors, as well as their swing, is indicated in elevation drawings.
- Vertical dimensions are typically given in elevations.
- Interior elevations are typically drawn to a larger scale and give a pictorial view of how an interior wall or feature is to be constructed.

REVIEW QUESTIONS

1. What are two types of elevation drawings?
2. Which elevation view will be facing east?
3. How are the four elevation views identified when there is no reference to north?
4. What is the importance to an electrician in knowing the exterior finish?

5. How are the foundation walls and footings shown on an elevation?
6. Describe the meaning of a $^6/_{12}$ roof pitch.
7. Describe how the swing of a window will be identified on an elevation.

ELEVATION EXERCISE

Use Problem 13–1 (page 216) to answer the following questions:

1. How many and what size of under-cabinet lights are to be installed in the kitchen?
2. How tall is the opening in the cabinets for the refrigerator?
3. How tall is the base cabinet in the living room elevation?
4. What is the height to the top of the backsplash in the utility room?

5. What is the finished wall material in the kitchen?
6. Describe the cooking appliances in the kitchen.
7. What are the height and width dimensions of the cabinet in the master bathroom?
8. What are the overall dimensions of the island in the kitchen?
9. What is the scale of the cabinet elevations?
10. How wide is the opening for the washer and dryer in the utility room?

Problem 13–1

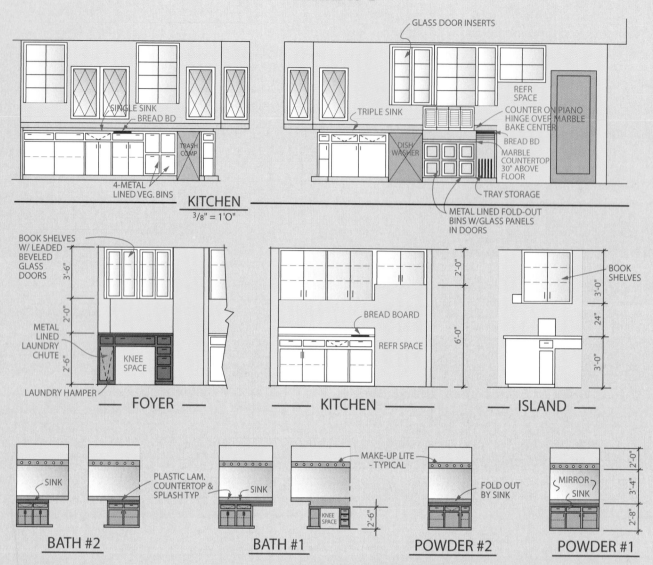

KITCHEN
3/8" = 1'0"

Use Problem 13–2 (page 217) to answer the following questions.

1. How many colors of brick are used on the building?

2. What are the dimensions of the garage door?

3. What is the distance from the first floor to the second floor?

4. What is the scale of the elevation?

5. What is shown with a hidden line?

6. Describe the garage door.

7. What is the distance between concrete piles?

8. What is above the elevation of the roof in the center of the building?

Problem 13–2

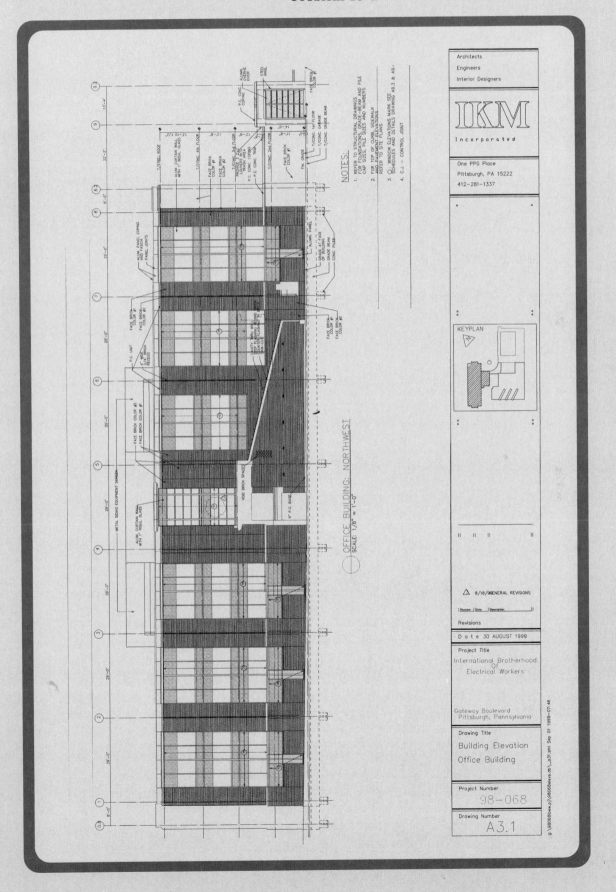

Architects
Engineers
Interior Designers

IKM
Incorporated

One PPG Place
Pittsburgh, PA 15222
412–281–1337

KEYPLAN

△ 8/18/99GENERAL REVISIONS

Revisions

Date 30 AUGUST 1999

Project Title
International Brotherhood
Of
Electrical Workers

Gateway Boulevard
Pittsburgh, Pennsylvania

Drawing Title
Building Elevation
Office Building

Project Number
98–068

Drawing Number
A3.1

14

Details and Sections

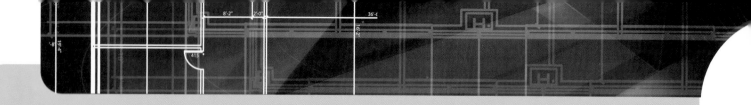

OVERVIEW

Site plans, floor plans, and elevations have an abundance of information but are limited in some aspects due to their scale and how they are viewed. Section and detail drawings can give important detailed information other drawings are unable to provide. This chapter covers the information found in section and detail drawings and relates it to the electrical industry.

OBJECTIVES

After completing this chapter, you should be able to:

• Describe the information given by a cutting-plane line and a ball note.

• Describe the information given in a wall section.

• List the four types of detail drawings.

• Apply the information found in a detail drawing to the construction site.

SECTIONS

Sections are scaled drawings giving a view of an object or part of a building that has been cut away to see the inside features. Using only plan and elevation views, it is difficult to determine materials and sizes of wall and framing members due to the drawing scale. Sections are used to give additional information about what materials are used and how a building is to be constructed.

A cutting-plane line will be drawn on an elevation or floor plan to indicate where the section has been taken. The cutting-plane line will have an arrow indicating which direction the cut is to be viewed, and will contain a letter or **ball note** to identify where the section has been drawn (see Figure 14–1). Sections that are drawn on the same page the section was taken from will be identified simply by letters (see Figure 14–2). Commercial and industrial drawings may have many pages of sections and

Ball Note

A symbol used to identify a section by number, page from which the section was taken, and page on which the section is drawn.

FIGURE 14–1 Cutting plane lines show where the imaginary cut is made as well as from which direction it is viewed.

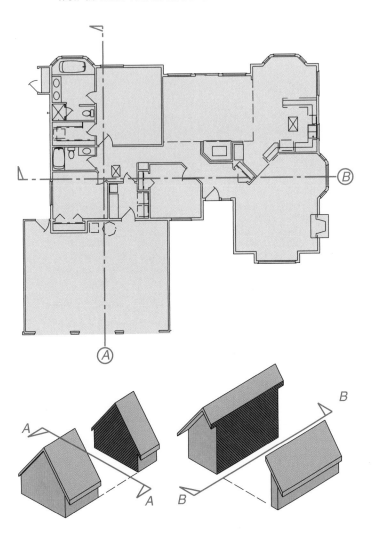

FIGURE 14-2 When a section is drawn on the same page that it was taken from, it is often simply identified by letters.

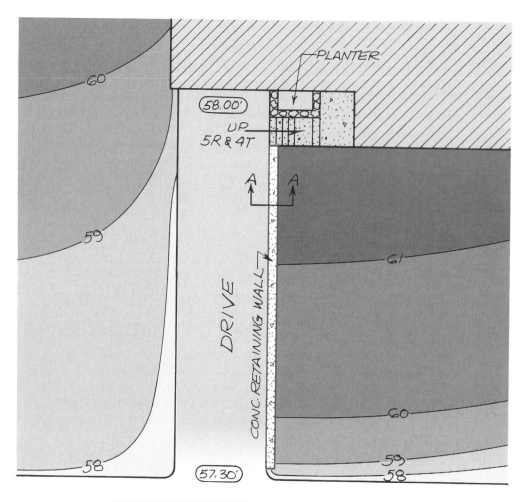

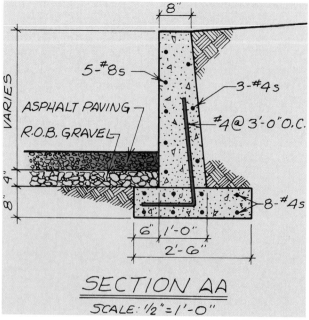

SECTION AA

SCALE: 1/2"=1'-0"

FIGURE 14-3 Arrows and ball notes are used to identify which section it is, which page it was taken from, and which page it is drawn on.

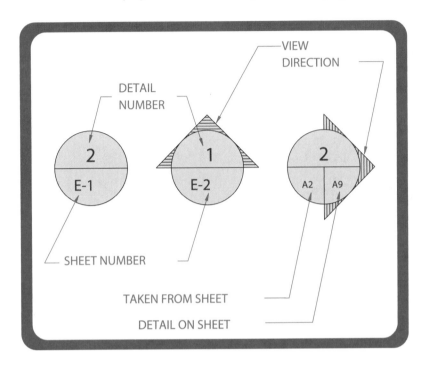

FieldNote!

An electrician will usually have to penetrate the exterior wall. It could be for various reasons: to bring power into the building, to feed an outdoor light, to feed parking lot lights, to accommodate an outdoor receptacle, and so on. A wall section can be used to see how the exterior wall can be penetrated and when the work should be performed. For example, if the wall is poured concrete, the electrician may want to be there before the pour to get a sleeve in the wall. If the wall is precast concrete, a core drill will be required. A lot of headaches can be saved by having the right tools and being in the right place at the right time.

details. A ball note is often used on commercial and industrial drawings to identify the section or detail and to indicate in which print it can be found (see Figure 14–3).

A very common section is a wall section. A wall section will contain information on what types and sizes of materials are used to build the wall. A typical wall section is a drawing that shows how a "typical" wall in the building is to be constructed. Dimensions and notes are used to convey all of the necessary information (see Figure 14–4).

A building section is a section view of the entire building. Building sections are used to give dimensions, to show relationships of parts of the building, and to show how the building is constructed (see Figure 14–5).

DETAIL DRAWINGS

Details are scaled drawings that are drawn to a large scale to show exactly how a feature is made or where it is located (see Figure 14–6). The detail will either be dimensioned or drawn to an accurate scale that will produce the desired feature with accuracy not possible from a print with a smaller scale. The scale usually runs from ¾ inch equals 1 foot to 3 inches equals 1 foot. Details may be drawn in plan, elevation, section, or isometric views.

Plan details clarify the layout of the building and give specific details. If the scale of the plan is too small to convey all necessary information, a detail of the portion of the plan that needs clarification will be drawn.

Plan Detail
A drawing in plan view drawn to a larger scale to provide additional information.

FIGURE 14–4   A typical wall section gives information about how a typical wall in a building is to be constructed.

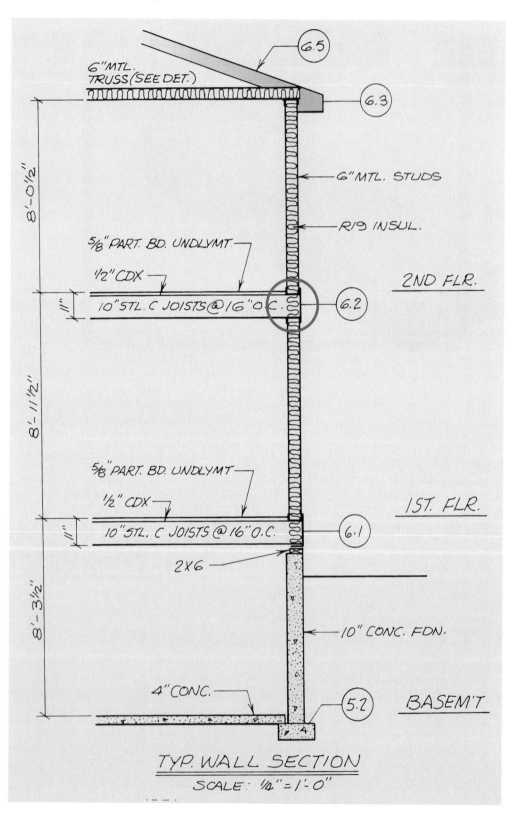

6" MTL. TRUSS (SEE DET.)

6.5

6.3

6" MTL. STUDS

R19 INSUL.

5/8" PART. BD. UNDLYMT.

1/2" CDX

10" STL. C JOISTS @ 16" O.C.

2ND FLR.

6.2

8'-0½"

11"

5/8" PART. BD. UNDLYMT.

1/2" CDX

10" STL. C JOISTS @ 16" O.C.

1ST. FLR.

6.1

2X6

8'-11½"

11"

10" CONC. FDN.

4" CONC.

8'-3½"

BASEM'T

5.2

TYP. WALL SECTION

SCALE: 1/2" = 1'-0"

FIGURE 14-5 A building section is a section view of the entire building.

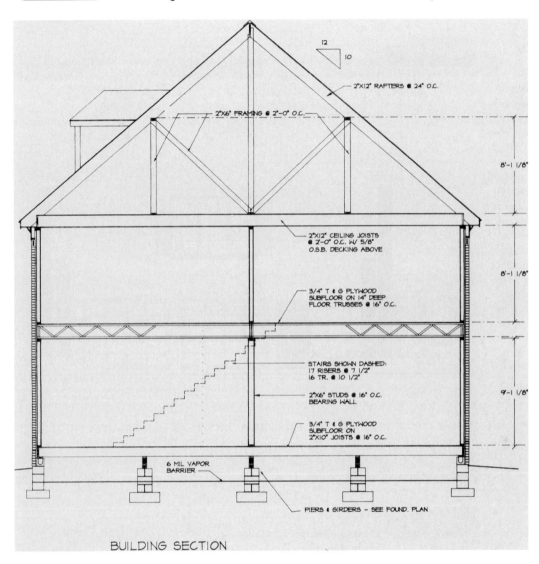

12
10

2"X12" RAFTERS @ 24" O.C.

2"X6" FRAMING @ 2'-0" O.C.

8'-1 1/8"

2"X12" CEILING JOISTS
@ 2'-0" O.C. W/ 5/8"
O.S.B. DECKING ABOVE

8'-1 1/8"

3/4" T & G PLYWOOD
SUBFLOOR ON 14" DEEP
FLOOR TRUSSES @ 16" O.C.

STAIRS SHOWN DASHED:
17 RISERS @ 7 1/2"
16 TR. @ 10 1/2"

2"X6" STUDS @ 16" O.C.
BEARING WALL

9'-1 1/8"

3/4" T & G PLYWOOD
SUBFLOOR ON
2"X10" JOISTS @ 16" O.C.

6 MIL VAPOR
BARRIER

PIERS & GIRDERS - SEE FOUND. PLAN

BUILDING SECTION

FIGURE 14-6

A detail drawing gives detailed information about how a specific portion of a building is to be constructed. This cable tray detail shows how the cable tray is to be hung from the ceiling. (Courtesy of Ulteig Engineers.)

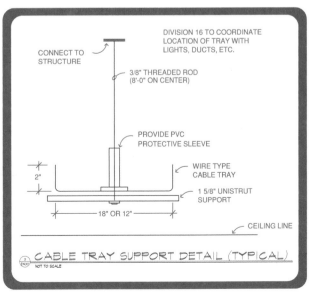

CONNECT TO
STRUCTURE

DIVISION 16 TO COORDINATE
LOCATION OF TRAY WITH
LIGHTS, DUCTS, ETC.

3/8" THREADED ROD
(8'-0" ON CENTER)

PROVIDE PVC
PROTECTIVE SLEEVE

WIRE TYPE
CABLE TRAY

1 5/8" UNISTRUT
SUPPORT

2"

18" OR 12"

CEILING LINE

CABLE TRAY SUPPORT DETAIL (TYPICAL)
NOT TO SCALE

Figure 14–7 shows a plan detail of a stairway. An enlarged plan view of the stairway has been drawn to provide additional information that could not be conveyed due to the small scale of the floor plan.

FIGURE 14–7   A plan detail gives a plan view of a portion of the building that is drawn at a larger scale to clarify and give additional information.

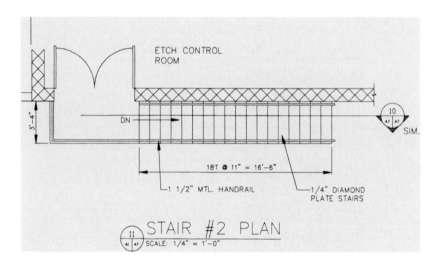

Elevation Detail

A drawing in elevation view drawn to a larger scale to provide additional information.

An **elevation detail** conveys how a portion of a building looks, gives dimensions for construction, and shows the relationships of features. Figure 14–8 shows an example of an elevation detail. The fireplace elevation gives dimensions and information about how the fireplace is to be constructed. Figure 14–9 shows an example of an electrical elevation detail. This detail gives placement and mounting instructions for an explosion-proof receptacle.

FIGURE 14–8   A fireplace elevation is a detail drawing that gives a view of how the fireplace is built as well as important dimensions.

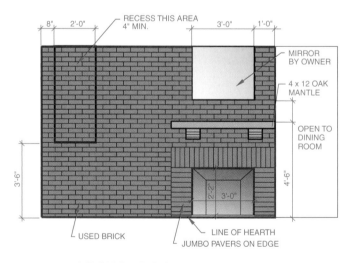

FIGURE 14–9 This detail of an explosion-proof receptacle gives details about how the receptacle is to be installed. It gives a vertical height dimension, as well as how the conduit seal, junction box, and straps are to be installed.

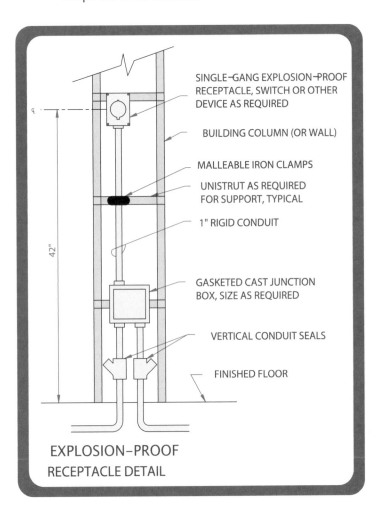

SINGLE–GANG EXPLOSION–PROOF RECEPTACLE, SWITCH OR OTHER DEVICE AS REQUIRED

BUILDING COLUMN (OR WALL)

MALLEABLE IRON CLAMPS

UNISTRUT AS REQUIRED FOR SUPPORT, TYPICAL

1" RIGID CONDUIT

GASKETED CAST JUNCTION BOX, SIZE AS REQUIRED

VERTICAL CONDUIT SEALS

FINISHED FLOOR

42"

EXPLOSION–PROOF RECEPTACLE DETAIL

Section Detail

A section drawing drawn to a larger scale to provide additional information.

Isometric Detail

An isometric drawing drawn to a larger scale to provide additional information.

Section details are the most commonly used detail. Section details are an enlarged drawing showing a view of an object or part of the building that has been cut away to see the inside. Figure 14–10 shows a section detail indicating how the wall/ceiling connection is to be constructed. Notes as well as dimensions are used to give detailed information.

Isometric detail drawings are drawn to help visualize parts of the building that may be difficult to understand from a plan or elevation view only. An example would be a piping isometric for a boiler (see Figure 14–12).

An electrician will use detail drawings for planning a job as well as during the construction process. Looking at a wall section can tell the electrician what type of construction is being used. This may determine what wiring method to use and whether the wiring will be flush or surface mounted. The kitchen details will be used to figure out where to locate receptacles and lighting so as to not interfere with the cabinetry. A concrete-encased raceway section will give information on how to install the raceways. It is very important for an electrician to understand the details and the information detail drawings and sections provide.

FieldNote!

An example of a section detail used by an electrician is the concrete-encased raceways shown in Figure 14–11. Figure 14–11 is an electrical site plan that has four sections drawn on the page. By looking at the site plan alone, the electrician can see where the concrete-encased raceways (duct banks) run but not how they are constructed. The sections views give the rest of the information necessary to install the raceways.

FIGURE 14–10 Detail drawings will use notes as well as dimensions to give detailed information about how the building is to be constructed.

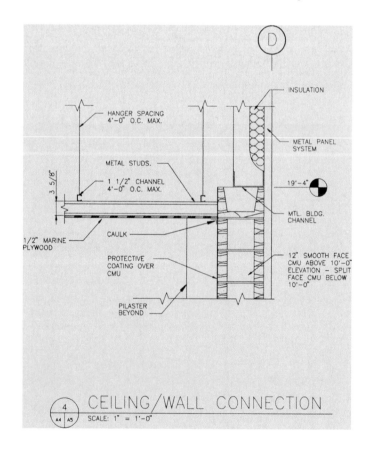

FIGURE 14-11 This site plan has several detail drawings drawn on page. The details are section views of the duct banks and indicate the orientation, size, and number of raceways that are to be installed.

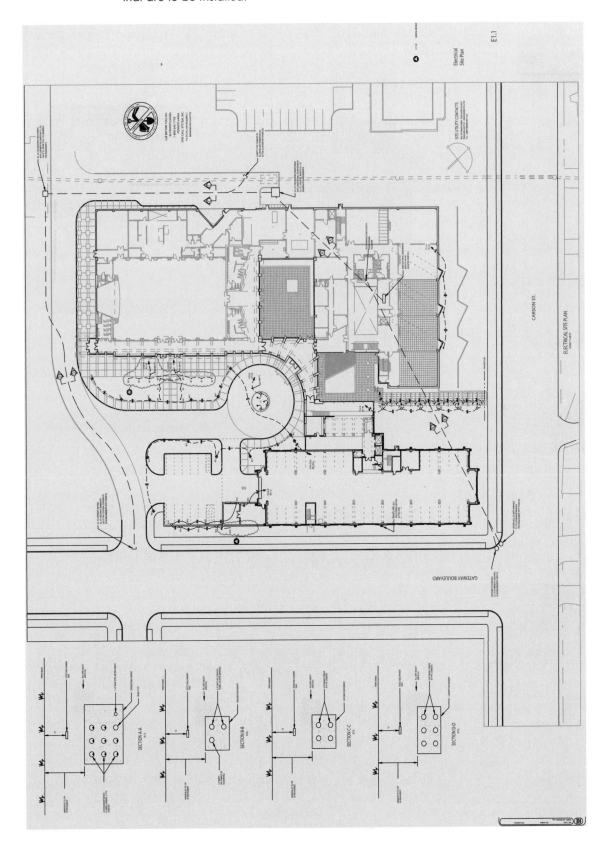

FIGURE 14–12   An equipment room isometric gives an isometric view of how the equipment and piping is to be laid out.

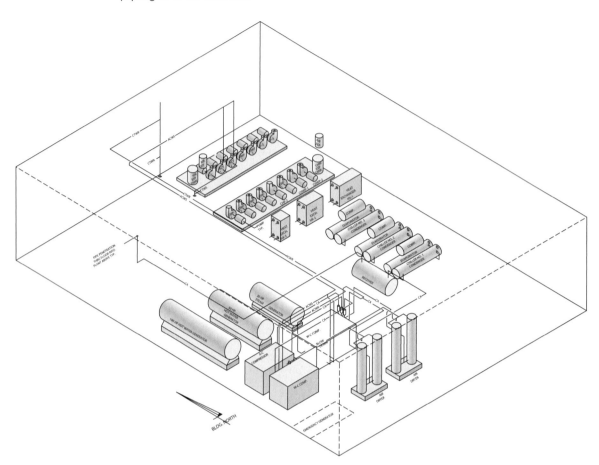

FieldNote!

Some details will contain more than one drawing. An example would be a fireplace detail (see Figure 14–13). A fireplace detail may contain plan, elevation, and section views.

FIGURE 14-13 This fireplace detail consists of plan, section, and elevation views.

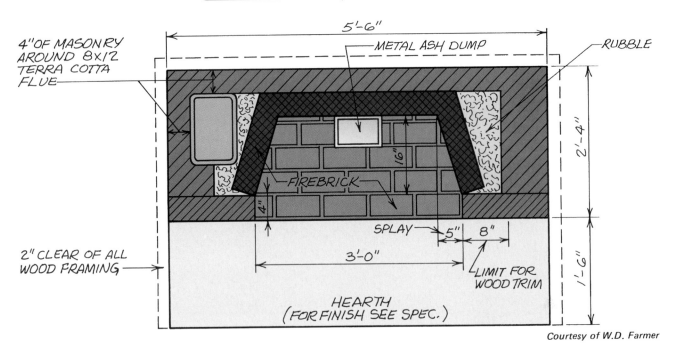

4" OF MASONRY AROUND 8x12 TERRA COTTA FLUE

METAL ASH DUMP

RUBBLE

5'-6"

2'-4"

FIREBRICK

16"

4"

SPLAY

5"

8"

3'-0"

LIMIT FOR WOOD TRIM

1'-6"

2" CLEAR OF ALL WOOD FRAMING

HEARTH (FOR FINISH SEE SPEC.)

Courtesy of W.D. Farmer

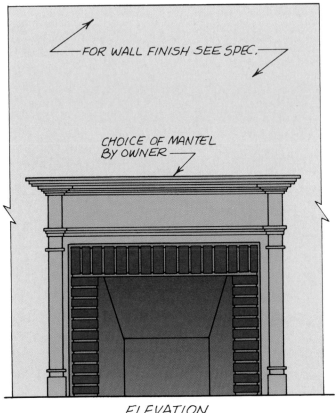

FOR WALL FINISH SEE SPEC.

CHOICE OF MANTEL BY OWNER

ELEVATION

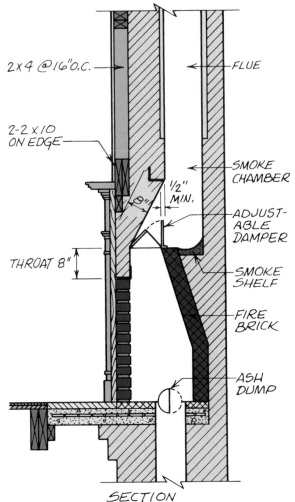

2x4 @16"O.C.

FLUE

2-2x10 ON EDGE

8"

1/2" MIN.

SMOKE CHAMBER

ADJUST-ABLE DAMPER

THROAT 8"

SMOKE SHELF

FIRE BRICK

ASH DUMP

SECTION

Courtesy of W.D. Farmer

SUMMARY

- Sections are scaled drawings giving a view of an object or part of a building that has been cut away to see the inside features.
- A cutting-plane line will indicate where a section has been taken, which direction the cut is being viewed, and where the section drawing is located.

- A typical wall section contains information on what sizes and types of materials are used to build the wall, as well as how it is constructed.
- Detail drawings are drawn to a larger scale for clarification.
- There are four types of detail drawings: plan, elevation, section, and isometric.

REVIEW QUESTIONS

1. What is a section drawing?
2. What does the arrow in a cutting-plane line indicate?
3. Describe the difference between a detailed wall section and a typical wall section.

4. What is a detail drawing?
5. What are four types of detail drawings?
6. Give an example of an elevation detail drawing.

DETAIL AND SECTION EXERCISE

Use Problem 14–1 to answer the following questions:

1. What is the size of the footing?
2. How far must the wood siding be above grade?
3. What is the scale of the drawing?
4. Describe the foundation wall.
5. How thick is the concrete slab in the basement?
6. Describe the wall insulation.

7. What is the dimension from the finished floor to the point where the wall meets the ceiling on the main level?
8. What is the finished wall material in the lower level?
9. What is the roofing material?
10. Describe the rafters.

Problem 14–1

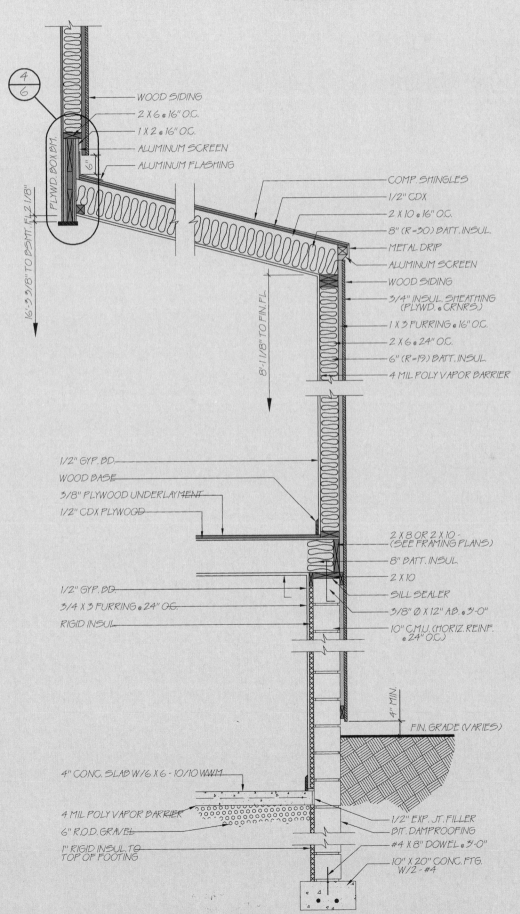

WOOD SIDING
2 X 6 • 16" O.C.
1 X 2 • 16" O.C.
ALUMINUM SCREEN
ALUMINUM FLASHING

PLYWD. BOX BM.

16'-5 3/8" TO BSMT. FL. 2 1/8"

6"

COMP. SHINGLES
1/2" CDX
2 X 10 • 16" O.C.
8" (R=30) BATT. INSUL.
METAL DRIP
ALUMINUM SCREEN
WOOD SIDING
3/4" INSUL. SHEATHING
(PLYWD. • CRNRS.)
1 X 3 FURRING • 16" O.C.
2 X 6 • 24" O.C.
6" (R=19) BATT. INSUL.
4 MIL POLY VAPOR BARRIER

8'-11 1/8" TO FIN. FL.

1/2" GYP. BD.
WOOD BASE
3/8" PLYWOOD UNDERLAYMENT
1/2" CDX PLYWOOD

2 X 8 OR 2 X 10 -
(SEE FRAMING PLANS)
8" BATT. INSUL.
2 X 10
SILL SEALER
3/8" Ø X 12" A.B. • 3'-0"
10" C.M.U. (HORIZ. REINF.
• 24" O.C.)

1/2" GYP. BD.
3/4 X 3 FURRING • 24" O.C.
RIGID INSUL.

4" MIN.

FIN. GRADE (VARIES)

4" CONC. SLAB W/6 X 6 - 10/10 WWM

4 MIL POLY VAPOR BARRIER
6" R.O.D. GRAVEL

1" RIGID INSUL. TO
TOP OF FOOTING

1/2" EXP. JT. FILLER
BIT. DAMPROOFING

#4 X 8" DOWEL • 5'-0"
10" X 20" CONC. FTG.
W/2 - #4

4
6

Use Problem 14–2 to answer the following questions:

1. What is the scale of the detail?

2. What height are receptacles to be when mounted above the backsplash of a counter?

3. What height are the switches to be installed?

4. How far from the door should the switch be mounted when the wall space is more than 36 inches wide?

5. At what height are below-counter receptacles to be mounted?

6. What height should a fire alarm station be mounted?

7. If a switch is to be mounted next to a door where the wall space is less than 36 inches, how is it to be installed?

Problem 14–2

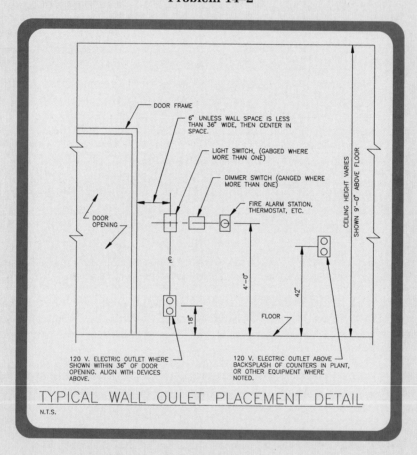

TYPICAL WALL OULET PLACEMENT DETAIL
N.T.S.

15

Schedules and Specifications

OUTLINE

OVERVIEW

This chapter covers schedules and specifications. Schedules and specifications give additional information about a job that cannot be found on blueprints. Schedules and specifications are an integral part of the construction process, from bidding to the completion of the project. This chapter discusses the information found in schedules and specifications and relates it to the electrical industry.

OBJECTIVES

After completing this chapter, you should be able to:

• Describe the information found on various schedules.

• Apply the information found on schedules to a job site.

• Describe the information found in specifications.

• Apply the information found in specifications to a job site.

SCHEDULES

A schedule is a chart that gives detailed information about a project that cannot be found by just looking at one of the drawings. Some standard types of schedules are architectural, mechanical, electrical, and various subsets of these. Schedules are an organized, systematic list of necessary materials and devices for a specific job. A schedule may also contain necessary data for voltage-load calculations. Schedules are valuable in estimating, ordering, and organizing the entire construction project. Various types of prints contain various types of schedules.

15.1 Room Schedules

Room schedules are used primarily on commercial and industrial prints and give pertinent information on the walls, ceilings, and floors of each room (see Figure 15–1). They provide mounting and routing information for electrical devices. The room schedule is valuable in determining building materials, and therefore mounting and run techniques.

FIGURE 15–1   Room schedule.

ROOM SCHEDULE

ROOM NO.	ROOM NAME	CURING	FLOOR	BASE	WALL		
					MTL	FINISH	
1.1	DOCK	C&S	C/1	--	CMU	LTX	
1.2	SUPPLY	C&S	C/1	--	CMU	LTX	
1.3	ELECTRICAL	C&S	C/1	--	CMU	LTX	
1.4	COMPACTOR	MC	STL.	--	CMU	LTX	
2.1	PRE-CLEAN	C&S	C/1	--	CMU	LTX	
3.1	PRIMARY IMAGE	MC	C/2	4" CV2	CMU	ENL	
3.2	SCREEN PREP	MC	*C/2	4" CV2	CMU	ENL	
3.3	AIR LOCK	MC	C/2	4" CV2	CMU	ENL	
4.1	ETCH	MC	C/3	4" CV3	CMU	EPX	
4.2	MEZZANINE CONTROL ROOM	MC	C/2	V	CMU	EPX	
5.1	SOLDER & PRINT	MC	C/2	4" CV/2	CMU	ENL	
6.1	CORRIDOR	C&S	C/1	--	CMU	ENL	
6.2	RESTROOM	C&S	C/1	--	CMU	ENL	
6.3	COMPACTOR	MC	STL.	--	CMU	LTX	
7.1	PUNCH PRESS	C&S	C/1	--	CMU	ENL	
8.1	INSPECTION & PKG	C&S	C/1	--	CMU	ENL	
8.2	ELECTRIC TEST	C&S	C/1	--	CMU	ENL	
9.1	MEZZANINE	C&S	C	--	MTL & CMU	ENL	
10.1	DIE STORE	C&S	C/1	--	CMU	ENL	
11.1	MACHINE SHOP	C&S	C/1	--	CMU	ENL	
11.2	STORAGE	C&S	C/1	--	CMU	ENL	
12.1	MAINTENANCE	C&S	C/1	--	CMU	ENL	

15.2 Window and Door Schedules

Window and door schedules are used on all types of prints (see Figure 15–2). They contain all pertinent information relating to the size, shape, material, and ordering information of doors and windows.

FIGURE 15–2 Door schedule.

DOOR SCHEDULE

DOOR NO.	SIZE WD - HT - THICK	EL	DOOR MATL	FRAME MATL	HDW SET	DETAILS HEAD	JAMB	SILL
1	PR 3'-0" X 7'-0"	F		STL	10	3B/A8 3B/A8 -		
2	3'-0" X 7'-0" X 1-3/4"	A	HM	HM	1	3B/A8 3B/A8 -		
3	8'-0" X 10'-0"	G	MTL	STL	10	2/A6 3C/A8 -		
4	8'-0" X 10'-0"	G	MTL	STL	10	2/A6 3C/A8 -		
5	8'-0" X 10'-0"	G	MTL	STL	10	2/A6 3C/A8 -		
6	10'-0" X 14'-0"	G	MTL	STL	10	2/A6SIM 3C/A8 -		
7	3'-0" X 7'-0" X 1-3/4"	C	HM	HM	1	3B/A8 3B/A8 -		
8	PR 4'-0" & 6'-0" = 10'-0" X 10'-0" HI	H	HM	STL	15	3B/A8 3B/A8 -		
9	PR 4'-0" & 6'-0" = 10'-0" X 8'-0" HI	H	WD/P	STL	15	3B/A8 3B/A8 -		
10	PR 4'-0" & 6'-0" = 10'-0" X 10'-0" HI	H	HM	STL	15	3B/A8 3B/A8 -		
11 A	3'-0" X 7'-0" X 1-3/4"	C	HM	HM	1	3B/A8 3B/A8 -		
11 B	3'-0" X 7'-0" X 1-3/4"	C	HM	HM	2	3B/A8 3B/A8 -		
12	3'-0" X 7'-0" X 1-3/4"	C	HM	HM	6	3B/A8 3B/A8 -		
13	3'-0" X 7'-0" X 1-3/4"	E	HM	HM	9	3B/A8 3B/A8 -		
14	PR 3'-0" X 7'-0"	--		STL	10	3B/A8 3B/A8 -		
15	8'-0" X 10'-0"	G	MTL	STL	10	2/A6 3C/A8 -		
16	3'-0" X 7'-0" X 1-3/4"	I	ALUM	ALUM	10			

15.3 Panelboard Schedules

Panelboard schedules are used on all types of construction (see Figure 15–3). They identify and describe the panel. They also list current, voltages, and loads. Panelboard schedules list and identify circuit branches. There is usually a space left on the schedule for notes.

15.4 Lighting Fixture Schedules

Lighting fixture schedules are used on all types of prints (see Figure 15–4). They identify fixture type (a letter that will correspond to a letter on the blueprints), manufacturer, catalog number, lamp number, mounting instructions, and other remarks. The lighting fixture schedule is extremely useful in estimating, ordering, and keeping a running inventory of materials on hand. There may be no lighting fixture schedule on a residential print. In this case, the architect will give the owner a lighting fixture allowance. If the owner is operating on a lighting allowance, the fixtures will be selected by the owner at a local lighting center.

FIGURE 15-3 Panelboard schedule.

PANELBOARD C SCHEDULE

VOLTAGE: 120/208 VOLT, 3 PH, 4 W
MAINS: 150 AMP MAIN BREAKER
MOUNTING: SURFACE
FEEDER: 2"C - 4#1/0, 1#6 GRD.

LOCATION: MACHINE SHOP
TIE TO BUS FEEDER #4

42 POLES

Note: In addition to the information shown, many municipalities now require Available Fault Current to be listed on the print prior to plan review.

CKT NO.	BREAKER	PHASE - LOADING (KVA)			CIRCUIT IDENTIFICATION	NOTES
		A	B	C		
		9.0	9.5	9.2		
1	30/3	1.2	1.2	1.2	RECEPTACLES	VERTICAL MILL (230 V)
2	30/3	1.0	1.0	1.0	RECEPTACLES	MONARCH LATHE (230 V)
3	30/3	1.0	1.0	1.0	RECEPTACLES	WELDER
4	30/1	.8	.8		RECEPTACLES	BELT SANDER (120 V)
5	20/1		.5		RECEPTACLES	BENCH GRINDER (120 V)
6	20/1			7.0	RECEPTACLES	
7	20/1	1.0			RECEPTACLES	
8	20/1		1.0		RECEPTACLES	
9	20/1			1.0	RECEPTACLES	

FIGURE 15-4 Light fixture schedule.

LIGHT FIXTURE SCHEDULE

TYPE	MANUFACTURER	CATALOG NUMBER	LAMPS	MOUNTING	REMARKS
A	LITHONIA	WA440A277	4-40W	CEILING	WRAPAROUND ACRYLIC LENS
			FLUOR	SURFACE	
B	KEYSTONE	2A440EXA277GPWS	4-40W	CEILING	2X4 GRID TROFFER
			FLUORESCENT	SURFACE	ACRYLIC LENS. AIR RETURN
C	KEYSTONE	2A440PWSGPW277S	4-40W	CEILING	2X4 GRID TROFFER
			FLUORESCENT	SURFACE	1/2"X1/2"X1/2" SILVER
					PARABOLIC LOUVER
					AIR RETURN
D	PHOENIX	DL-300D60-3	1-300W	WALL	DUAL ARM DOCK
			R40	7'-0" AFF	LIGHT, 62" LENGTH W/ VERT
					ADJUSTMENT, WIREGUARD, &
					BUILT-IN SWITCH

15.5 Control Circuit Schedules

Control circuit schedules are used most often in commercial and industrial construction (see Figure 15–5). Control circuit schedules list lighting or motor control information by room, circuit designation, and contactor or connections.

FIGURE 15–5 Lighting control schedule.

LIGHTING CONTROL SCHEDULE				
LIGHTING CONTROL DESCRIPTION		CONTACTOR(S) BEING CONTROLLED	PANEL-CIRCUIT BEING SERVED	ROOM NAME
LCC #1 (MASTER) (SEE DETAIL)	SWITCH#1	A,B,C,D,E,F,G,H	AA - 1-19,21-29,31,32,34	SOLDER MASK, PUNCH PRESS, INSPECTION AND PACKING, MACHINE SHOP, MAINTENANCE
	SWITCH#2	I,J,K,L,M,N,O	BB - 1-10,12-21,38	PRIMARY IMAGE, PRE-CLEAN, DOCK, SCREEN PREP.
	SWITCH#3	P	BB - 20-24	OFFICE AREA, TOILETS, BREAK ROOM
LS #2		A	AA - 1,2,5,6,9	PUNCH PRESS
LS #3		B	AA - 3,4,7,8,11	PUNCH PRESS
LS #4		C	AA - 10,12	MACHINE SHOP
LS #5		D	AA - 17,19	MAINTENANCE OFFICE AND SHOP
LS #6		E	AA - 13,14,28,32	PACK INSPECTION
LS #7		F	AA - 15,16,26,34	PACK INSPECTION
LS #8		G	AA - 21,22,25,31	SOLDER MASK AND PRINTING
LS #9		H	AA - 23,24,27,29	SOLDER MASK AND PRINTING
LS #10		I	BB - 1,2,5	PRIMARY IMAGE
LS #11		J	BB - 3,4,6,7	PRIMARY IMAGE
LS #12		K	BB - 13,15,35	SCREEN PREP
LS #13		L	BB - 17,19	DOCK

15.6 Connected Load Schedules

Connected load schedules provide information on voltage, current, and power requirements (see Figure 15–6). These schedules are used in all construction types. The connected load schedule is used to separate circuits and balance loads when this information is required by local utility companies.

15.7 Equipment Schedules

Equipment schedules are used to determine the voltage, current, and power requirements for motors and other electrical equipment on construction jobs (see Figure 15–7). Equipment schedules also list wire sizes and run information for the various machinery and equipment.

FIGURE 15-6   Connected load schedule.

CONNECTED LOAD:

LIGHTING --------------- 12.2 KW

MISCELLANEOUS ----------- 42.0 KW

WATER HEATING ---------- 7.5 KW

HVAC -------------------- 27.0 KW

TOTAL: 88.7 KW

VOLTAGE 120/208
PHASE 3
WIRE 4

FIGURE 15-7   Equipment schedule.

PARTIAL EQUIPMENT SCHEDULE

NO.	Designation	HP/KW	Voltage	Remarks
AC1	Compressor	110 HP	460/3/60	
AC2	Compressor	200 HP	460/3/60	
EH1	Heater	1.5 KW	120/1/60	
CT1	Tower Pump	15 HP	460/3/60	

FieldNote!

Specifications will often require an electrician to have a certain amount of field experience installing certain types of systems. Aspects of the job that typically require field experience are controls such as motor controls, fire alarms, and security. This is to prevent damage to expensive equipment, as well as other costly and time-consuming mistakes.

SPECIFICATIONS

Most jobs, from small residential projects to massive industrial projects, have specifications. Small projects (such as residential construction) may detail the specifications in the drawings, whereas a large commercial project may contain hundreds of pages of specifications.

Specifications are a list of detailed job requirements, under which all work must be performed. The specifications will detail how the job is to be bid and what requirements the contractor must meet to be able to bid on the job. The specifications are responsible for spelling out what types of materials must be used on the project and will detail various installation procedures, including on occasion how much work experience a person must have to perform a job.

The specifications will also detail the responsibility of the contractor after the job is finished (for example, any warranty that must be provided). If material substitutions or installation techniques must be changed, written approval should be received from the architect, engineer, or owner. The contractor could be held financially or legally responsible for any changes made without proper permission.

All contractors, general or subcontractors, bid jobs based on specifications. When jobs are bid, the contractors are held responsible for completing the job per blueprints and specifications. If there is a conflict between the prints and the specifications, clarification should be sought from the architect, engineer, and/or owner—not from individuals in the field.

These specifications were previously divided into 16 divisions. Some divisions are general, and others are specific. General requirements cover areas such as site work and may affect multiple crafts. Specific specifications are aimed at one craft, such as plumbing, mechanical, or electrical. Table 15–1 lists the 16 divisions commonly used in specifications. Division 16 is the electrical category that covers the majority of electrical work.

TABLE 15–1 Specification categories

SPECIFICATION CATEGORIES	
Division 1	General Requirements
Division 2	Site Work
Division 3	Concrete
Division 4	Masonry
Division 5	Metals
Division 6	Wood and Plastics
Division 7	Thermal and Moisture Protection
Division 8	Doors and Windows
Division 9	Finishes
Division 10	Specialties
Division 11	Equipment
Division 12	Furnishings
Division 13	Special Construction
Division 14	Conveying System
Division 15	Mechanical
Division 16	Electrical

In 2004, the Construction Specifications Institute (CSI) implemented a new format called MasterFormat—which increased the number of divisions to 49. The MasterFormat is a standard format to be used to organize project specifications. CSI was hoping that by 2007 the new format would be in widespread use, designers and workers having had time to adjust from the 16-division specifications. Time will only tell if the industry adopts and supports the new MasterFormat. Figure 15–8 shows a listing of divisions under the new MasterFormat.

Some key differences of the new MasterFormat are as follows. First, the new expanded system is able to cover commercial, industrial, and process specifications. Next, there is room for expansion built into the system—which allows for emerging technologies, change, and expansion. Finally, you will look for information in a different place from that of the 16-division system. For example, divisions 20 through 29 cover and expand the information formerly in divisions 15 and 16. Under the new format, there are several divisions the electrician will look to for information.

NOTE: For more information about the new MasterFormat, contact CSI at http://www.csinet.org.

Contractors and field supervisors should be familiar with specification articles that pertain to their work or that may directly affect that work. It is also important that a tradesperson review his or her craft specifications as well as related specifications. This will help in understanding the overall scope of the job. It will also help prevent costly rework.

Appendix A of this text has a partial set of commercial specifications. Included in Appendix A are a table of contents, a portion of Division 1 (General Requirements), and a portion of Division 16 (Electrical). Appendix B of this text provides a partial set of industrial specifications. Included in Appendix B are a table of contents and a portion of Division 16. Both sets of specifications have the old 16-division format.

FIGURE 15-8

The Construction Specifications Institute has created a new *MasterFormat*™ that contains 49 divisions. The new format addresses the new developments in the construction industry and has room for expansion as industry continues to change. This format is intended to gradually replace the old 16 division format. (The Numbers and Titles used in this textbook are from *MasterFormat*™ 2004, published by The Construction Specifications Institute (CSI) and Construction Specifications Canada (CSC), and are used with permission from CSI. For those interested in a more in-depth explanation of *MasterFormat*™ 2004 and its use in the construction industry visit www.csinet.org/ MasterFormat or contact: The Construction Specifications Institute (CSI); 99 Canal Center Plaza, Suite 300; Alexandria, VA 22314; 800-689-2900; 703-684-0300; http://www.csinet.org.

MasterFormat™ 2004 Edition – Numbers & Titles   Division Numbers & Titles
November 2004

Division Numbers and Titles

PROCUREMENT AND CONTRACTING REQUIREMENTS GROUP
Division 00   Procurement and Contracting Requirements

SPECIFICATIONS GROUP

GENERAL REQUIREMENTS SUBGROUP
Division 01   General Requirements

FACILITY CONSTRUCTION SUBGROUP
Division 02   Existing Conditions
Division 03   Concrete
Division 04   Masonry
Division 05   Metals
Division 06   Wood, Plastics, and Composites
Division 07   Thermal and Moisture Protection
Division 08   Openings
Division 09   Finishes
Division 10   Specialties
Division 11   Equipment
Division 12   Furnishings
Division 13   Special Construction
Division 14   Conveying Equipment
Division 15   Reserved
Division 16   Reserved
Division 17   Reserved
Division 18   Reserved
Division 19   Reserved

FACILITY SERVICES SUBGROUP
Division 20   Reserved
Division 21   Fire Suppression
Division 22   Plumbing
Division 23   Heating, Ventilating, and Air Conditioning
Division 24   Reserved
Division 25   Integrated Automation
Division 26   Electrical
Division 27   Communications
Division 28   Electronic Safety and Security
Division 29   Reserved

SITE AND INFRASTRUCTURE SUBGROUP
Division 30   Reserved
Division 31   Earthwork
Division 32   Exterior Improvements
Division 33   Utilities
Division 34   Transportation
Division 35   Waterway and Marine Construction
Division 36   Reserved
Division 37   Reserved
Division 38   Reserved
Division 39   Reserved

PROCESS EQUIPMENT SUBGROUP
Division 40   Process Integration
Division 41   Material Processing and Handling Equipment
Division 42   Process Heating, Cooling, and Drying Equipment
Division 43   Process Gas and Liquid Handling, Purification, and Storage Equipment
Division 44   Pollution Control Equipment
Division 45   Industry-Specific Manufacturing Equipment
Division 46   Reserved
Division 47   Reserved
Division 48   Electrical Power Generation
Division 49   Reserved

Div Numbers - 1

SUMMARY

- A schedule is a chart that gives detailed information about a project that cannot be found by looking at the construction drawings. Some common schedules are:

  - Room schedule
  - Window and door schedule
  - Light fixture schedule
  - Control schedule

- Connected load schedule
- Equipment schedule

- Specifications are a detailed list of job requirements, under which all work must be performed.
- Specifications are used throughout the construction process, from bidding on a job to providing a warranty for the job.

REVIEW QUESTIONS

1. What information would be found on a light fixture schedule?

2. What information would be found on a room schedule?

3. What information would be found on an equipment schedule?

4. What are specifications?

5. Under the old format, what division of specifications is dedicated to electrical?

6. Under the new MasterFormat, how many divisions are there?

7. What role do specifications play when bidding on a job?

RELATED QUESTIONS

Use the commercial specifications found in Appendix A to answer the following questions.

1. Who should be in attendance at a progress meeting?

2. What depth of earth cover is required for concrete-encased PVC ducts housing secondary service conductors?

3. What type conductors are to be used in exterior underground raceways?

4. What is the color code to be used for 277/480 V conductors?

5. How must the distribution panelboard switching and protective devices be marked?

6. How much experience must the electrician installing the combination motor starters have had with installing similar systems?

7. What fire alarm control panel is to be installed?

8. What length of warranty must be provided on the labor and material of the television distribution system?

16

Residential Print Reading

O U T L I N E

OVERVIEW

This is the first of three chapters in area-specific print reading. The chapter covers residential plans, elevations, details, and estimating. The purpose of this chapter is to work through, utilize, and analyze an actual set of residential prints.

OBJECTIVES

After completing this chapter, you should be able to:

• Describe the process of studying a residential print.

• Recognize and identify the various prints used on a residential project.

• Utilize plans and elevations to determine device placement and electrical service location.

• Utilize plans to perform device take-off and practice the cost-estimation process.

• Visualize a residence by looking at a set of residential construction drawings.

RESIDENTIAL PRINT-READING ANALYSIS

Residential construction plans consist of a cover sheet, plot plan, floor plans, elevation, details, sections, and schedules. Prints are reproduced by various methods, such as computer-aided design (CAD) plots, xerographic reproduction, blue-line printing, and photo offset printing. Regardless of the reproduction technique, the scale of the reproduced drawing should be carefully checked before scaling the drawing.

As stated, the purpose of this chapter is to work through and analyze an actual set of residential prints. The print set is titled *Franklin Residence,* which is a house located in the Fairway subdivision, Knoxville, Tennessee. The plans are actual plans, but the owner's name and location have been changed. The house was completed in September of 1992. The CD at the back of this text contains CAD files as well as PDF files of the Franklin residence.

Due to the reduction of some of the figures in this chapter, you should consult the residential print set while reading the chapter. Regardless of your craft, the process of print reading is generally the same. It is important to understand all phases of print reading to do a professional and accurate job in the construction trades. The electrical worker should understand the entire print, but will obviously concentrate on the floor plan and elevations containing electrical information.

The first step in print reading is to examine the cover sheet for any pertinent information and a list of the available prints. The cover sheet on residential prints may be excluded. If the cover sheet is present, it may contain a pictorial of the residence and a list of drawings included (Figure 16–1).

FieldNote!

Most residential construction is a **design build** job for the electrical contractor. "Design build" means that the contractor designs and lays out all electrical for the job and performs the installation. There are no electrical engineers or designers involved. The contractor will typically meet with the homeowner or general contractor to find out their specific needs and to make suggestions. He or she will then design and build the job based on that information as well as code requirements. This is not as common for commercial and industrial jobs due to their complexity. However, it does occur.

Design Build
A project in which the electrical contractor designs and lays out all electrical for the job and performs the installation.

FIGURE 16–1 Cover sheet.

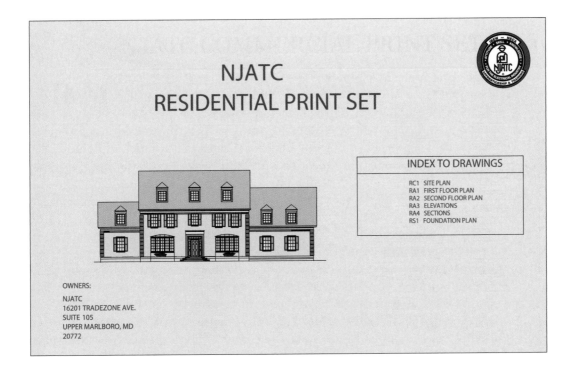

NJATC
RESIDENTIAL PRINT SET

INDEX TO DRAWINGS

RC1 SITE PLAN
RA1 FIRST FLOOR PLAN
RA2 SECOND FLOOR PLAN
RA3 ELEVATIONS
RA4 SECTIONS
RS1 FOUNDATION PLAN

OWNERS:

NJATC
16201 TRADEZONE AVE.
SUITE 105
UPPER MARLBORO, MD
20772

FieldNote!

When electrical contractors estimate a residential job, they typically include everything required to complete the installation except for the light fixtures. This is due to the fact that light fixtures are as much a decoration as a light source and homeowners are often very particular about which light they want installed. To ensure that the owners get the fixture they want, the electrical contractor or general contractor will have a **lighting allowance.**

This will be a set amount of money set aside for the homeowner to purchase light fixtures. If the homeowner picks out fixtures that cost more than the allotted lighting allowance, they will have to pay the extra out of pocket. If they spend less, the balance will be subtracted from the total bill. The amount of money in the lighting allowance will be an educated guess by the contractor based on the number and type of light fixtures in the house.

NOTE: Recessed light fixtures are typically included in an electrical contractor's bid because they are normally purchased by the electrical contractor.

Lighting Allowance
A set amount of money set aside for a homeowner to purchase light fixtures.

16.1 Plans and Views

The next step in residential print reading is to examine the plot plan to determine the orientation of the house and possible electrical service information. The Franklin residence is located in an area with all services underground. The electrical transformer and underground hookup are represented by the square box inside the Franklin driveway circle and just off Walnut Circle (Figure 16–2).

The next and most important step for the electrician is to examine the floor plan (or plans, in multilevel structures). The Franklin residence has two floors, marked A1 and A2 for the first- and second-floor plans, respectively. The letter *A* refers to architectural. On commercial and industrial projects, the prints will be divided into *A* for architectural, *M* for mechanical, *E* for electrical, *P* for plumbing, and so on. The Franklin residence has the electrical information shown on prints A1 and A2, which are shown in Figures 16–3 (see page 252) and 16–4 (see page 253). It is common to have the electrical information on residential architectural prints.

Another consideration is the elevation views. In residential construction, they serve to show the cutting planes for the floor plans—as shown in the Franklin residence print A3 by cutting-plane lines. Elevations also give an overall pictorial view of the structure and show locations of exterior lighting. The Franklin residence print elevations do not show exterior lighting fixtures. The exterior fixtures are covered by the **lighting allowance** allotted to Mr. Franklin and are placed and wired according to local codes (see Figure 16–5, page 254).

Two final considerations on the Franklin residential print are C1 (which is the foundation plan) and A4, which are the framing-section detail drawings. The only concern of the electrician with these two prints is in the materials he or she must run wires or conduit through. The Franklin residence plan has underground service. The type of building material and the local code determine how the runs are to be made in the prints shown in Figures 16–6 (see page 255) and 16–7 (see page 255).

As mentioned previously, the most important prints to the electrician are the floor plans. There are two considerations on residential floor plans. First, one must consider devices, locations, and runs. Second, one must consider the relationship of the first and second floors to each other for the purpose of efficient wire runs. On commercial and industrial prints, schedules, lists of materials, and devices are usually supplied. It is common in residential construction for the electrical contractor to have to figure materials, circuits, loads, and all other common schedules. This figuring must be done in compliance with the latest local codes. Knowledge of code and careful counting of devices and figuring of circuit loads make it possible to develop usable residential schedules.

Devices are shown on the plans by graphic symbology, as discussed in Chapter 7. When determining individual circuits and circuit loads, the devices are separated into various rooms. Depending on the size of each room and the number of lights or receptacles, some circuits may cover more than one room. Sometimes, as in kitchens, there will be several circuits in one room. Large appliance circuits are individual circuits, and receptacle and lighting circuits must follow local code requirements. Figure 16–8 (see page 256) contains a detail of the kitchen in the Franklin residence. According to your local code, how many circuits would be in this kitchen? Name them.

FIGURE 16-2 Site plan.

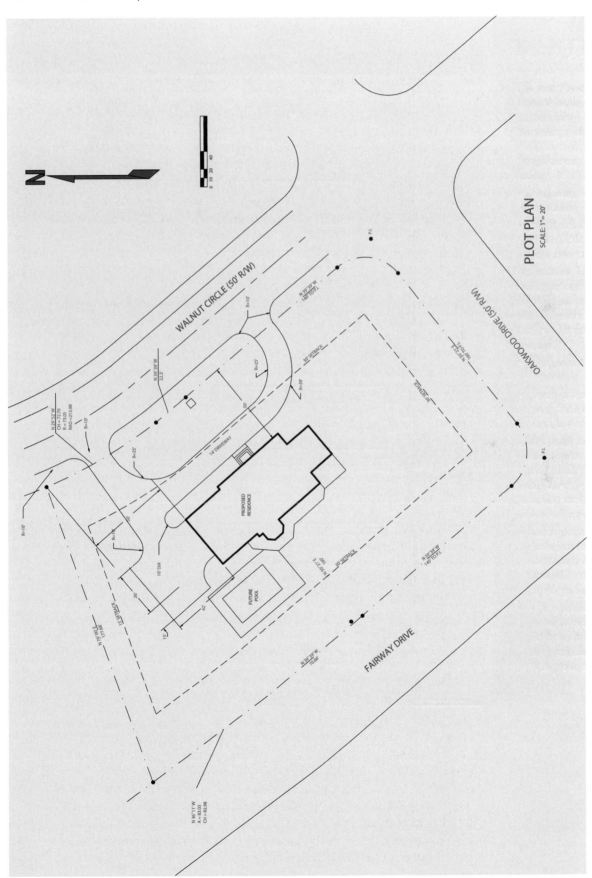

PLOT PLAN
SCALE: 1" = 20'

FIGURE 16–3 First-floor plan.

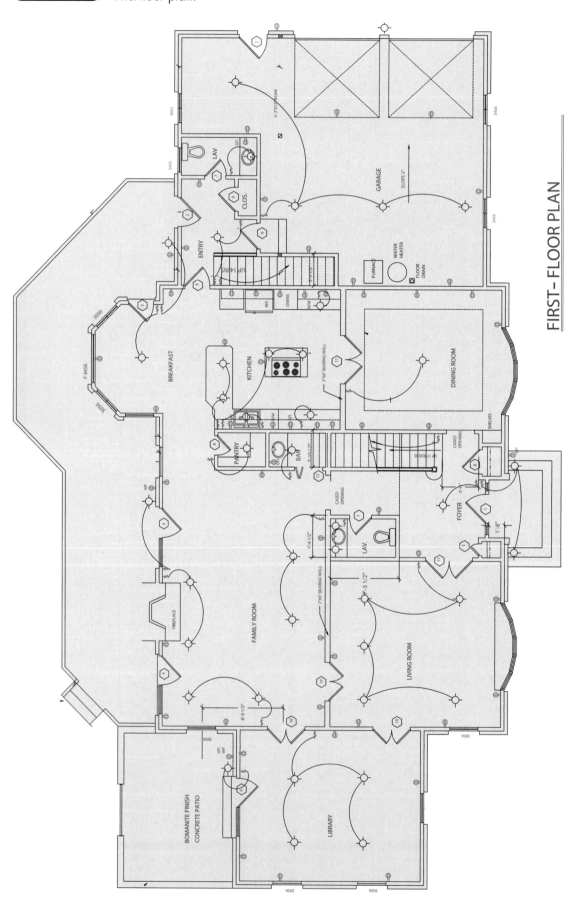

FIRST–FLOOR PLAN

FIGURE 16-4 Second-floor plan.

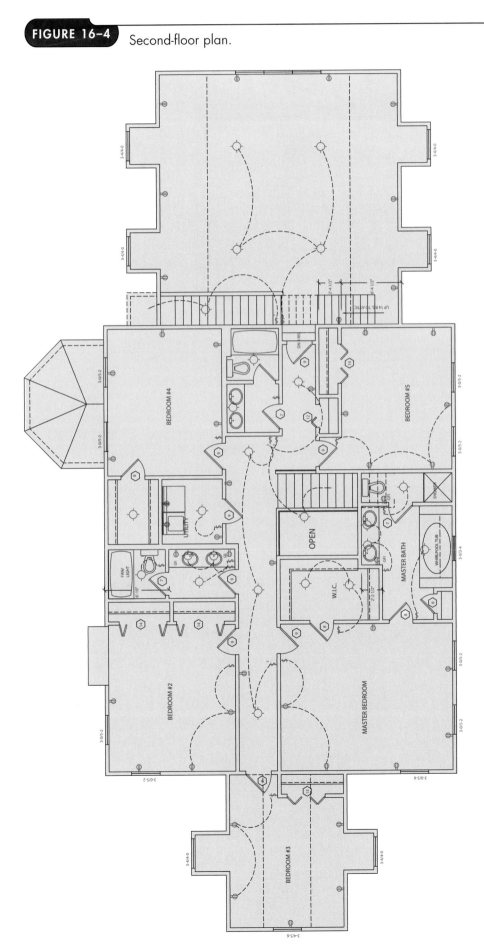

SECOND – FLOOR PLAN

FIGURE 16-5 Elevations.

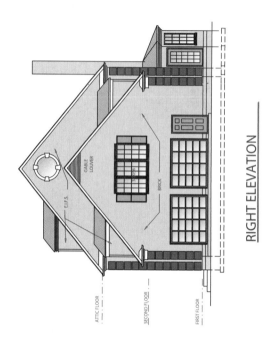

RIGHT ELEVATION
SCALE: 1/8" = 1'-0"

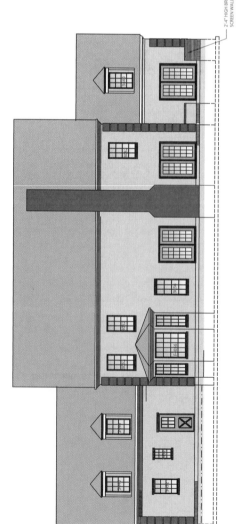

REAR ELEVATION
SCALE: 1/8" = 1'-0"

FIGURE 16–6 Foundation plan.

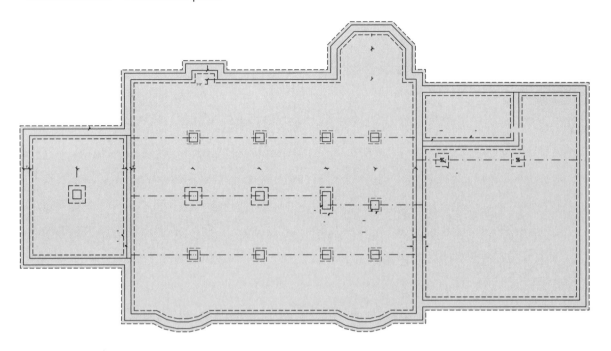

FIGURE 16–7 Sections.

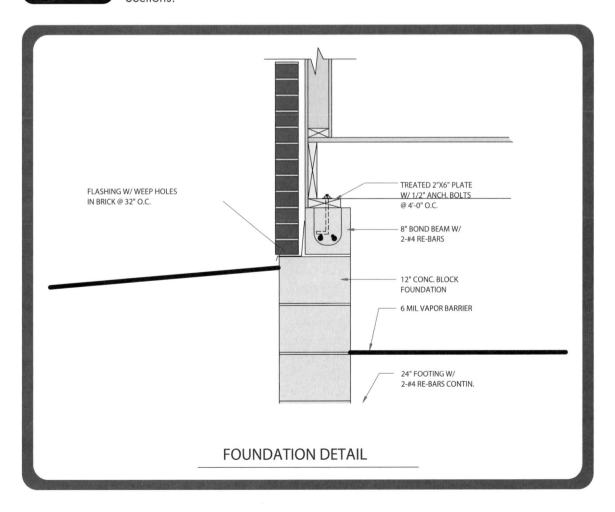

FLASHING W/ WEEP HOLES
IN BRICK @ 32" O.C.

TREATED 2"X6" PLATE
W/ 1/2" ANCH. BOLTS
@ 4'-0" O.C.

8" BOND BEAM W/
2-#4 RE-BARS

12" CONC. BLOCK
FOUNDATION

6 MIL VAPOR BARRIER

24" FOOTING W/
2-#4 RE-BARS CONTIN.

FOUNDATION DETAIL

FIGURE 16–8   Kitchen plan detail.

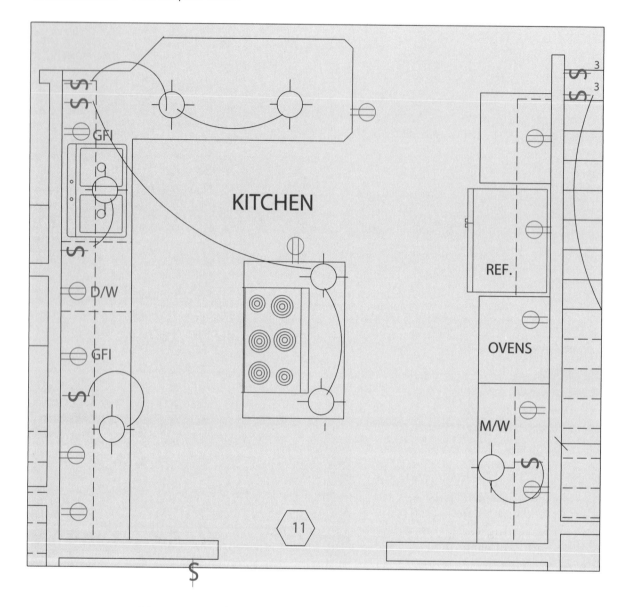

Figure 16–9 shows one possible answer based on the National Electrical Code and common wiring practices. Your local code restrictions may change the answer slightly.

16.2 Estimating

Estimating

The process of calculating what it will cost to build or complete a project.

Estimation Sheets

Sheets designed to aid in estimating a project.

Estimating may be an important concern. With access to complete electrical schedules, **estimation sheets,** and some practice, estimating can become less difficult. The various electrical schedules were discussed in Chapter 15. An understanding of those schedules should be of great help in tallying costs and installation times. Some contractors use a set figure (such as $30 per residential box opening) to estimate costs. Some contractors use a set figure based on the type of device. Others tally costs, times, and other expenses required to complete the job.

FIGURE 16-9 Circuit listing.

CIRCUIT LIST		
1	LIGHTING	110V
2	APPLIANCE 1	110V GFI
3	APPLIANCE 2	110V GFI
4	DISHWASHER	110V
5	DISPOSAL	110V
6	MICROWAVE	110V
7	REFRIGERATOR	110V
8	OVEN	220V
9	RANGE	220V

A cost-estimating sheet can be developed to improve estimating ability. This sheet should list all rooms or sections of the building on the side bar. On the top bar, the individual devices or wire runs should be listed. The squares formed by this chart would then contain the number of each device needed in each room. The bottom of the chart will have a total column. Multiplying the number of each total column by the unit cost will give a good cost estimation of materials (Figure 16–10). Counting the devices room by room will reduce the chances of missing devices while counting.

Estimating the amount of time necessary to complete a job is more difficult. With experience, a company estimator can develop an average time for installing various devices. This average time can be multiplied by the number of each device to acquire the total work hours for installation of each device (Figure 16–11). Before an estimator can estimate cost per box or job costs, he or she must know the total **hourly labor cost.** The total hourly labor cost is affected by many variables. These variables are local collective bargaining agreements, state taxes, local taxes, insurance regulations, and local assessments. Some specific examples are Social Security, local pensions, health and welfare, National Electric Benefit Fund, annuities, industry fund, and apprenticeship training (JATC).

After wage costs have been figured, other costs must be considered. These costs include direct job costs, indirect job costs, and profit. **Direct job costs** include all items that would be attributed directly to the project. Some of these are temporary buildings, vehicle expenses, specialty tools, rentals, on-the-job supervision, and so on. **Indirect job costs** include items generally thought of as **overhead.** Some of these are the electrical contractor's shop, the wages and benefits of office staff in the shop, cost of the estimator's salary and benefits, and all of the items to keep a contractor in business but that are not directly project related.

Hourly Labor Cost
The amount an employer has to pay for each hour an employee works, including wages, insurance, benefits, and so on.

Direct Job Costs
Any cost directly associated to a job, such as wages and materials.

Indirect Job Costs
Employers costs that are not directly associated with a specific job, such as office staff and shop expenses.

Overhead
Costs associated with running a business, including indirect job costs.

FieldNote!

When estimating the electrical for a house, many contractors go room by room counting devices, using a sheet similar to the estimation sheet shown in Figure 16–10. The contractor will have an assigned value for each device. This value will include everything necessary to install that device. For example, a single pole switch may have a cost of $25 per switch. That price includes the box, switch, plate, cable, wire nuts, straps, and labor for the complete installation. Each device could have a different value based on the amount of materials and labor required for the installation. Although not every switch or receptacle takes the same amount of time and materials, the cost used is an average value. Some may require more time and materials, while others may require less.

FIGURE 16–10   Estimation sheet.

SAMPLE FIXTURE ESTIMATING SHEET

Room ↓ DEV. →	S	S₃	S₄	⊖	⊖ GFCI	⊖ R	F			
Kitchen										
Utility										
Dining										
Bed 1										
Bed 2										
Bath 1										
Bath 2										
Halls										
Closets										
Garage										
Exterior										
TOTAL #										
UNIT COST										
TOTAL COST										

FIGURE 16–11   Time chart.

	NUMBER	UNIT TIME	TOTAL TIME
Receptacles			
Switches			
220 Outlets			
Light Fixtures			
Recessed Lights			
Panel Box			

Total Time		
Cost/Hour	*	
LABOR COST	=	

Profit is the item that is normally misunderstood. There is a common misconception that contractors have a large profit margin. In reality, a large profit margin figured on a job may equate to loss of the bid and therefore loss of jobs. A named profit and overhead margin on bid jobs may range from 5% to 20% based on the size of the job and its location. Large jobs tend to have smaller profit margins.

Once all costs have been assessed, it is the estimator's job to decide the cost of a labor unit of 1 hour—including all costs described previously. This is done by placing percentage figures on direct costs, indirect costs, and profit. These percentage figures are taken against the total labor and materials costs for the job. Many contractors use a cost-estimating manual or program such as the National Electrical Contractors Association (NECA) labor unit manual or RSMEANS Electrical Cost Data to help determine costs until they develop their own system.

Total job cost is essentially figured by adding materials, wages, direct labor costs, direct and indirect job costs, and profit. This is not an easy task, and takes practice. Figure 16–12 is a generic example of how to figure total job costs.

NOTE: Before starting any work or estimating, it is extremely important to check for print dates and **change notes.** Always make sure the print in hand has the most current date. After checking the date, check the cover sheet and all other sheets for change notes. Change notes explain changes or revisions on the prints. Overlooking these notes may cause costly rework (Figure 16–13).

Profit
The amount of money left from a project after all costs have been paid.

Total Job Cost
The amount a job costs, including direct costs, indirect costs, and profit.

Change Notes
Notes that explain changes or revisions on prints.

FIGURE 16–12 Total job cost.

Total Labor	+
Total Materials	+
Direct Costs	+
Indirect Costs	+
Profit	+
TOTAL JOB COST	=

FIGURE 16–13 Change notes.

PANELBOARD CHANGE NOTES:

1. PROVIDE AND INSTALL IN SEPARATE DEDICATED CABINET WITH LOCKABLE CONTINUOUS HINDED DOOR AND SHALL CONTAIN: 6-POLE 20 AMP CONTACTOR "A" TO CONTROL CIRCUITS AA-1, 2, 5, 6, 9; 6-POLE 20 AMP CONTACTOR "B" TO CONTROL CIRCUITS AA-3, 4, 7, 8, 11; 2-POLE 20 AMP CONTACTOR "C" TO CONTROL CIRCUITS AA-10, 12; 2-POLE 20 AMP CONTACTOR "D" TO CONTROL CIRCUITS AA-17; 6-POLE 20 AMP CONTACTOR "E" TO CONTROL CIRCUITS AA-13,14, 28, 32, 36; 4-POLE 20 AMP CONTACTOR "F" TO CONTROL CIRCUITS AA-15, 16, 30, 34; 4-POLE 20 AMP CONTACTOR "G" TO CONTROL CIRCUITS AA-21, 22, 25, 31; 4 POLE 20 AMP CONTACTOR "H" TO CONTROL CIRCUITS AA-23, 24, 27, 29; 2-POLE 20 AMP CONTACTOR "R" TO CONTROL CIRCUIT AA-19.

2. PROVIDE AND INSTALL IN SEPARATE DEDICATED CABINET WITH LOCKABLE CONTINUOUS HINDED DOOR AND SHALL CONTAIN: 4-POLE 20 AMP CONTACTOR "I" TO CONTROL CIRCUITS BB-1, 2, 5; 4-POLE 20 AMP CONTACTOR "J" TO CONTROL CIRCUITS BB-3, 4, 7; 4-POLE 20 AMP CONTACTOR "K" TO CONTROL CIRCUITS BB-13, 15, 35; 2-POLE 20 AMP CONTACTOR "L" TO CONTROL CIRCUITS BB-17, 19; 4-POLE 20 AMP CONTACTOR "M" TO CONTROL CIRCUITS BB-14, 16, 18; 4-POLE 20 AMP CONTACTOR "N" TO CONTROL CIRCUITS BB-8, 11, 12, 25; 2-POLE 20 AMP CONTACTOR TO CONTROL CIRCUITS BB-9, 10; 6-POLE 20 AMP CONTACTOR "P" TO CONTROL CIRCUITS BB-21, 22, 23, 24; 2-POLE CONTACTOR "Q" TO CONTROL CIRCUIT BB-20.

3. PROVIDE AND INSTALL IN SEPARATE DOOR COVER; 8-POLE 20 AMP CONTACTOR TO CONTROL CIRCUITS CC-1, 2, 3, 4, 5. CONTACTOR TO BE CONTROLLED BY PHOTO-CELL.

4. PROVIDE AND INSTALL IN SEPARATE DOOR OVER; 4-POLE 20 AMP CONTACTOR TO CONTROL CIRCUIT CC-6. CONTACTOR TO BE CONTROLLED BY PHOTO-CELL AND ALSO BY TIME SWITCH.

SUMMARY

- The electrical portion of residential blueprints is often designed and drawn by the electrical contractor.
- In residential construction, it is often up to the electrician performing the work to plan the runs and circuitry.

- It is important for an electrician to be able to visualize how the completed job will look to properly plan and install the electrical devices, lights, and circuitry.
- Estimating a residential job is a systematic process of counting devices and equipment and applying the appropriate cost.

REVIEW QUESTIONS

1. What is the first step in residential print reading?
2. Looking at the first-floor plan and plot plan, where would you locate the electrical service box?

   NOTE: This house has an underground feed designated by the square box on Walnut Circle shown on the plot plan.

3. What is the purpose of elevation views?
4. Why would the electrician consult the foundation detail plan?
5. What are the most important plans to the electrician? Why?

   NOTE: If the prints in the book are too small to read for some of the following questions, a CD version is available for viewing or printing at a larger size.

6. What is the overall length of the house?
7. Looking at the first- and second-floor plans (see Figure 16–3, page 252, and Figure 16–4, page 253), how many three-way switch systems are used? Where?
8. How many receptacles are shown in the kitchen detail (see Figure 16–8, page 256)?
9. Based on the circuit list shown (see Figure 16–9, page 257), how many 110 volt and how many 220 volt circuits are required in this kitchen?
10. How many lights are used in the garage area?

11. How many lights are used in the family room and how many switches are used to control these lights?
12. Where is the switch that controls the back door outside light?
13. Where is the pantry light switch located?
14. Bedroom number 3 on the second floor is above what room on the first floor?
15. Where is the laundry or utility room on this project?
16. How many bathrooms are listed on this project?
17. Why are two receptacles shown above the garage doors on the first-floor plan?
18. Where is the water heater located?
19. Roughly how long would the 220 volt wire run be from the breaker box if located in the back left corner of the garage to the range in the kitchen?
20. No outside spot lights are shown on this project. Where would you locate them?

   NOTE: At this point, it would be best to print out a set of full-size floor plans to do print-reading, take-off, and cost-estimating practice with your instructor. You or your instructor should get current device prices and determine labor costs for purposes of a cost-estimating exercise. You can use the estimating sheet examples in the text or create your own. Each student should perform a take-off and cost estimate and then compare with others and the instructor.

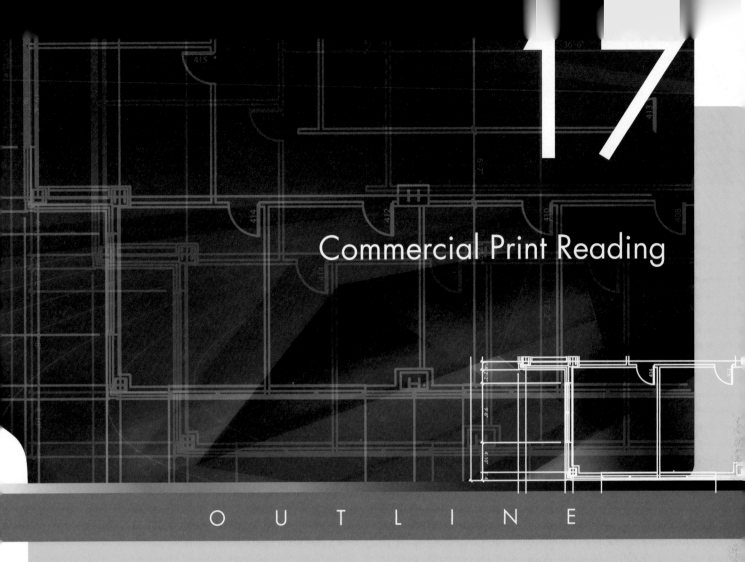

Commercial Print Reading

17

OVERVIEW

This chapter covers commercial print and specification reading. Specifications are an important part of commercial and industrial prints. The purpose of this chapter is to work through a partial set of commercial prints and specifications. The specifications are in this text and the prints are available on CD. The print and specifications for this exercise are for the IBEW training center, conference center, and offices recently constructed in Pittsburgh, Pennsylvania.

OBJECTIVES

After completing this chapter, you should be able to:

• Describe the process of studying a set of commercial construction drawings.

• Visualize a commercial building by looking at a set of commercial construction drawings.

• Read and gather necessary information from a given set of commercial specifications.

• Gather required information from a variety of commercial plans and elevations.

• Read and gather required data from feeder riser diagrams.

• Read and gather required data from legends and schedules.

• Describe the layout of the electrical on the job by looking at construction drawings.

COMMERCIAL PRINT ANALYSIS

Commercial print sets may include a cover sheet, plot plan, elevations, various floor and ceiling plans, details, sections, and schedules. These prints tend to be much more extensive than residential prints. The overall size of commercial projects is usually larger in scope than residential projects.

The specifications section previously consisted of 16 divisions. As explained in Chapter 15, the Construction Standards Institute (CSI) has upgraded the MasterFormat to include commercial, industrial, and process specifications. This new MasterFormat contains 49 numbered divisions.

The purpose of this chapter is to work through and analyze an actual set of commercial prints. The new commercial print set is the IBEW Local 5 office and training center constructed in 1999. The title sheet lists more than 200 prints, of which 12 are supplied on the CD at the back of this text. Due to the reduction of some of the figures in this chapter, you should consult the commercial prints while reading the chapter.

This chapter references only a selection of the supplied drawings as a basic introduction to commercial print reading. When possible, a print or partial print section is included to help explain the text. In some cases, an entire print reduced to page size would be illegible. In these cases, a small section of the entire print is used for discussion purposes. The best way to read this chapter is with access to the master full-size or half-size commercial print set.

FIGURE 17-1 Cover sheet.

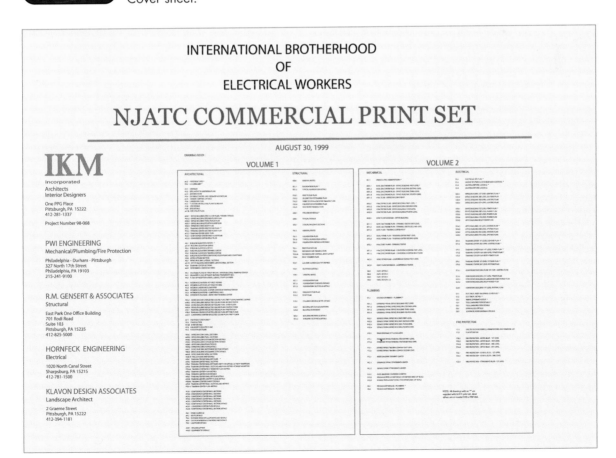

TechTips!

Riser diagrams show the electrical relationship of the power distribution system (see Figure 17–3). Examples of some of the equipment shown on a riser diagram are switchboards, panelboards, disconnects, transformers, and busways. The riser diagram will have lines running between pieces of equipment representing a feeder. The feeder will either have a note next to it or a designation letter or number.

If it has a designation letter or number it will correspond with the feeder schedule (see Figure 17–4). The feeder schedule will indicate the conduit type and size and the number and size of the conductors in the conduit. If a note is used instead of a designation letter or number, it will have the necessary information about the raceway and conductors. The material following walks through a portion of the riser diagram shown in Figure 17–3.

NOTE: The feeder schedule in Figure 17–4 will have some of the information.
- Power from the utility transformer comes into the main switchboard, which is 277/480 V.
  - Feed comes in on eight 4 inch conduits.
  - Each conduit contains four 500 kcmil conductors.
  - One spare 4 inch conduit is to be installed.

(continued)

Riser Diagram

A diagram that shows the relationship of the electrical distribution system.

It is obvious that there is no perfect way or proper order in which to read a print set. However, several prints must be consulted before starting a construction job. It is imperative at this point to have the necessary sections of the job specifications, as discussed in Chapter 15. This chapter provides a logical example of how to work through a print set. It also explains the basic components and content of various prints and schedules.

17.1 Commercial Specifications

The first step in commercial print reading is to examine the building specifications, as listed in Appendix A of this book. Next, consult the cover sheet for dates and print designations. Make sure the print set has the most current possible date. The cover sheet should list all included drawings by number and title. The commercial office building print set has a drawing list complete with print numbers that appear at the lower right-hand corner of each individual sheet. (This cover sheet is shown in Figure 17–1.)

17.2 Commercial Plans and Elevations

Next, the site plan E1.1 should be examined. The plan lists general and specific electrical notes. It shows the overall commercial complex layout and placement of conduit and cables throughout the project. Whenever possible, a visit to the actual site will help give a feel for the project and will help in interpreting drawings. The electrical legend and site plan on this print fail to show cable television and telephones systems, which may be listed in specifications or contracted to other contractors.

Special care should be taken on any project when trenching between buildings or from external utility sources. All utility companies should be contacted before digging. Your company could be held liable for any damage resulting from improper digging practices. *Call before you dig!* A condensed version of print E1.1 is shown in Figure 17–2 to illustrate the principles previously described.

17.3 Feeder Riser Diagrams and Panelboard Schedules

To get an overview of this commercial project, consult the site plan E1.1, the riser diagram E5.1, and the panelboard schedule E1.2. With attention focused on the feeder riser diagram, you can see the electrical feeds and origins. The riser diagram also indicates conduit size, number of conductors, breaker sizes, and designations. The riser diagram E5.1 is in the print set, and a partial riser is shown in Figure 17–5.

The panelboard schedules in drawing E1.2 will give specific information on panel designations, mounting requirements, voltages, and branch circuits. Transformer and disconnect information may also be listed on this schedule sheet (Figure 17–6).

In examining the prints previously listed, consideration should be given to feed location and size, switch-gear location, site lighting, and other power and electrical requirements. Looking at the site plan and checking on plumbing, sewer, and underground feed locations is important. Coordination of digging related to footer depths and utility feed depths is imperative. Specific local and state codes may affect entry depths of various utilities.

FIGURE 17-2 Site plan.

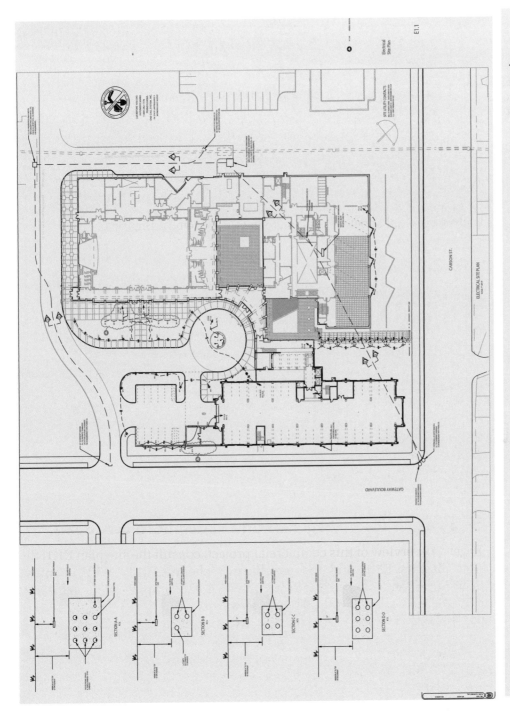

TechTips! *(cont'd)*

- The main switchboard feeds several items, one of which is transformer TT-1 (which steps the voltage down to 120/208 V. (TT-1 stands for "transformer, training facility, #1.")
  - Fed with two 3 inch conduits.
  - Each conduit contains three 350 kcmil conductors and a #1⁄0 ground.
- Transfomer TT-1 feeds distribution board DPBT1L. (DPBT1L stands for "distribution panelboard, training facility, 1st floor, lower voltage.")
  - Fed with three 3-1⁄2 inch conduits.
  - Each conduit contains four 500 kcmil conductors and a #1⁄0 ground.
- Distribution board DPBT1L feeds panels T1LA, T1LB, T1LC, T1LD, T1LF, T2LA, and T2LA. (T1LA stands for "training acility, 1st floor, lower voltage, panel A.")
  - Each panel on the first level is fed with a 1-1⁄2 inch conduit with four #2 conductors and a #6 ground inside.
  - Each panel on the second level is fed with a 2 inch conduit with four #1⁄0 conductors and a #2 ground inside.

FIGURE 17-3 Partial riser diagram.

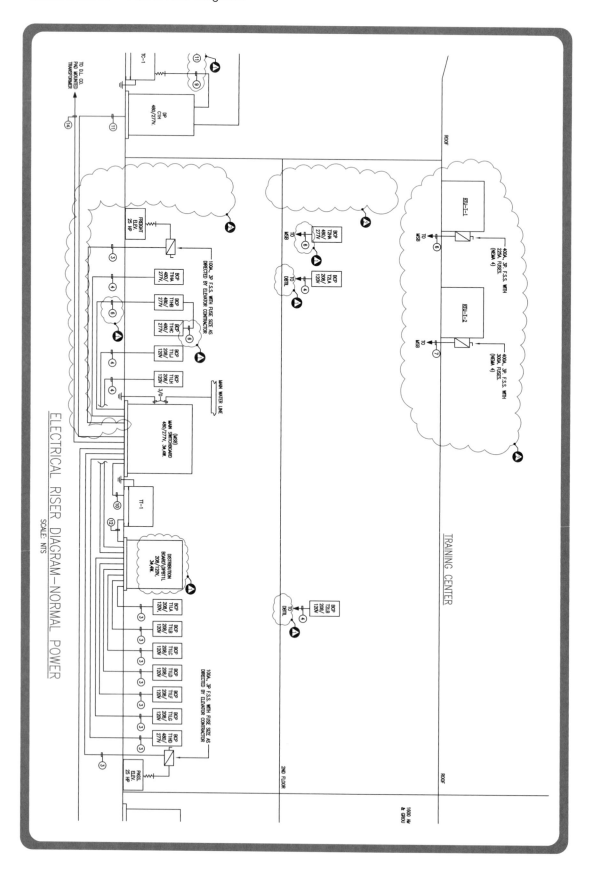

FIGURE 17–4 Feeder schedule.

FEEDER SCHEDULE

FEEDER DESIGNATION	WIRE QUANTITY & SIZE	GROUND WIRE QUANTITY & SIZE	CONDUIT SIZE
1	3–#4	1–#6	1 1/4"
2	3–#2	1–#6	1 1/2"
3	4–#2	1–#6	1 1/2"
4	4–#1/0	1–#2	2"
5	3–#2/0	1–#2	2"
6	4–#4/0	1–#2	2 1/2"
7	4–350 MCM	1–#1/0	3"
8	4–500 MCM	1–#1/0	3 1/2"
9	3–500 MCM	1–#1/0	3 1/2"
10	2 RUNS 3–350 MCM	1–#1/0	3"
11	2 RUNS 4–500 MCM	1–#1/0	3 1/2"
12	3 RUNS 4–500 MCM	1–#1/0	3 1/2"
13	5 RUNS 4–500 MCM	1–#1/0	3 1/2"
14	8 RUNS 4–500 MCM + 1 SPARE 4"	—	4"

FIGURE 17–5 Riser diagram.

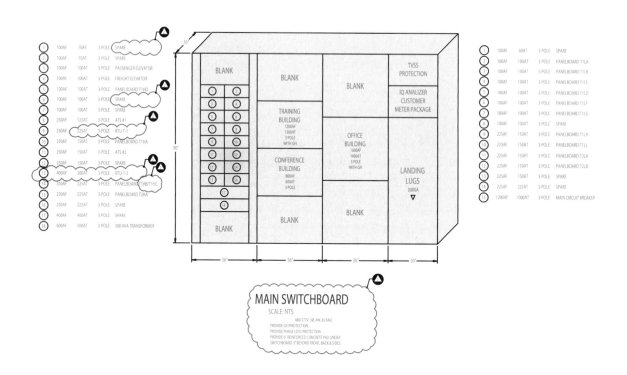

MAIN SWITCHBOARD
SCALE: NTS

480/277V, 3Ø, 4W, 65 KAIC
PROVIDE GFI PROTECTION
PROVIDE PHASE LOSS PROTECTION
PROVIDE 6" REINFORCED CONCRETE PAD UNDER
SWITCHBOARD 3" BEYOND FRONT, BACK & SIDES

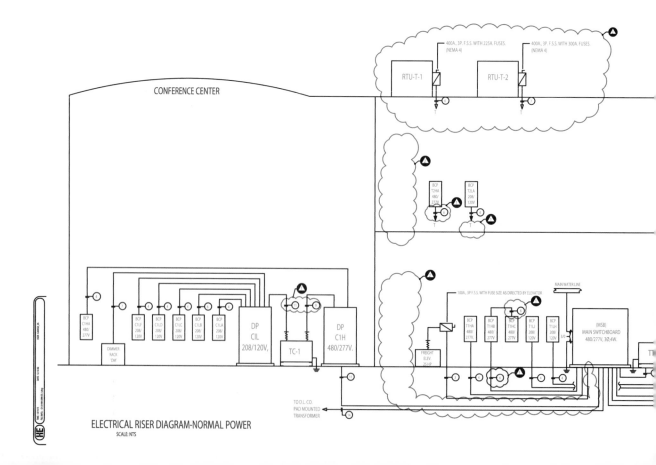

ELECTRICAL RISER DIAGRAM-NORMAL POWER
SCALE: NTS

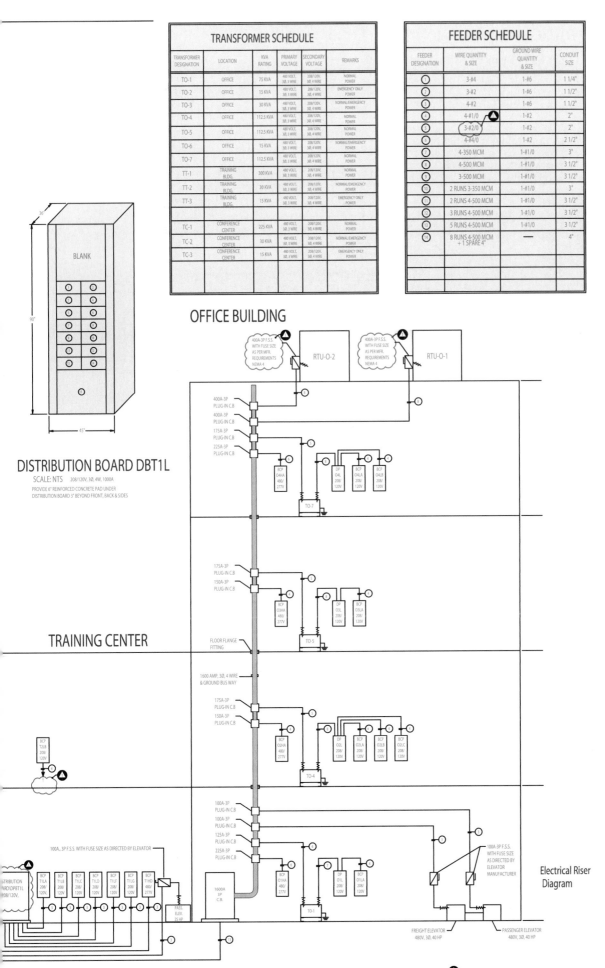

TRANSFORMER SCHEDULE

TRANSFORMER DESIGNATION	LOCATION	KVA RATING	PRIMARY VOLTAGE	SECONDARY VOLTAGE	REMARKS
TO-1	OFFICE	75 KVA	480 VOLT, 3Ø, 3 WIRE	208/120V, 3Ø, 4 WIRE	NORMAL POWER
TO-2	OFFICE	15 KVA	480 VOLT, 3Ø, 3 WIRE	208/120V, 3Ø, 4 WIRE	EMERGENCY ONLY POWER
TO-3	OFFICE	30 KVA	480 VOLT, 3Ø, 3 WIRE	208/120V, 3Ø, 4 WIRE	NORMAL/EMERGENCY POWER
TO-4	OFFICE	112.5 KVA	480 VOLT, 3Ø, 3 WIRE	208/120V, 3Ø, 4 WIRE	NORMAL POWER
TO-5	OFFICE	112.5 KVA	480 VOLT, 3Ø, 3 WIRE	208/120V, 3Ø, 4 WIRE	NORMAL POWER
TO-6	OFFICE	15 KVA	480 VOLT, 3Ø, 3 WIRE	208/120V, 3Ø, 4 WIRE	NORMAL/EMERGENCY POWER
TO-7	OFFICE	112.5 KVA	480 VOLT, 3Ø, 3 WIRE	208/120V, 3Ø, 4 WIRE	NORMAL POWER
TT-1	TRAINING BLDG.	300 KVA	480 VOLT, 3Ø, 3 WIRE	208/120V, 3Ø, 4 WIRE	NORMAL POWER
TT-2	TRAINING BLDG.	30 KVA	480 VOLT, 3Ø, 3 WIRE	208/120V, 3Ø, 4 WIRE	NORMAL/EMERGENCY POWER
TT-3	TRAINING BLDG.	15 KVA	480 VOLT, 3Ø, 3 WIRE	208/120V, 3Ø, 4 WIRE	EMERGENCY ONLY POWER
TC-1	CONFERENCE CENTER	225 KVA	480 VOLT, 3Ø, 3 WIRE	208/120V, 3Ø, 4 WIRE	NORMAL POWER
TC-2	CONFERENCE CENTER	30 KVA	480 VOLT, 3Ø, 3 WIRE	208/120V, 3Ø, 4 WIRE	NORMAL/EMERGENCY POWER
TC-3	CONFERENCE CENTER	15 KVA	480 VOLT, 3Ø, 3 WIRE	208/120V, 3Ø, 4 WIRE	EMERGENCY ONLY POWER

FEEDER SCHEDULE

FEEDER DESIGNATION	WIRE QUANTITY & SIZE	GROUND WIRE QUANTITY & SIZE	CONDUIT SIZE
①	3-#4	1-#6	1 1/4"
②	3-#2	1-#6	1 1/2"
③	4-#2	1-#6	1 1/2"
④	4-#1/0	1-#2	2"
⑤	3-#2/0	1-#2	2"
⑥	4-#4/0	1-#2	2 1/2"
⑦	4-350 MCM	1-#1/0	3"
⑧	4-500 MCM	1-#1/0	3 1/2"
⑨	3-500 MCM	1-#1/0	3 1/2"
⑩	2 RUNS 3-350 MCM	1-#1/0	3"
⑪	2 RUNS 4-500 MCM	1-#1/0	3 1/2"
⑫	3 RUNS 4-500 MCM	1-#1/0	3 1/2"
⑬	5 RUNS 4-500 MCM	1-#1/0	3 1/2"
⑭	8 RUNS 4-500 MCM + 1 SPARE 4"	—	4"

BLANK

DISTRIBUTION BOARD DBT1L

SCALE: NTS   208/120V, 3Ø, 4W, 1000A

PROVIDE 6" REINFORCED CONCRETE PAD UNDER
DISTRIBUTION BOARD 3" BEYOND FRONT, BACK & SIDES

OFFICE BUILDING

TRAINING CENTER

Electrical Riser
Diagram

△ 12-REVISED MECHANICAL SYSTEM

△ 8-10-98 GENERAL REVISION

E5.1

FIGURE 17–6 Panelboard schedule.

PANELBOARD SCHEDULE

PANEL DESIGNATION	MOUNTING	MAINS		SIZE OF BUS	TOTAL NUMBER OF POLES	VOLTAGE
		M.L.O. OR M.C.B.	LOCATION			
NEO1HA	SURFAC	60A-3P M.C.B.	BOTTOM	225A	24	480/277V.,3Ø,4W.
NEO1HB	SURFAC	100A.-3P. M.C.B.	BOTTOM	100A	42	480/277V.,3Ø,4W.
NEO1LB	SURFAC	100A.-3P M.C.B.	TOP	100A	24	208/120V.,3Ø,4W.
EO1LA	SURFAC	60A-3P M.C.B.	BOTTOM	100A	18	208/120V.,3Ø,4W.
O1HA	SURFAC	225A M.L.O.	TOP	225A	42	480/277V.,3Ø,4W.
DPO1L	SURFAC	300A.-3P. M.C.B.	TOP	400A	24	208/120V.,3Ø,4W.
O1LA	SURFAC	225A M.L.O.	TOP	225A	42	208/120V.,3Ø,4W.
O2HA	SURFAC	225A M.L.O.	TOP	225A	42	480/277V.,3Ø,4W.
DPO2L	SURFAC	400A.-3P. M.C.B.	TOP	400A	24	208/120V.,3Ø,4W.
O2LA	FLUSH	225A M.L.O.	TOP	225A	42	208/120V.,3Ø,4W.
O2LB	SURFAC	225A M.L.O.	TOP	225A	42	208/120V.,3Ø,4W.
O2LC	FLUSH	225A M.L.O.	TOP	225A	42	208/120V.,3Ø,4W.
NEO3HA	SURFAC	100A M.L.O.	TOP	100A	24	480/277V.,3Ø,4W.
NEO3LB	SURFAC	60A.-3P. M.C.B.	TOP	100A	24	208/120V.,3Ø,4W.
O3HA	SURFAC	225A M.L.O.	TOP	225A	42	480/277V.,3Ø,4W.
DPO3L	SURFAC	400A.-3P. M.C.B.	TOP	400A	24	208/120V.,3Ø,4W.
O3LA	FLUSH	225A M.L.O.	TOP	225A	42	208/120V.,3Ø,4W.
O4HA	SURFAC	225A M.L.O.	TOP	225A	42	480/277V.,3Ø,4W.
DPO4L	SURFAC	400A.-3P. M.C.B.	TOP	400A	24	208/120V.,3Ø,4W.
O4LA	FLUSH	225A M.L.O.	TOP	225A	42	208/120V.,3Ø,4W.
O4LB	FLUSH	225A M.L.O.	TOP	225A	42	208/120V.,3Ø,4W.

BRANCH CIRCUIT BREAKERS	REMARKS	INTERRUPTING RATING
20A.-1P., 4-1P. SPACES		14000
)A.-1P., 7-20A.-3P., 2-30A.-3P., 1-50A.-3P., 6-1P. SPACES		14000
20A.-1P., 6-1P. SPACES		10000
)A.-1P.		10000
20A.-1P., 5-20A.-3P., 1-30A.-3P.		14000
)A.-3P.,1-100A.-3P., 2-150A.-3P., 12-1P. SPACES		10000
20A.-1P.	▲4	10000
20A.-1P., 1-20A.-3P., 3-30A.-1P.		14000
)0A.-3P., 3-150A.-3P., 12-1P. SPACES		10000
20A.-1P.		10000
20A.-1P., 1-30A.-2P.,1-50A-2P	▲3 ▲6	10000
20A.-1P., 4-1P. SPACES		10000
20A.-1P., 10-1P. SPACES		14000
20A.-1P., 10-1P. SPACES	▲4	10000
20A.-1P., 1-20A.-3P., 1-30A.-1P., 1-40A.-1P., 9-1P. SPACES		14000
)0A.-3P., 2-150A.-3P., 15-1P. SPACES	▲6	10000
20A.-1P.,1-50A-2P.,2-1P-SPACES	▲4	10000
20A.-1P., 4-20A.-3P., 2-30A.-1P., 2-1P. SPACES		14000
)A.-3P., 1-50A.-3P., 2-150A.-3P., 12-1P. SPACES		10000
20A.-1P.	▲6	10000
20A.-1P.,1-50A-2P		10000

continued

FIGURE 17–6 (continued)

PANEL DESIGNATION	MOUNTING	MAINS		SIZE OF BUS	TOTAL NUMBER OF POLES	VOLTAGE
		M.L.O. OR M.C.B.	LOCATION			
NEC1HA	SURFAC	60A.-3P. M.C.B.	BOTTOM	100A	24	480/277V.,3Ø,4W.
NEC1HB	SURFAC	100A.-3P. M.C.B.	BOTTOM	100A	42	480/277V.,3Ø,4W.
NEC1LB	SURFAC	100A.-3P. M.C.B.	BOTTOM	100A	24	208/120V.,3Ø,4W.
EC1LA	SURFAC	60A.-3P. M.C.B.	BOTTOM	100A	18	208/120V.,3Ø,4W.
DPC1H	SURFAC	800A.-3P. M.C.B.	BOTTOM	800A	42	480/277V.,3Ø,4W.
C1HA	FLUSH	225A M.L.O.	BOTTOM	225A	42	480/277V.,3Ø,4W.
DPC1L	SURFAC	800A.-3P. M.C.B.	BOTTOM	800A	30	208/120V.,3Ø,4W.
C1LA	FLUSH	225A M.L.O.	BOTTOM	225A	42	208/120V.,3Ø,4W.
C1LB	FLUSH	225A M.L.O.	BOTTOM	225A	42	208/120V.,3Ø,4W.
C1LC	SURFAC	225A M.L.O.	BOTTOM	225A	42	208/120V.,3Ø,4W.
C1LD	SURFAC	225A M.L.O.	BOTTOM	225A	42	208/120V.,3Ø,4W.
C1LF	FLUSH	225A M.L.O.	BOTTOM	225A	42	208/120V.,3Ø,4W.
NET1HBA	SURFAC	100A M.L.O.	BOTTOM	100A	42	480/277V.,3Ø,4W.
NET1HA	SURFAC	225A M.L.O.	BOTTOM	225A	30	480/277V.,3Ø,4W.
NET1HB	SURFAC	225A M.L.O.	BOTTOM	225A	24	480/277V.,3Ø,4W.
NETILB	SURFAC	100A.-3P. M.C.B.	BOTTOM	100A	42	208/120V.,3Ø,4W.
NET2LB	SURFAC	100A. M.L.O.	BOTTOM	100A	30	208/120V.,3Ø,4W.
ETIHA	SURFAC	60A.-3P. M.C.B.	TOP	100A	24	480/277V.,3Ø,4W.

BRANCH CIRCUIT BREAKERS	REMARKS	INTERRUPTING RATING
A-1P, 8-1P SPACES		14000
A-1P, 6-20A.-3P., 1-30A.-3P., 3-1P SPACES		14000
A-1P, 12-1P SPACES		10000
A-1P, 10-1P SPACES		10000
-3P, 1-60A-3P, 2-100A-3P, 2-150A-3P, 1-250A-3P,1-350A-3P, 1-3P SPACE		22000
A-1P, 1-30A-3P, 1-50A-3P, 16-1P SPACES		22000
-3P, 1-100A-3P, 4-150A-3P, 1-225A-3P, 1-300A-3P, 2-3P SPACE		10000
A-1P, 6-1P SPACES	◇1	10000
A-2P, 1-50A-2P, 4-20A-3P, 1-30A-3P, 2-50A-3P, 1-60A-3P., 6-1P. SPACES	◇2	10000
A-1P, 1-30A-1P, 9-1P SPACES		10000
A-1P, 6-1P SPACES		10000
A-1P, 10-1P SPACES		10000
A.-3P., 2-30A.-3P., 17-20A.-1P; 5-30A.-1P., 8-1P. SPACES		14000
A.-3P., 1-40A.-3P., 4-20A.-1P; 7-1P. SPACES		14000
A.-3P., 3-100A.-3P., 9-1P. SPACES		14000
0A.-1P, 5-30A.-1P, 1-60A.-3P, 1-40A.-1P		10000
0A.-1P, 4-40A.-1P, 6-1P. SPACES		10000
A.-3P., 2-40A.-3P., 15-1P. SPACES		14000

17.4 Legends and Schedules

Next, study the architectural floor plans listed for the training center and office building as shown on the print list. Use the actual prints for practice exercises. These plans have exact dimensions and should be consulted rather than scaling the lighting and power prints. It is always more accurate to trust a given dimension than to trust a scaled dimension. The first-floor plan also contains wall sections. This wall section is important due to penetration considerations of electrical devices when passing through walls. The wall section gives necessary device-mounting information for the electrical worker (Figure 17–7).

FIGURE 17–7 Wall section.

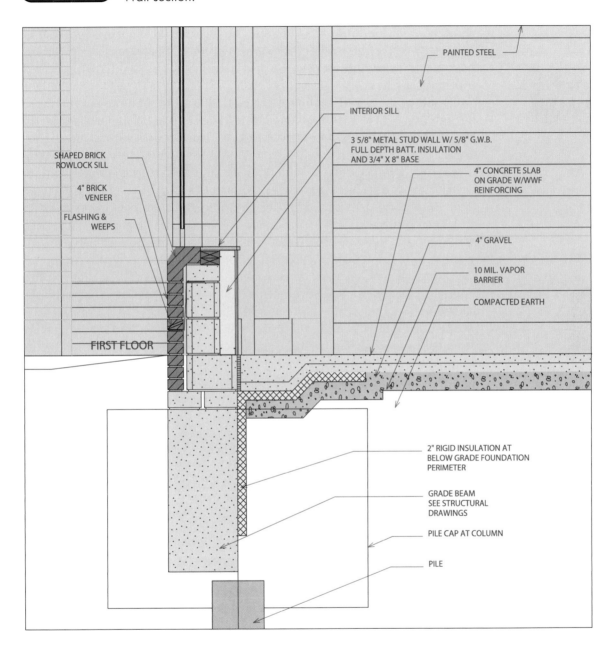

PAINTED STEEL

INTERIOR SILL

3 5/8" METAL STUD WALL W/ 5/8" G.W.B.
FULL DEPTH BATT. INSULATION
AND 3/4" X 8" BASE

4" CONCRETE SLAB
ON GRADE W/WWF
REINFORCING

4" GRAVEL

10 MIL. VAPOR
BARRIER

COMPACTED EARTH

SHAPED BRICK
ROWLOCK SILL

4" BRICK
VENEER

FLASHING &
WEEPS

FIRST FLOOR

2" RIGID INSULATION AT
BELOW GRADE FOUNDATION
PERIMETER

GRADE BEAM
SEE STRUCTURAL
DRAWINGS

PILE CAP AT COLUMN

PILE

WALL SECTION
SCALE: 3/4" = 1'-0"

The next step is to consider lighting, power, and system requirements. Plans required for the exercises in this book are found on the print CD. These plans illustrate lighting fixtures, switches, electrical circuitry, and panel connection numbers as seen in the partial lighting plan E02.1 (Figure 17–8). While reading the lighting plans, it is important to refer to the lighting schedule on plan E1.3 to determine specific lighting-fixture information (Figure 17–9).

At this point, the reflected ceiling plan AT6.1 should be considered with the lighting plans. This ceiling plan indicates sizes of materials and tile types (Figure 17–10). Commercial prints will also have a variety of telecommunications prints and details. Print E5.4 is one of the telecommunications prints and detail sheets provided. A sample of the telecommunications details form (print E5.4) is shown in Figure 17–11.

It is important to review all of the other mechanical prints, such as HVAC and plumbing. These prints will help make run determinations. After reviewing all other mechanical prints, a coordinated effort between the various crafts may save time, materials, and money for all parties involved. The electrical worker often has last choice of runs over HVAC and piping. It may be possible to make minor mechanical run deviations that will save the electrical worker time and material. The first-level HVAC plan M01.1 is an illustration of why other craft prints should be consulted before making electrical runs. HVAC hanger and duct locations must be considered before making electrical runs (Figure 17–12).

NOTE: Larger-size complete drawings of all prints listed in this chapter are found in the Commercial Print Set on the companion CD. The NJATC print CD has commercial prints in multiple formats. The NJATC print CD contains far more prints than listed in the text.

FIGURE 17-8   Partial lighting plan.

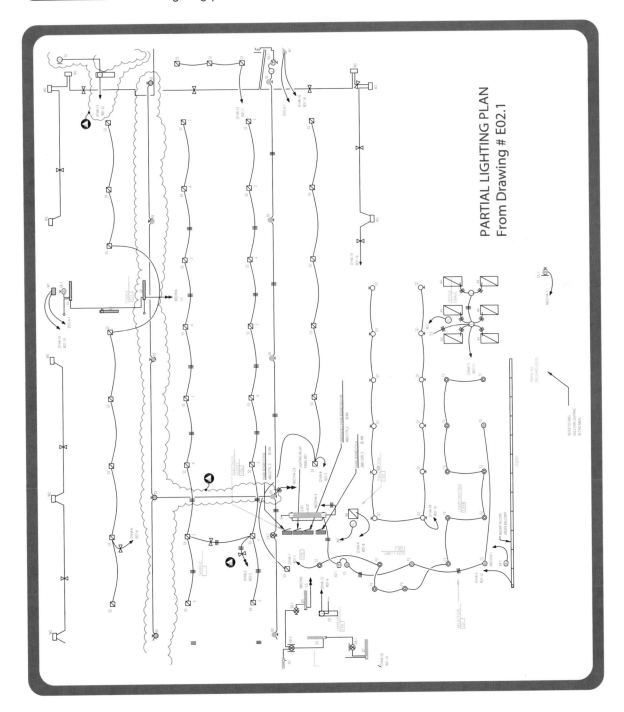

PARTIAL LIGHTING PLAN
From Drawing # E02.1

FIGURE 17-9 Fixture schedule.

R3	1-150 W PAR 38 METAL HALIDE & 1-100W. QUARTZ	1	ENCAPSULATED HPF	277	RECESSED	KIRLIN #RR40710-58-49-66-LP-45
T1	—	—	—	—	PENDANT MOUNTED CEILING	STAFF #5800 SERIES 3 CCT. TRACK SATIN ALUMINUM FINISH
U1	1-150W T-7 METAL HALIDE	1	HPF	277	PENDANT	LOUIS POULSEN #ARP763 WHITE/WHITE
V1	6-F40SP35 BIAX & 1-21 WATT 2DSP35 FLUORESCENT	4	ELECTRONIC	277	PENDANT- AIRCRAFT CABLE	LITECONTROL #P1D336BX40/21-FP-ELB-277
X1	2-F32 WATT TT SP35 COMPACT FLUORESCENT	1	ELECTRONIC	120	SURFACE WALL	DAC #D1017-2F32-277-CTBS-SCF
Y1	TO BE DETERMINED	—	—	120	SUSPENDED	FIXTURE TO BE SELECTED LATER ALLOWANCE $1,200.00 EACH.
Z1	2-F32 T8 SP35 FLUORESCENT	1	ELECTRONIC	277	SURFACE MOUNTED ON CONTINUOUS TRUNK CHANNEL SUSP. FROM CLG.	ZUMTOBEL #ZX-L-N2/32-4-S2/ZX-ZAC-70/ ZX-70-1/4-20/2X-ACF

7 1/2" DIAMETER RECESSED ADJUSTABLE METAL HALIDE DOWNLIGHT WITH ALZAK REFLECTOR, ENCAPSULATED HPF BALLAST, 3 YEAR WARRANTY WITH EMERGENCY ONLY SOCKET. U.L. LISTED & IBEW LABEL.	OFFICE BUILDING
PENDANT MOUNTED 3-CIRCUIT LIGHTING TRACK WITH ALL NECESSARY FITTINGS FOR CONFIGURATION INDICATED. U.L. LISTED & IBEW LABEL.	
24" DIAMETER x 31 1/2" HIGH PENDANT MOUNTED CHANDELIER. CAST ALUMINUM HOUSING LENGTH OF PENDANTS TO BE DETERMINED TO ACHIEVE +5'-0" MOUNTING HEIGHT ABOVE 2ND FLOOR LEVEL. UL LISTED & IBEW LABEL.	
33" DIAMETER SUSPENDED DIRECT/INDIRECT FLUORESCENT LIGHTING FIXTURE WITH 6 LAMPS UP & 1 LAMP DOWN. PERFORATED SPUN STEEL HOUSING, WHITE FINISH, ELECTRONIC BALLASTS, CENTER 1/8" ACRYLIC DIFFUSER. U.L. LISTED, & IBEW LABEL.	TRAINING CENTER
WALL MOUNTED TWIN FLUORESCENT INDUSTRIAL FIXTURE 18" LONG X 5" HIGH X 8" DEEP WITH GLASS RIBBED GLOBES & PERFORATED METAL SLEEVE ENCLOSURE, ELECTRONIC BALLAST. UL LISTED & IBEW LABEL.	
DECORATIVE CHANDELIER	BOARD ROOM 4TH FLOOR OFFICE BUILDING
CONTINUOUS DIRECT LOUVERED LINEAR FLUORESCENT LIGHTING SYSTEM CONSISTING OF PENDANT MOUNTED TRUNK CHANNEL WITH 4' FLUORESCENT WHITE LOUVERED FIXTURES AS INDICATED WITH ELECTRONIC BALLAST AND ALL NECESSARY FITTINGS AND CABLES & POWER FEEDS FOR A COMPLETE INSTALLATION. REFER TO DWGS. FOR CONFIGURATION & LOCATIONS OF NORMAL & EMERGENCY FIXTURES. UL LISTED & IBEW LABEL.	TRAINING CENTER REFER TO PLANS FOR LAYOUT OF TRUNK CHANNEL.

FIGURE 17–9   *(continued)*

AA	1-70 WATT PAR 30 METAL HALIDE MASTER COLOR M98	1	ELECTRONIC	120	LIGHTING TRACK T1	STAFF #909-3-MP070-TN
BB	2-26 WATT T4 SP35 COMPACT FLUORESCENT	1	ELECTRONIC	120	LIGHTING TRACK T1	STAFF #13226-3-TN
CC	2-39 WATT T5 SP35 COMPACT FLUORESCENT	1	ELECTRONIC	120	LIGHTING TRACK T1	STAFF #135-3-TN
DD	50 WATT MR16	1	ELECTRONIC	120/12	RECESSED CEILING	ZUMTOBEL #MT4660-S1-415-WH
FF	1-250WATT METAL HALIDE	1	HPF	277	PENDANT 'J' HOOK	HOLOPHANE #RA250MH-27-H-31-B-6-25LH-6-C3
UC-3	1-F21 T5 SP35 FLUORESCENT	1	ELECTRONIC	120	UNDER COUNTER	ALKCO #LINCS100FS35/RSW/ES
UC-4	1-F21 T5 SP35 FLUORESCENT	1	ELECTRONIC	120	UNDER COUNTER	ALKCO #LINCS1OOFS46/RSW/ES
EX-1	LED INTEGRAL	—	—	277	SURFACE OR RECESSED CEILING	YORKLITE #SOL-AC-CR-1C-R-XX SOL-AC-SC-1C-R-XX
EX-2	LED INTEGRAL	—	—	277	SURFACE OR RECESSED CEILING	YORKLITE #SOL-AC-CR-2C-R-X SOL-AC-SC-2C-R-X
EX-3	LED INTEGRAL	—	—	277	SURFACE WALL	YORKLITE #SOL-AC-5W-1C-R-XX
EX-4	LED INTEGRAL	—	—	277	SURFACE WALL END MOUNT	YORKLITE #SOL-AC-5E-LC-R-XX
EX-5	LED INTEGRAL	—	—	277	SURFACE PENDANT CEILING MOUNTED	YORKLITE #SOL-AC-SC-2C-R-XX-P
S3	1-175 WATT COATED METAL HALIDE ED/BT-28	1	HPF	277	RECESSED	KIRLIN #RS51456-11-45-46-58

Partial Fixture Schedule From Drawing # E1.3

ALUMINUM DIE CAST HID TRACK FIXTURE WITH INTEGRAL BALLAST & SATIN ALUMINUM FINISH. U.L. LISTED & IBEW LABEL.	
13" LONG ALUMINUM FLUORESCENT WALLWASH TRACK FIXTURE WITH ELECTRONIC BALLAST & SATIN ALUMINUM FINISH. U.L. LISTED & IBEW LABEL.	
18" ALUMINUM FLUORESCENT WALLWASH TRACK FIXTURE WITH ELECTRONIC BALLAST & SATIN ALUMINUM FINISH. U.L. LISTED & IBEW LABEL.	
4" DIAMETER RECESSED LOW VOLTAGE ADJUSTABLE DOWNLIGHT WITH FACETED REFLECTOR, THERMAL PROTECTION, COMPLETE WITH 120 TO 12 VOLT TRANSFORMER. U.L. LISTED & IBEW LABEL.	
	TRAINING CENTER PIPE LAB
INDUSTRIAL HID LOW BAY LUMINAIRE WITH HIGH POWER FACTOR BALLAST, GLASS PRISMATIC REFLECTOR WITH SAFETY HOOK CHAIN & MATCHING RECEPTACLE.	
3' UNDERCOUNTER LIGHT WITH EXTRUDED ALUMINUM HOUSING, CLEAR LINEAR PRISIM LENS END CAPS, CONNECTORS FOR ROW MOUNT, ROCKER SWITCH, ETCHED SILVER FINISH & ELECTRONIC BALLAST, U.L. LISTED, & IBEW LABEL.	
4' UNDERCOUNTER LIGHT WITH EXTRUDED ALUMINUM HOUSING, CLEAR LINEAR PRISIM LENS END CAPS, CONNECTORS FOR ROW MOUNT, ROCKER SWITCH, ETCHED SILVER FINISH & ELECTRONIC BALLAST, U.L. LISTED, & IBEW LABEL.	
SINGLE FACE EDGE GLO LED EXIT SIGN WITH RED LETTERS ON CLEAR BACKGROUND PROPER HOUSING CEILING (SURFACE OR RECESSED) SHALL BE FURNISHED IN AREA USED. FINISHED AREAS SHALL UTILIZE A RECESSED HOUSING & UNFINISHED OR EXPOSED AREAS SHALL UTILIZE SURFACE HOUSING WITH CHEVRONS WHERE INDICATED. U.L. LISTED, & IBEW LABEL.	
DOUBLE FACE EDGE GLO LED EXIT SIGN WITH RED LETTERS ON CLEAR BACKGROUND CEILING (SURFACE) MOUNT SHALL BE FURNISHED IN AREA USED. FINISHED AREAS SHALL UTILIZE A RECESSED HOUSING & UNFINISHED OR EXPOSED AREAS SHALL UTILIZE SURFACE HOUSING WITH CHEVRONS WHERE INDICATED. U.L. LISTED, & IBEW LABEL.	
SINGLE FACE EDGE GLO LED EXIT SIGN WITH RED LETTERS ON CLEAR BACKGROUND WALL (SURFACE) SHALL BE FURNISHED IN AREA USED. FINISHED AREAS SHALL UTILIZE A RECESSED HOUSING & UNFINISHED OR EXPOSED AREAS SHALL UTILIZE SURFACE HOUSING WITH CHEVRONS WHERE INDICATED. U.L. LISTED, & IBEW LABEL.	
DOUBLE FACE EDGE GLO LED EXIT SIGN WITH RED LETTERS ON CLEAR BACKGROUND PROPER HOUSING WALL (SURFACE OR RECESSED) END MOUNT SHALL BE FURNISHED IN AREA USED. FINISHED AREAS SHALL UTILIZE A RECESSED HOUSING & UNFINISHED OR EXPOSED AREAS SHALL UTILIZE SURFACE HOUSING WITH CHEVRONS WHERE INDICATED. U.L. LISTED, & IBEW LABEL.	
DOUBLE FACE EDGE GLO LED EXIT SIGN WITH RED LETTERS ON CLEAR BACKGROUND PROPER HOUSING CEILING (SURFACE) SHALL BE FURNISHED IN AREA USED. FINISHED AREAS SHALL UTILIZE A RECESSED HOUSING & UNFINISHED OR EXPOSED AREAS SHALL UTILIZE SURFACE HOUSING WITH CHEVRONS WHERE INDICATED. U.L. LISTED, & IBEW LABEL.	
14" SQUARE x 8" DEEP RECESSED HID FIXTURE WITH ALUMINUM HOUSING, DOUBLE GASKETING TEMPERED GLASS SPREAD LENS, HIGH POWER FACTOR BALLAST, FUSED PRIMARY, WET LOCATION, UL & IBEW LABEL.	

FIGURE 17–10 Reflected ceiling.

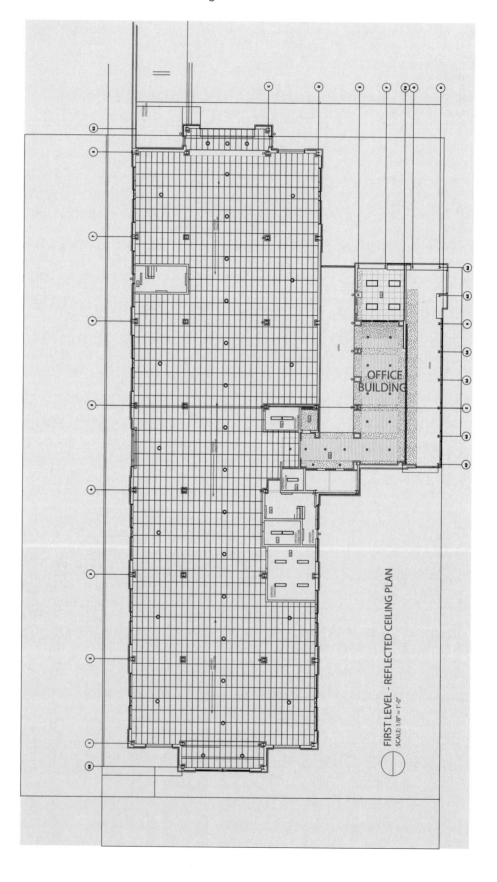

OFFICE BUILDING

FIRST LEVEL - REFLECTED CEILING PLAN
SCALE 1/8" = 1'-0"

FIGURE 17–11 Telecommunication details.

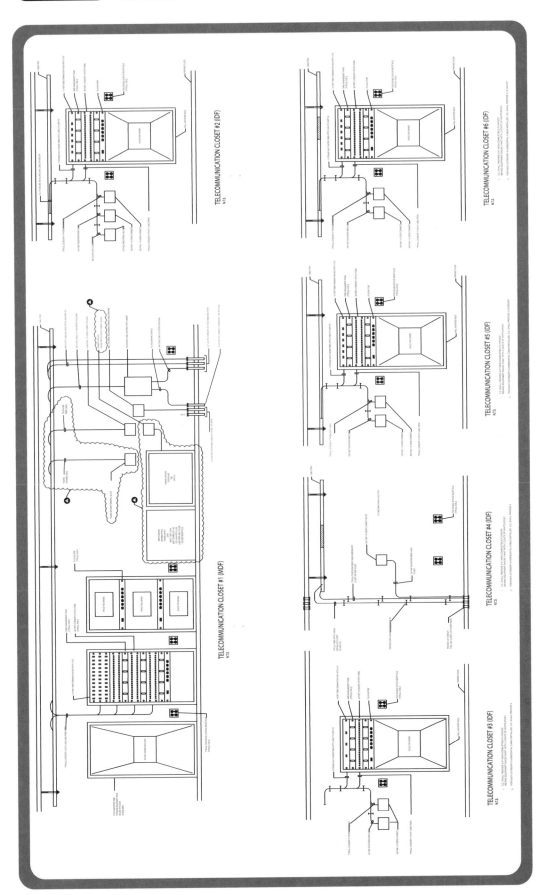

FIGURE 17-12 HVAC plan.

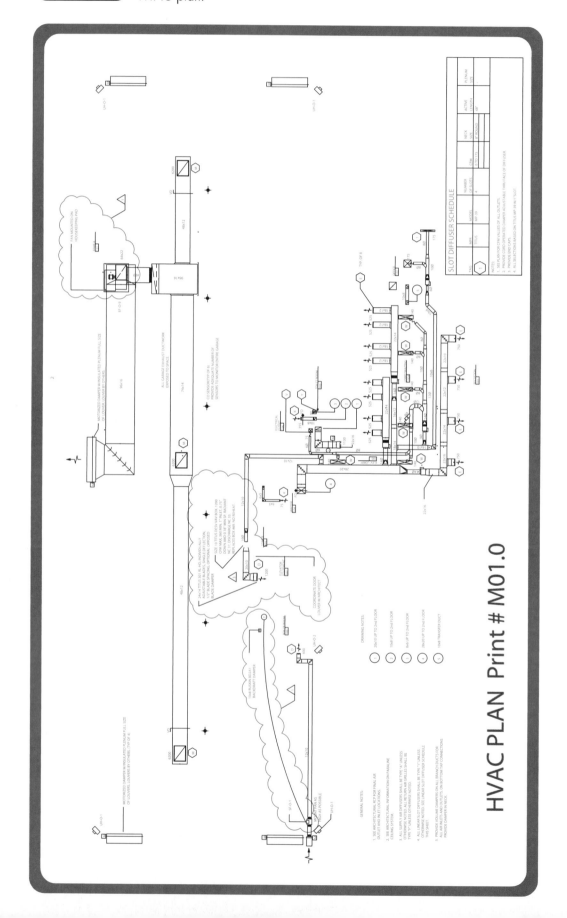

HVAC PLAN Print # M01.0

SUMMARY

- An electrician must be comfortable using and exploring commercial plans, specifications, elevations, legends, and schedules.
- To be able to visualize and completely understand a commercial set of construction drawings, a person must spend time looking at the various drawings and specifications.
- Becoming proficient at reading construction drawings takes time and practice.

REVIEW QUESTIONS

1. What are print specifications?

   NOTE: Use the commercial specifications in the book and the CD print set for the following questions.

2. What are divisions 15 and 16?

3. What division of the commercial specifications gives information relative to motors and starters?

4. In the area of interior wiring installation, do the branch circuits' "hose runs" need separate neutrals?

5. What types of connectors are used on splices in conduits with No. 8 wire and smaller?

6. Of what material are exposed outlet boxes on this project?

7. What Underwriters Laboratory (UL) and National Electrical Manufacturers Association (NEMA) manufacturers standard must switchboards meet?

8. What is the allowable sound level for dry k-rated transformers of 51 to 150 KVA?

9. What section explains main service grounding?

10. What type of fire alarm control is used on this project?

11. Who furnishes and installs the motor starters for HVAC equipment?

12. What three prints will give the electrician a good overview of the project?

NOTE: At this point, it would be best to print out a set of full-size prints for questions 13 through 19 and for actual print-reading practice.

13. What is the location of the D.L. Duquesne Light Company pad-mounted transformer that feeds the project's main switchboard?

14. What information is available on the riser diagram?

15. What three types of electrical plans are available for each floor of each building on a complete print set?

16. Why does the electrician care about wall sections, stair details, and other architectural considerations?

17. What is the importance of a reflected ceiling plan?

18. Why is it important to review and understand HVAC and piping prints?

19. How many telecommunications closets are required on this job?

NOTE: This chapter has generally covered commercial print and specification reading. It would be best at this point if you and your instructor worked through the full-size prints on the CD and developed more practice exercises.

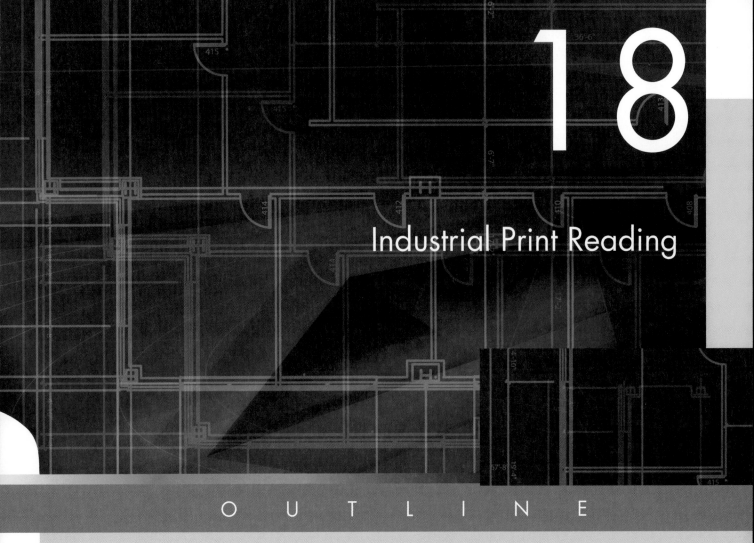

18

Industrial Print Reading

OVERVIEW

This chapter covers industrial print and specification reading. Specifications are a very important part of industrial prints. The purpose of this chapter is to work through a partial set of industrial prints and specifications. The specifications are in this text and the prints are available on CD. The prints and specifications for these exercises are for a manufacturing plant.

OBJECTIVES

After completing this chapter, you should be able to:

• Describe the process of studying a set of industrial construction drawings.

• Visualize an industrial building by looking at a set of industrial construction drawings.

• Read and gather necessary information from a given set of industrial specifications.

• Gather required information from a variety of industrial plans and elevations.

• Read and gather required data from feeder riser diagrams.

• Read and gather required data from legends and schedules.

• Describe the layout of the electrical on the job by looking at the construction drawings.

INDUSTRIAL PRINT ANALYSIS

Like commercial prints, industrial print sets may include a cover sheet, plot plan, elevations, various floor and ceiling plans, details, sections, schedules, and specifications. Industrial prints may also contain process flow diagrams and machine wiring diagrams. Industrial prints are larger and more extensive than residential prints and most commercial prints.

The specifications section previously consisted of 16 divisions. As explained in Chapter 15, the Construction Specifications Institute has upgraded the MasterFormat to include commercial, industrial, and process specifications. This new MasterFormat contains 49 numbered divisions.

There are several differences between commercial and industrial prints and specifications. A look at section 16 of the commercial and industrial print specifications (Appendixes A and B, respectively) will illustrate this point. Industrial projects generally have more detailed specifications than commercial prints. More heavy-duty materials may be used, such as rigid instead of electrical metallic tubing conduit. Hazardous environment concerns such as explosion-proof devices are evident. Advanced electronics and motor control are common in industrial projects.

Another difference between commercial and industrial projects is the actual type of building involved. Office buildings, stores, malls, and warehouses are generally commercial projects. Chemical plants, refineries, manufacturing and construction plants, resource and development facilities, and water treatment plants are industrial projects.

The purpose of this chapter is to work through an actual set of industrial prints. The industrial print set is titled *Manufacturing Plant* and is an actual set of prints from an unnamed component-manufacturing plant. The project prints were issued on 20 July 1990, and the building is now completed. The title sheet lists 88 prints, of which 20 are supplied on the print CD.

This chapter refers to a selection of those prints. In some cases, only a partial section of each print is displayed in this text. This is due to the readability lost when reducing large prints to book size. It is best to have access to the master full-size or half-size print set when reading this chapter.

NOTE: If any of the print reductions shown are difficult to read, consult the industrial print set on the CD.

It is obvious that there is no perfect way of reading or order in which to read a set of prints. There are, however, several prints that must be consulted before starting construction. This chapter provides a logical example of how to work through an industrial print set and explains the basic components and content of prints and schedules.

18.1 Cover Sheet

The first step in reading industrial prints is to examine the cover sheet for dates and print revisions. Make sure your print has the most current date possible. The cover sheet should list all included drawings by number and title. The manufacturing plant print set contains such a print list. A reduced copy of the cover sheet is shown in Figure 18–1.

FIGURE 18-1   Cover sheet.

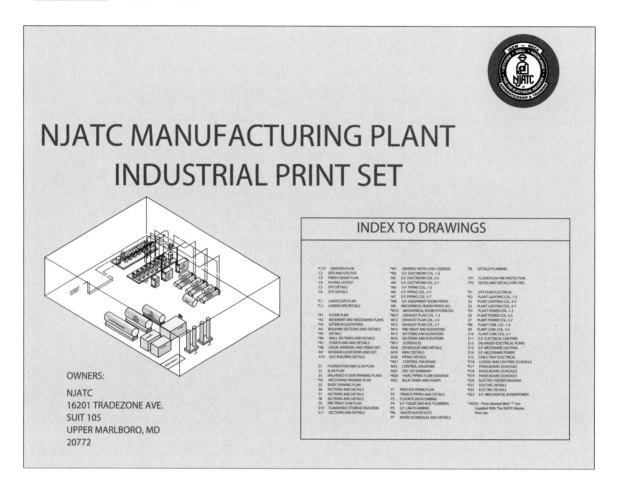

18.2 Industrial Specifications

Next, examine the general and specific specifications, as supplied in Appendix B. These specifications contain important details not found in the print set. The specifications should be consulted throughout the print-reading process.

Next, the site plan E1 should be examined for information regarding power, telephone, security, and lighting. A partial section of this electrical site plan is shown in Figure 18–2. Shown are a riser and underground telephone information, as well as transformer locations and security camera locations.

18.3 Feeder Riser Diagram and Panelboard Schedule

Next, for a project overview examine the feeder riser diagram E20 simultaneously with the panelboard schedules (E17 through E19) and the one-line power distribution diagram on the electrical site plan. The schedules give specific information depicted in a block diagram on the feeder riser print. For example, the MSB2 information from the feeder riser diagram is located on the E17 panelboard schedule (Figure 18–3). Time should be taken at this point to examine the riser diagrams and schedule for the necessary information they contain.

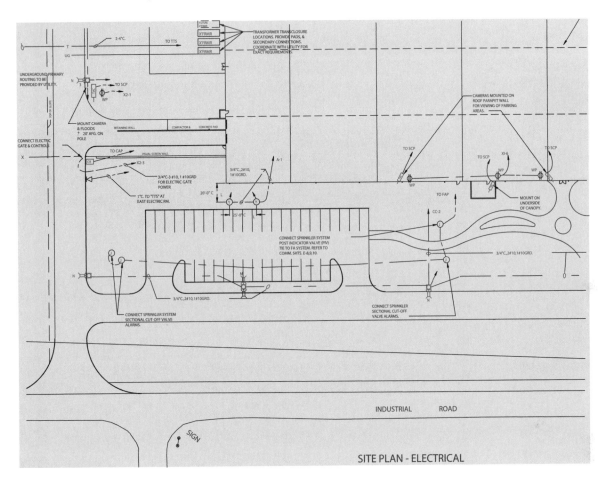

FIGURE 18-2 Site plan.

SITE PLAN - ELECTRICAL

18.4 Plans and Elevations

Next, study the architectural floor plans and elevations A1 and A2. By looking at the plans and elevations, you can get a feel for the physical construction of the entire building. The floor plans are also important on this print set, for two reasons. The first is that floor plans contain actual detailed dimensions that are more accurate than scaling the electrical plans in the print set. The second reason is that this print set uses match-line sections and key section plans. These help relate the various prints to the overall project.

The industrial floor plan A1 is divided into six vertical sections and three horizontal sections by column letters and numbers. The reason for this division is reproduction clarity. Looking at the sections of this building in three parts presents larger and clearer prints. The print CD contains a sample of the prints required for the electrical worker to complete this construction project. Only enough related craft prints are supplied to work through column sections 1 through 3. The reason for only following columns 1 through 3 is to keep the master print set down to a usable size.

FIGURE 18-3 Schedule and riser.

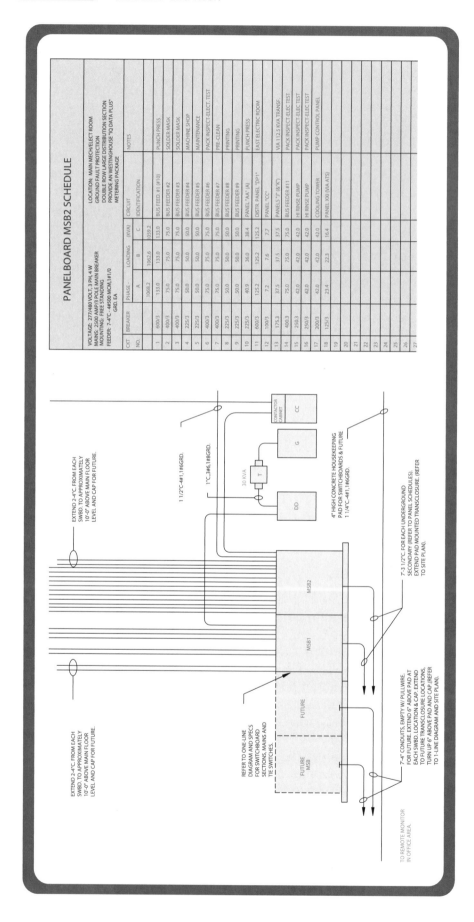

PANELBOARD MSB2 SCHEDULE

VOLTAGE: 277/480 VOLT, 3 PH, 4 W
MAINS: 2500 AMP/3 POLE MAIN BREAKER
MOUNTING: FREE STANDING.
FEEDER: 7-4"C.-4#500 MCM, 1#1/0
GRD. EA.

LOCATION: MAIN MECH/ELECT ROOM
GROUND FAULT PROTECTION
DOUBLE ROW LARGE DISTRIBUTION SECTION
PROVIDE AN WESTINGHOUSE "IQ DATA PLUS"
METERING PACKAGE

CKT NO.	BREAKER	PHASE LOADING (KVA) A	B	C	CIRCUIT IDENTIFICATION	NOTES
		1068.2	1062.6	1059.2		
1	600/3	133.0	133.0	133.0	BUS FEED #1 (#10)	PUNCH PRESS
2	400/3	75.0	75.0	75.0	BUS FEEDER #2	SOLDER MASK
3	400/3	75.0	75.0	75.0	BUS FEEDER #3	SOLDER MASK
4	225/3	50.0	50.0	50.0	BUS FEEDER #4	MACHINE SHOP
5	225/3	50.0	50.0	50.0	BUS FEEDER #5	MAINTENANCE
6	400/3	75.0	75.0	75.0	BUS FEEDER #6	PACK/INSPECT-ELECT. TEST
7	400/3	75.0	75.0	75.0	BUS FEEDER #7	PRE-CLEAN
8	225/3	50.0	50.0	50.0	BUS FEEDER #8	PRINTING
9	225/3	50.0	50.0	50.0	BUS FEEDER #9	PRINTING
10	225/3	40.9	36.0	38.4	PANEL "AA" (A)	PUNCH PRESS
11	600/3	125.2	125.2	125.2	DISTR. PANEL "DP1"	EAST ELECTRIC ROOM
12	100/3	7.2	7.6	7.7	PANEL "CC"	
13	175.3	37.5	37.5	37.5	PANELS "J" (&"K")	VIA 112.5 KVA TRANSF.
14	400/3	75.0	75.0	75.0	BUS FEEDER #11	PACK/INSPECT-ELEC TEST
15	250/3	42.0	42.0	42.0	HI RINSE PUMP	PACK/INSPECT-ELEC TEST
16	250/3	42.0	42.0	42.0	HI RINSE PUMP	PACK/INSPECT-ELEC TEST
17	200/3	42.0	42.0		COOLING TOWER	PUMP CONTROL PANEL
18	125/3	23.4	22.3	16.4	PANEL XXI (KVA ATS)	
19						
20						
21						
22						
23						
24						
25						
26						
27						

Print M2 is HVAC columns 1 through 3, M5 is piping columns 1 through 3, E2 is lighting columns 1 through 3, E5 is power columns 1 through 3, and E8 is communications 1 through 3. To partially explain this column-designation process, Figure 18–4 shows a reduced view of the floor plan and Figure 18–5 shows a column 1 through 3 section of the HVAC information contained on master print M-2. Note the key plan section at the lower right in print M2. This shows how the section you are viewing actually fits within the overall project. The process of dividing plans into columns is common on industrial and commercial print sets.

Next, lighting and power plans E2 and E5 should be examined. These plans depict lighting, electrical, and power requirements. Pay particular attention to notes and references to feeder riser diagrams—as in Figure 18–6, which shows a particular plan section of columns 1 through 3 power plan from sheet E5.

In addition, pay attention to notes referencing panel schedules such as print sheet E2. A section of print E2 shows notes and drawings that refer to panel schedules AA and BB found on print E18. Figure 18–7 shows part of print E2 with notes. Figure 18–8 shows schedules AA and BB from print E18, which is titled *Panelboard Schedules.* These panelboard schedules give specific detail and hookup information not available on the lighting or power prints.

When reading the lighting prints, you must examine the lighting schedules located on print E16. This schedule lists room location panel, circuit information, and manufacturers' numbers for fixtures (Figure 18–9).

With this industrial print set there are details of small areas such as the mechanical room and electrical room print, which have enlarged floor plans for greater detail (as on prints A2 and M8). These prints may also have details to help explain placement of devices, such as shown in Figure 18–10 (the partial section taken from print M8).

NOTE: If any of the prints shown in Figures 18–3 through 18–10 are difficult to read, consult the industrial print set on the CD.

18.5 Detail Sheets

There are a variety of important electrical detail sheets. These sheets are usually detailed elevations with precise measurements needed for installation, as in prints E21 and E22. An example of an electrical detail is the placement and sealing of an explosion-proof receptacle, as shown on print E22 (Figure 18–11). In the case of this print set, column grounding details are also on the detail sheets. Sheet E21 lists grounding details (Figure 18–12).

It is necessary at this time to consult architectural details, as found on print A6. This print contains structural wall sections and details. Structural sections enable the electrical worker to decide how to pass runs through various wall materials. Stair sections help in determining placement and mounting of lighting and conduit runs. Figure 18–13 shows a typical stair and wall section from the master print set.

Fire walls are an example of when architectural prints and specifications need to be consulted. Various special code requirements exist when passing runs through fire walls. Another special concern is open-air plenums, as used

FIGURE 18-4 Main floor plan.

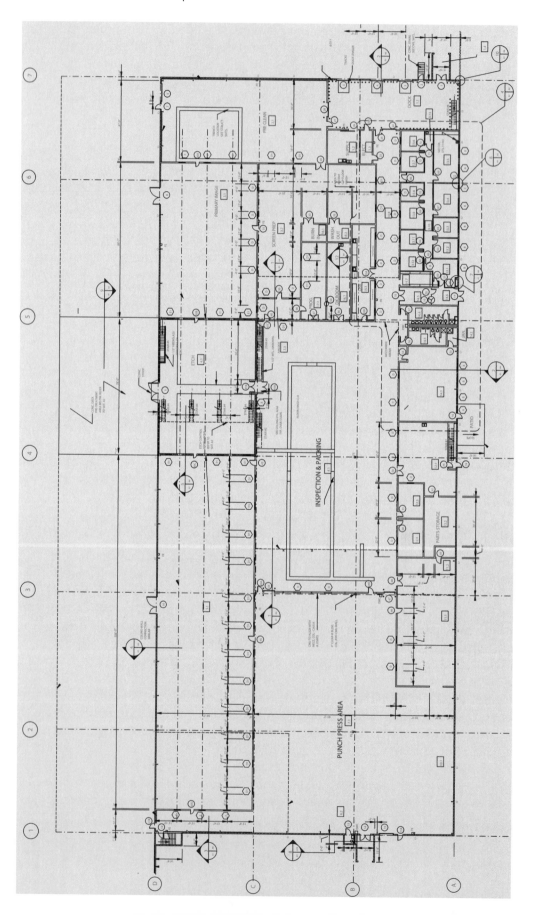

FIGURE 18–5   HVAC plan.

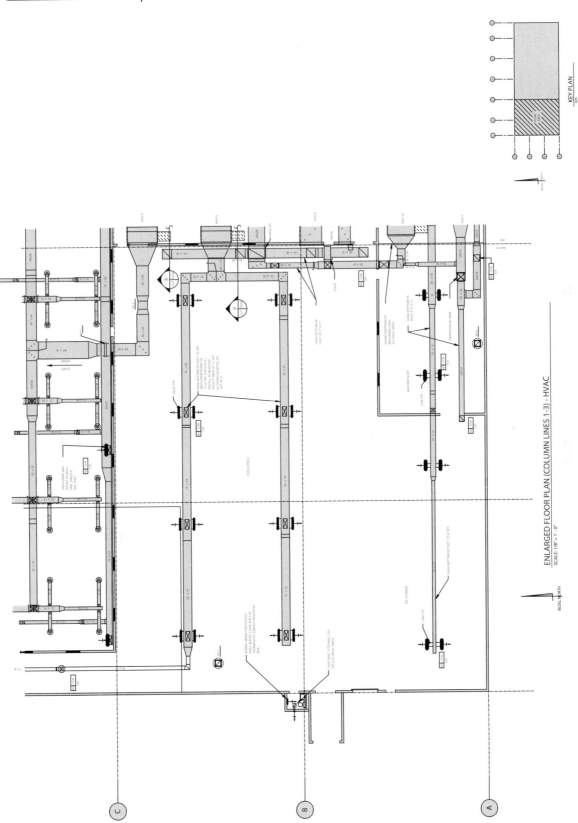

FIGURE 18-6 Plant power.

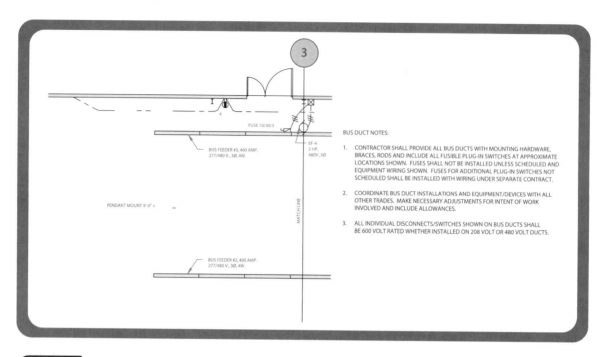

BUS DUCT NOTES:

1. CONTRACTOR SHALL PROVIDE ALL BUS DUCTS WITH MOUNTING HARDWARE, BRACES, RODS AND INCLUDE ALL FUSIBLE PLUG-IN SWITCHES AT APPROXIMATE LOCATIONS SHOWN. FUSES SHALL NOT BE INSTALLED UNLESS SCHEDULED AND EQUIPMENT WIRING SHOWN. FUSES FOR ADDITIONAL PLUG-IN SWITCHES NOT SCHEDULED SHALL BE INSTALLED WITH WIRING UNDER SEPARATE CONTRACT.

2. COORDINATE BUS DUCT INSTALLATIONS AND EQUIPMENT/DEVICES WITH ALL OTHER TRADES. MAKE NECESSARY ADJUSTMENTS FOR INTENT OF WORK INVOLVED AND INCLUDE ALLOWANCES.

3. ALL INDIVIDUAL DISCONNECTS/SWITCHES SHOWN ON BUS DUCTS SHALL BE 600 VOLT RATED WHETHER INSTALLED ON 208 VOLT OR 480 VOLT DUCTS.

FIGURE 18-7 Plant lighting.

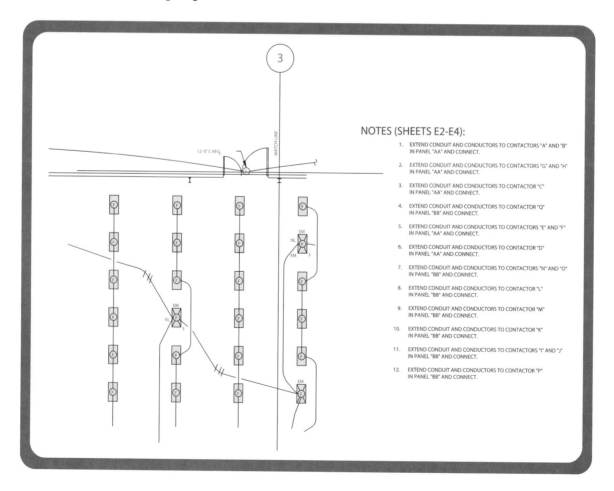

NOTES (SHEETS E2-E4):

1. EXTEND CONDUIT AND CONDUCTORS TO CONTACTORS "A" AND "B" IN PANEL "AA" AND CONNECT.

2. EXTEND CONDUIT AND CONDUCTORS TO CONTACTORS "G" AND "H" IN PANEL "AA" AND CONNECT.

3. EXTEND CONDUIT AND CONDUCTORS TO CONTACTOR "C" IN PANEL "AA" AND CONNECT.

4. EXTEND CONDUIT AND CONDUCTORS TO CONTACTOR "Q" IN PANEL "BB" AND CONNECT.

5. EXTEND CONDUIT AND CONDUCTORS TO CONTACTORS "E" AND "F" IN PANEL "AA" AND CONNECT.

6. EXTEND CONDUIT AND CONDUCTORS TO CONTACTOR "D" IN PANEL "AA" AND CONNECT.

7. EXTEND CONDUIT AND CONDUCTORS TO CONTACTORS "N" AND "O" IN PANEL "BB" AND CONNECT.

8. EXTEND CONDUIT AND CONDUCTORS TO CONTACTOR "L" IN PANEL "BB" AND CONNECT.

9. EXTEND CONDUIT AND CONDUCTORS TO CONTACTOR "M" IN PANEL "BB" AND CONNECT.

10. EXTEND CONDUIT AND CONDUCTORS TO CONTACTOR "K" IN PANEL "BB" AND CONNECT.

11. EXTEND CONDUIT AND CONDUCTORS TO CONTACTORS "I" AND "J" IN PANEL "BB" AND CONNECT.

12. EXTEND CONDUIT AND CONDUCTORS TO CONTACTOR "P" IN PANEL "BB" AND CONNECT.

FIGURE 18-8   Panel schedule.

PANELBOARD AA SCHEDULE

VOLTAGE: 277/480 VOLT, 3 PH, 4 W.
MAINS: 225 AMP 3 POLE MLO
MOUNTING: SURFACE
FEEDER: 2-1/2 C*-4#4/0, 1#2 GRD.
LOCATION: PUNCH PRESS NOTE #1
42 POLES

CKT NO.	BREAKER	PHASE A	B	C	CIRCUIT IDENTIFICATION	NOTES
		44.8	41.6	33.2		
1	20/1	4.0			LIGHTING	
2	20/1		3.4		LIGHTING	
3	20/1			1.6	LIGHTING	
4	20/1	2.8			LIGHTING	
5	20/1		3.4		LIGHTING	
6	20/1			3.4	LIGHTING	
7	20/1	3.8			LIGHTING	
8	20/1		3.4		LIGHTING	
9	20/1			1.2	LIGHTING	
10	20/1	2.2			LIGHTING	
11	20/1		1.2		LIGHTING	
12	20/1			2.2	LIGHTING	
13	20/1	1.6			LIGHTING	
14	20/1		3.0		LIGHTING	
15	20/1			1.8	LIGHTING	
16	20/1	2.6			LIGHTING	
17	20/1		3.8		LIGHTING	
18	20/1			2.4	LIGHTING	
19	20/1	2.4			LIGHTING	
20	50/3	10.0	10.0	10.0	PANEL "AA"	VIA 30 KVA TRANSF.
21	20/1			2.2	LIGHTING	
22	20/1	2.8			LIGHTING	
23	20/1		2.8		LIGHTING	
24	20/1			2.2	LIGHTING	
25	20/1	2.8			LIGHTING	
26	20/1		2.8		LIGHTING	
27	20/1			2.4	LIGHTING	
28	20/1	3.0			LIGHTING	
29	20/1		2.8		LIGHTING	
30	20/1			1.7	LIGHTING	
31	20/1	1.4			LIGHTING	
32	20/1		2.0		LIGHTING	
33	20/1			1.0	LIGHTING	
34	20/1	2.6			LIGHTING	
35	20/1		1.0		SPARE	
36	20/1			1.0	SPARE	
37	20/1	1.0			SPARE	
38	20/1		1.0		SPARE	
39	20/1			1.0	SPARE	
40	20/1	1.0			SPARE	
41						
42						

PANELBOARD A SCHEDULE

VOLTAGE: 120/208 VOLT, 3 PH, 4 W.
MAINS: 100 AMP 3-POLE MAIN BREAKER
MOUNTING: SURFACE
FEEDER: 1-1/4"C.-4#3, 1#8 GRD.
LOCATION: PUNCH PRESS
42 POLES

CKT NO.	BREAKER	PHASE A	B	C	CIRCUIT IDENTIFICATION	NOTES
		6.0	7.0	7.0		
1	20/2		7.0	7.0	LIGHTING	WEST SIGN FLOODS
2	20/1	.5	.5		LIGHTING	
3	20/1			1.5	L/V	
4	20/1			1.0	WALL HEATER	
5	20/2	1.0	1.0		SPARE	
6	20/1		.5		VACUUM PUMP	
7	20/1			.5	ROLL-UP DOOR	
8	20/1	1.0			COMPACTOR DR. CONTROL	
9	20/1		1.0		RECEPTACLES	
10	20/1	1.0			RECEPTACLES	
11	20/1		1.0		RECEPTACLES	
12	20/1			1.0	RECEPTACLES	
13	20/1	.5			RECEPTACLES	
14	20/1		.5		ROLL-UP DOOR	
15	20/1			1.0	ROLL-UP DOOR	
16	20/1	1.0			SPARE	
17	20/1		1.0		SPARE	
18	20/1			1.0	SPARE	
19	20/1				SPARE	
20	20/1				PROVISIONS	
21	20/1				PROVISIONS	
22	20/1				PROVISIONS	
23	20/1				PROVISIONS	
24	20/1				PROVISIONS	
25	20/1				PROVISIONS	
26	20/1				PROVISIONS	
27	20/1				PROVISIONS	
28	20/1				PROVISIONS	
29	20/1				PROVISIONS	
30	20/1				PROVISIONS	
31	20/1				PROVISIONS	
32	20/1				PROVISIONS	
33	20/1				PROVISIONS	
34	20/1				PROVISIONS	
35	20/1				PROVISIONS	
36	20/1				PROVISIONS	
37	20/1				PROVISIONS	
38	20/1				PROVISIONS	
39						
40						
41						
42						

PANELBOARD BB SCHEDULE

VOLTAGE: 277/480 VOLT, 3 PH, 4 W.
MAINS: 225 AMP 3-POLE MLO
MOUNTING: SURFACE
FEEDER: 2"C.-4"4/0, 1#2 GRD. EA
LOCATION: EAST ELEC. ROOM NOTE #2
42 POLES

CKT NO.	BREAKER	PHASE A	B	C	CIRCUIT IDENTIFICATION	NOTES
		41.8	38.0	40.0		
1	20/1	1.8			LIGHTING	
2	20/1		2.0		LIGHTING	
3	20/1			2.2	LIGHTING	
4	20/1	2.2			LIGHTING	
5	20/1		1.8		LIGHTING	
6	20/1			2.2	LIGHTING	
7	20/1	2.2			LIGHTING	
8	20/1		2.4		SPARE	
9	20/1			1.0	LIGHTING	
10	20/1	2.8			SPARE	
11	20/1		1.0		SPARE	
12	20/1			1.4	LIGHTING	
13	20/1	3.2			LIGHTING	
14	20/1		2.8		LIGHTING	
15	20/1			2.8	LIGHTING	
16	20/1	2.8			LIGHTING	
17	20/1		2.4		LIGHTING	
18	20/1			2.0	LIGHTING	
19	20/1	1.0			LIGHTING	
20	20/1		2.4		LIGHTING	
21	20/1	2.4			LIGHTING	
22	20/1			3.6	LIGHTING	
23	20/1		2.6		LIGHTING	
24	20/1			2.4	LIGHTING	
25	20/1	3.0			LIGHTING	
26	20/1		2.4		LIGHTING	
27	20/1			3.6	LIGHTING	
28	20/1	3.0			LIGHTING	
29	20/1		3.4		LIGHTING	
30	20/1			3.4	LIGHTING	
31	20/1	3.4			LIGHTING	
32	20/1		1.2		LIGHTING	
33	20/1			1.0	LCC#1	
34	20/1	1.0			LIGHTING	
35	20/1		1.6		LIGHTING	
36	50/3	10.0	10.0	10.0	PANEL "B"	VIA 30 KVA TRANSF.
37	20/1			3.4	LIGHTING	
38	20/1	2.0			LIGHTING	
39	20/1				SPARE	
40	20/1		1.0		SPARE	
41	20/1	1.0			SPARE	
42	20/1			1.0	SPARE	

FIGURE 18–9   Lighting schedule.

LIGHTING CONTROL SCHEDULE

LIGHTING CONTROL DESCRIPTION		CONTACTOR(S) BEING CONTROLLED	PANEL-CIRCUIT BEING SERVED	ROOM NAME
LCC #1 (MASTER) (SEE DETAIL)	SWITCH#1	A,B,C,D,E,F,G,H	AA - 1-19,21-29,31,32,34	SOLDER MASK, PUNCH PRESS, INSPECTION AND PACKING, MACHINE SHOP, MAINTENANCE
	SWITCH#2	I,J,K,L,M,N,O	BB - 1-10,12-21,38	PRIMARY IMAGE, PRE-CLEAN, DOCK, SCREEN PREP.
	SWITCH#3	P	BB - 20-24	OFFICE AREA, TOILETS, BREAK ROOM
LS #2		A	AA - 1,2,5,6,9	PUNCH PRESS
LS #3		B	AA - 3,4,7,8,11	PUNCH PRESS
LS #4		C	AA - 10,12	MACHINE SHOP
LS #5		D	AA - 17,19	MAINTENANCE OFFICE AND SHOP
LS #6		E	AA - 13,14,28,32	PACK INSPECTION
LS #7		F	AA - 15,16,26,34	PACK INSPECTION
LS #8		G	AA - 21,22,25,31	SOLDER MASK AND PRINTING
LS #9		H	AA - 23,24,27,29	SOLDER MASK AND PRINTING
LS #10		I	BB - 1,2,5	PRIMARY IMAGE
LS #11		J	BB - 3,4,6,7	PRIMARY IMAGE
LS #12		K	BB - 13,15,35	SCREEN PREP
LS #13		L	BB - 17,19	DOCK
LS #14		M	BB - 14,16,18	PRE-CLEAN
LS #15		N	BB - 8,12,38	ETCH
LS #16		O	BB - 10	ETCH
LS #17		P	BB - 21,22,23,24	OFFICE AREA
LS #18		Q	BB - 20	TOILETS, BREAK ROOM
LS #19		R	AA - 19	MAINTENANCE OFFICE

LCC #1 NOTES:

1.   PROVIDE AND INSTALL PLASTIC ENGRAVED NAME PLATES ADJACENT
     TO EACH SWITCH INDICATING ROOM/ROOMS BEING CONTROLLED.

LIGHT   FIXTURE   SCHEDULE

TYPE	MANUFACTURER	CATALOG NUMBER	LAMPS	MOUNTING	REMARKS
A	LITHONIA	WA440A277	4-40W	CEILING	WRAPAROUND ACRYLIC LENS
			FLUOR.	SURFACE	
B	KEYSTONE	2A440EXA277GPWS	4-40W	CEILING	2X4 GRID TROFFER
			FLUORESCENT	SURFACE	ACRYLIC LENS. AIR RETURN
C	KEYSTONE	2A440PWSGPW277S	4-40W	CEILING	2X4 GRID TROFFER
			FLUORESCENT	SURFACE	1/2"X1/2"X1/2" SILVER
					PARABOLIC LOUVER
					AIR RETURN
D	PHOENIX	DL-300D60-3	1-300W	WALL	DUAL ARM DOCK
			R40	7'-0" AFF	LIGHT, 62" LENGTH W/ VERT
					ADJUSTMENT, WIREGUARD, &
					BUILT-IN SWITCH
E	METALUX	DIM-296277	2-75W	PENDANT	INDUSTRIAL REFLECTOR
			FLUORESCENT	16'-0" AFF	8'-0" LENGTH
F	KEYSTONE	2SG440WFEXA277V	4-40W	RECESSED	2X4 GRID TROFFER
		GOLD LAMPS	GOLD FLUOR.		ACRYLIC LENS
			LAMPS		
G	LITHONIA	DD296277	2-75W	CEILING	MOLDED IMPACT-RESISTANT
			FLUORESCENT	SURFACE	WHITE PLASTIC HOUSING
					AND CLIPS. 8'-0" LENGTH
H	METALUX	DIM-296277	2-75W	CEILING	INDUSTRIAL REFLECTOR
			FLUORESCENT	SURFACE	8'-0" LENGTH
I	LITHONIA	DD296A277	2-75W	CEILING	MOLDED IMPACT-RESISTANT
			GOLD FLUOR	SURFACE	WHITE PLASTIC HOUSING
			LAMPS		AND CLIPS. 8'-0" LENGTH
J	NOT USED				
	AS LIGHT FIXTURE				
K	HI-TEK	TWH200S277DMB	1-200W	WALL, AFG AS	WALL PACK
			HPS	NOTED ON DWGS.	BRONZE FINISH
L	KEYSTONE	2SG440WFEXA277V	4-40W	RECESSED	2X4 GRID TROFFER
		4 - RED LAMPS	FLUORESCENT		ACRYLIC LENS
M	McGRAW-EDISON	TWO-CS7262277BZCA40	2-400W	30' SQUARE	RECTANGULAR HOUSING
	K W INDUST. INC.	SSP30-5.0-7BZ	HPS	STEEL POLE	BRONZE FINISH
					SEE DETAIL
N	McGRAW-EDISON	CS7262277BZCA40	1-400W	30' SQUARE	RECTANGULAR HOUSING
	K W INDUST. INC.	SSP30-5.0-7BZ	HPS	STEEL POLE	BRONZE FINISH

FIGURE 18–10 Partial mechanical plan.

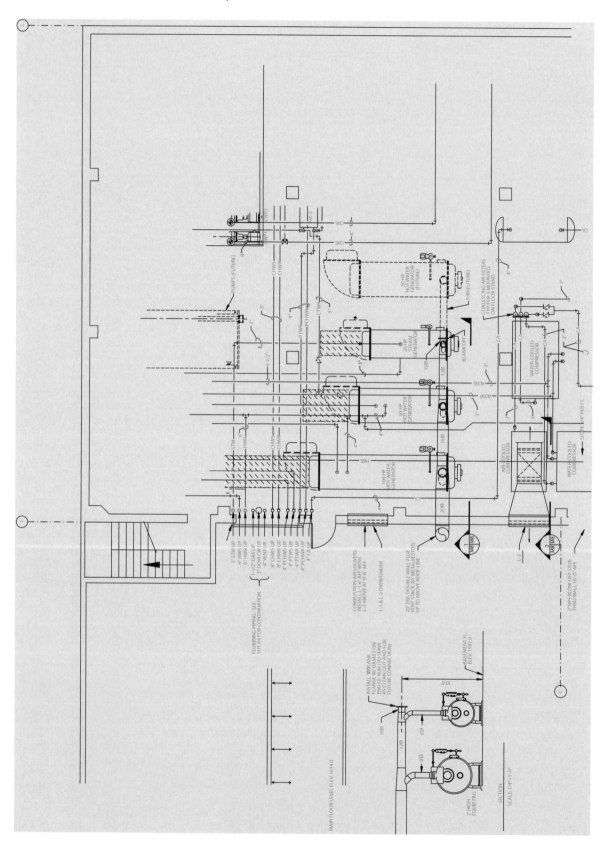

FIGURE 18–11 Explosion-proof receptacle detail.

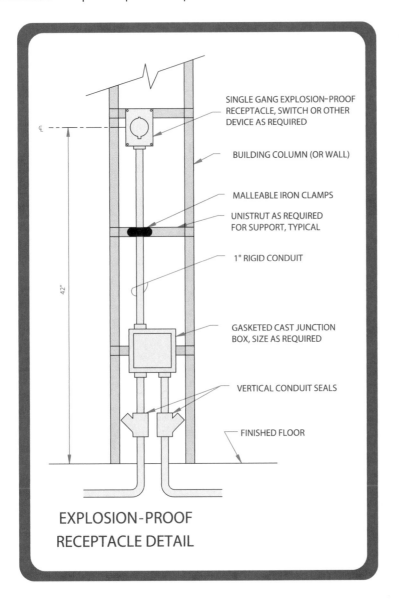

SINGLE GANG EXPLOSION-PROOF
RECEPTACLE, SWITCH OR OTHER
DEVICE AS REQUIRED

BUILDING COLUMN (OR WALL)

MALLEABLE IRON CLAMPS

UNISTRUT AS REQUIRED
FOR SUPPORT, TYPICAL

1" RIGID CONDUIT

GASKETED CAST JUNCTION
BOX, SIZE AS REQUIRED

VERTICAL CONDUIT SEALS

FINISHED FLOOR

42"

EXPLOSION-PROOF
RECEPTACLE DETAIL

in HVAC return systems. There are specifications and code requirements that must be followed when passing through plenums.

As discussed previously, it is important to mention the necessity for understanding other craft symbols and prints. The electrical worker will save time and rework by coordinating runs with the other crafts. There is a tendency to make electrical runs wait until mechanical runs are completed. This may be avoided by looking over mechanical prints carefully and determining runs accordingly. Electrical runs may be made earlier in the construction process based on proper understanding of related mechanical prints.

Structural prints and specifications are important in determining mounting techniques. Beam type and considerations such as fire treatment make a difference in mounting devices. If an area is to be fire treated, it is usually best to coordinate with the insulators and mount the conduit before fire treatment is sprayed.

FIGURE 18–12   Grounding details.

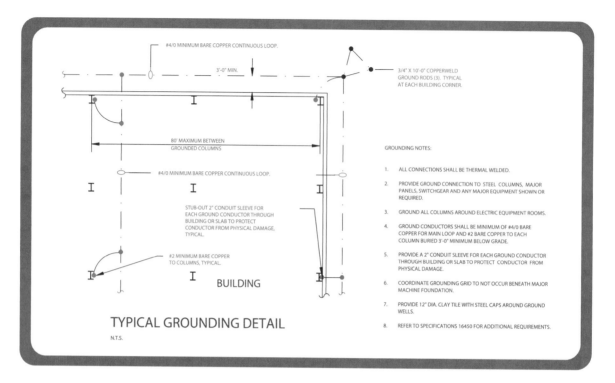

#4/0 MINIMUM BARE COPPER CONTINUOUS LOOP.

3'-0" MIN.

3/4" X 10'-0" COPPERWELD
GROUND RODS (3). TYPICAL
AT EACH BUILDING CORNER.

80' MAXIMUM BETWEEN
GROUNDED COLUMNS

#4/0 MINIMUM BARE COPPER CONTINUOUS LOOP.

STUB-OUT 2" CONDUIT SLEEVE FOR
EACH GROUND CONDUCTOR THROUGH
BUILDING OR SLAB TO PROTECT
CONDUCTOR FROM PHYSICAL DAMAGE,
TYPICAL.

#2 MINIMUM BARE COPPER
TO COLUMNS, TYPICAL.

BUILDING

TYPICAL GROUNDING DETAIL

N.T.S.

GROUNDING NOTES:

1. ALL CONNECTIONS SHALL BE THERMAL WELDED.

2. PROVIDE GROUND CONNECTION TO STEEL COLUMNS, MAJOR
    PANELS, SWITCHGEAR AND ANY MAJOR EQUIPMENT SHOWN OR
    REQUIRED.

3. GROUND ALL COLUMNS AROUND ELECTRIC EQUIPMENT ROOMS.

4. GROUND CONDUCTORS SHALL BE MINIMUM OF #4/0 BARE
    COPPER FOR MAIN LOOP AND #2 BARE COPPER TO EACH
    COLUMN BURIED 3'-0" MINIMUM BELOW GRADE.

5. PROVIDE A 2" CONDUIT SLEEVE FOR EACH GROUND CONDUCTOR
    THROUGH BUILDING OR SLAB TO PROTECT CONDUCTOR FROM
    PHYSICAL DAMAGE.

6. COORDINATE GROUNDING GRID TO NOT OCCUR BENEATH MAJOR
    MACHINE FOUNDATION.

7. PROVIDE 12" DIA. CLAY TILE WITH STEEL CAPS AROUND GROUND
    WELLS.

8. REFER TO SPECIFICATIONS 16450 FOR ADDITIONAL REQUIREMENTS.

Another type of print in the master industrial set is very useful in this area. Print M10 is the equipment room isometric, which shows pictorial isometric layouts of the mechanical room. These pictorial layouts can help give the electrical worker the feel and layout of the mechanical equipment room (Figure 18–14).

Finally, there are specific wiring and control diagrams used by electrical and mechanical workers on the job. One of these diagrams is called the control (process control) diagram. Print M21 shows control diagrams for various air-handling units on this industrial job (Figure 18–15).

The goal of this chapter and of Chapters 16 and 17 is to help improve the understanding of a rational process of print reading. As you can see, print reading is somewhat difficult and time consuming. Understanding of print reading is critically important in completing jobs to specifications. It is hoped that this text has been of some help in learning to read blueprints. The only way to become proficient in print reading is to read prints. On-the-job training and practice sessions with other electrical workers and journeypersons will improve your skills greatly.

FIGURE 18-13   Sections.

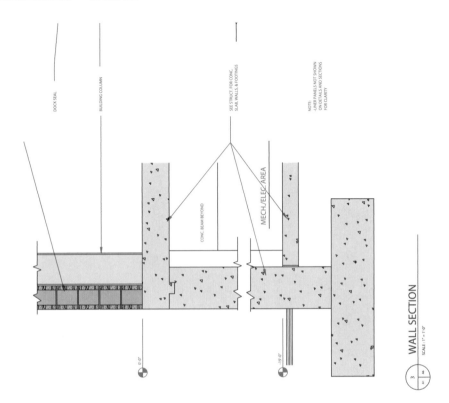

DOCK SEAL

BUILDING COLUMN

SEE STRUCT. FOR CONC.
SLAB, WALLS, & FOOTINGS

NOTE:
- LINER PANELS NOT SHOWN
ON DETAILS AND SECTIONS
FOR CLARITY

CONC. BEAM BEYOND

MECH./ELEC. AREA

0'-0"

-19'-0"

WALL SECTION

3   SCALE: 1" = 1'-0"

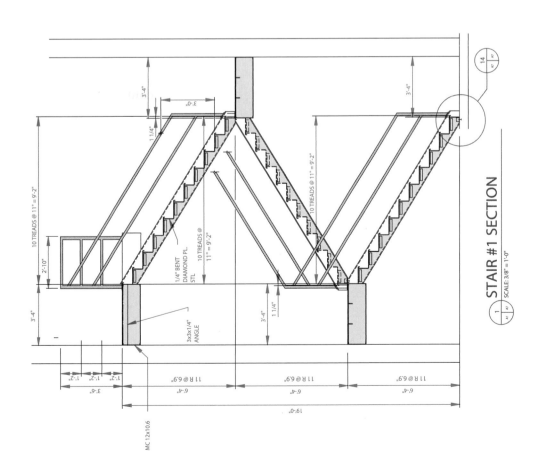

3'-4"

3'-0"

3'-4"

14

1 1/4"

10 TREADS @ 11" = 9'-2"

10 TREADS @ 11" = 9'-2"

10 TREADS @
11" = 9'-2"

1/4" BENT
DIAMOND PL
STL

2'-0"

3'-4"

1 1/4"

3'-4"

3x3x1/4"
ANGLE

11 R @ 6.9"   11 R @ 6.9"   11 R @ 6.9"

6'-4"   6'-4"   6'-4"

1'-2" 1'-2" 1'-2" 1'-2"

3'-6"

19'-0"

MC 12x10.6

STAIR #1 SECTION

1   SCALE: 3/8" = 1'-0"

FIGURE 18–14 Equipment room isometric.

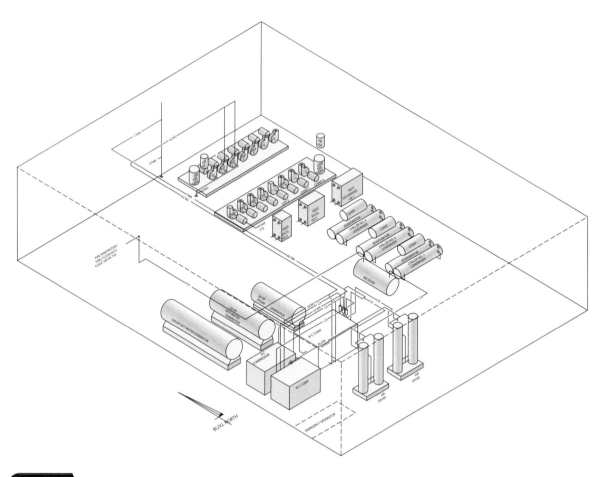

FIGURE 18–15 Control diagram.

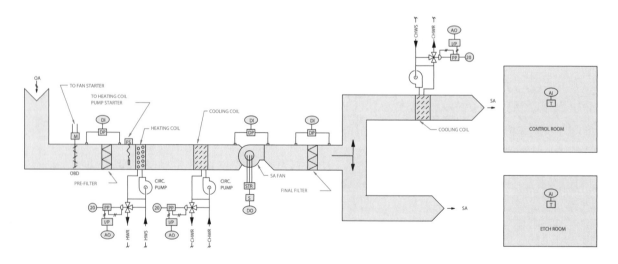

CONTROL DIAGRAM - AHU-3

SUMMARY

- An electrician must be comfortable using and exploring industrial plans, specifications, elevations, legends, and schedules.

- To be able to visualize and completely understand an industrial set of construction drawings, a person must spend time looking at the various drawings and specifications.

- Becoming proficient in reading construction drawings takes time and practice.

REVIEW QUESTIONS

NOTE: Some questions that would be applicable for both commercial and industrial prints were only used in the commercial chapter.

1. What division and section covers submittals?

2. What division and section covers compressed air dryers?

3. What division is most important to the electrician?

4. What division and section would supply information on switches and disconnects?

5. What division and section covers grounding practices?

6. Who should be present for all electrical acceptance tests?

7. What type of conduit is used in raceways on this project?

8. What type of boxes are used for telephone, televisions, and computers?

9. What type of manual motor starters are used on this job?

10. What are the specifications of the project's standby generator?

11. What type of receptacle is used on this project?

12. What is the maximum ground resistance to earth for ground rods and fields?

13. What is the input/output capacity required for the security alarm system on this project?

14. What is the mounting height for telephone outlets on this project?

15. List two common differences between commercial and industrial prints.

    NOTE: Using the set of full-size prints on the CD will make questions 16 through 30 easier to complete.

16. List several items shown on the industrial site plan that were not on the commercial site plan.

17. What type of information is available from the industrial riser diagram?

18. Why does Figure 18–3 in the text show the riser and partial panelboard together?

19. What type of information is available from the electrical floor plans?

20. What is the function of the key plans located on the drawings?

21. What is the reason for reviewing the HVAC enlarged floor plan, print M2, and Figure 18–5?

22. What information is shown on lighting plan E2?

23. What information is shown on print power plan E5?

24. What information is shown on print E16, the legend and lighting schedule?

25. Print M8, or Figure 18–10, is an enlarged view of the mechanical room. What information is available from this print?

26. What print contains industrial grounding details?

27. What is the use of print M10, mechanical room system isometric, or Figure 18–14?

28. What is the purpose of print M21, HVAC control diagrams, or Figure 18–15?

29. How high is the disconnect mounting on a typical battery-charging station detail?

30. If you are reading an electrical floor plan and find a symbol that looks like a rectangle with a circle in its center and a letter *C* within the circle, what does this symbol represent?

    NOTE: This chapter has generally covered industrial print and specification reading. It would be best at this point if you and your instructor worked through the full-size prints on the CD and developed more practice exercises.

COMMERCIAL SPECIFICATIONS

NOTE: This section contains a partial set of commercial specifications taken from the IBEW Local 5 commercial building specifications. This reduced commercial specification set is sufficient for the lessons in commercial print and specification reading. It contains all information generally necessary for an electrical worker.

TABLE OF CONTENTS FOR COMMERCIAL SPECIFICATIONS

* These documents are included for reference only and are part of the bidding documents as if included herein. These documents may be reviewed at the architect's office or purchased from the Pittsburgh chapter of the AIA.

Division 1—General Requirements

01010	Summary of Work
01020	Allowances
01030	Alternates
01060	Regulatory Requirements
01090	Definitions and Standards
01200	Project Meetings
01300	Submittals
01320	Construction Schedule
01400	Quality Control Services (*not included in this submission*)
01500	Temporary Facilities
01631	Products and Substitutions
01700	Field Engineering
01770	Project Closeout

Division 2—Site Work

02230	Site Clearing
02300	Earthwork
02465	Auger Cast Grout Piles
02530	Trench Drain
02740	Asphalt Concrete Paving
02750	Portland Cement Concrete Paving
02800	Site Improvements
02810	Underground Irrigation System (Alternate No. 2)
02900	Landscaping

Division 3—Concrete

03100	Concrete Formwork
03200	Concrete Reinforcement
03300	Cast-in-Place Concrete

Division 4—Masonry

04200	Unit Masonry

Division 5—Metals

05120	Structural Steel
05210	Steel Joists
05310	Steel Deck
05400	Light Gauge Metal Framing
05500	Metal Fabrications
05511	Manufactured Steel Stair Systems
05521	Pipe and Tube Railings
05530	Gratings
05810	Expansion Joint Cover Assemblies

Division 6—Wood and Plastics

06100	Rough Carpentry
06402	Finish Carpentry and Architectural Woodwork

Division 7—Thermal and Moisture Protection

07111	Composite Sheet Waterproofing
07200	Insulation
07270	Firestopping
07412	Steel Wall Panels
07415	Preformed Metal Standing Seam Roofing
07532	EPDM Single-Ply Membrane Roofing—Fully Adhered
07680	Sheet Metal Work
07700	Roof Hatch
07900	Joint Sealers

Division 8—Doors and Windows

08100	Steel Doors and Frames
08210	Wood Doors
08305	Access Doors
08345	Rolling Counter Shutters
08360	Sectional Overhead Doors
08400	Aluminum Curtainwall, Storefront, and Entrances
08700	Hardware (*not included in this submission*)
08721	Automatic Door Operator Systems
08800	Glass and Glazing
08960	Sloped Glazing Assemblies

Division 9—Finishes

09250	Gypsum Board Assemblies
09265	Gypsum Board Shaftwall Assemblies
09310	Ceramic Tile
09510	Acoustical Ceilings
09546	Linear Metal Ceilings
09550	Prefinished Wood Flooring
09650	Resilient Flooring
09680	Carpet
09705	Troweled Epoxy Flooring System
09900	Painting

Division 10—Specialties

10100	Markerboards
10155	Toilet Compartments
10210	Architectural Louvers
10270	Access Flooring
10522	Fire Extinguishers, Cabinets, and Accessories
10620	Operable Walls
10800	Toilet Accessories

Division 11—Equipment

11132	Projection Screens
11160	Loading Dock Equipment
11172	Waste Compactor
11400	Food Service Equipment

Division 12—Furnishings

12500	Window Treatment
12680	Recessed Foot Grilles

Division 13—Special Construction

Not Used

Division 14—Conveying Systems

14240	Hydraulic Elevators

Division 15—Mechanical

15000	Specific Mechanical Requirements
15004	Adjusting and Balancing
15005	Coordination with Testing and Balancing of HVAC Systems
15011	Identification of Mechanical Equipment and Piping Systems
15080	Water Piping Specialties
15090	Pipe Guides and Anchors
15091	Non-Fire Rated Sleeves and Seals
15092	Fire Rated Penetration Systems
15093	Counter Flashing—Mechanical Equipment
15094	Pipe Hangers and Supports
15141	Centrifugal Pumps
15160	Flexible Connectors
15170	Thermometers and Gauges
15174	Expansion Tanks
15178	Underground Fuel-Oil Storage Tank
15190	Motors and Starters
15191	Adjustable Frequency AC Motor Drives
15200	Vibration Control
15201	Concrete Equipment Foundations, Pads, and Curbs
15202	Roof Curbs and Supports for Mechanical Equipment
15250	Insulation for Mechanical Piping Systems
15251	Insulation—Heating and Cooling Equipment
15252	Insulation—Breeching, Mufflers, and Exhaust Pipes
15258	Ductwork Insulation
15342	Chemical Water Treatment
15350	Natural Gas Piping System
15401	Domestic Water System
15402	Domestic Water Pressure-Booster System
15405	Sanitary and Storm Water Drainage System
15421	Drains

Division 16—Electrical

DIVISION 1—GENERAL REQUIREMENTS

SECTION 01010

Summary of Work

Part 1. General

1.1 Related Documents

    A. Drawings and general provisions of the Contract, including General and Supplementary Conditions and other Division 1 Specification Sections, apply to work of this section.

1.2 Project Identification

   A. New Headquarters Complex, International Brotherhood of Electrical Workers, Carson Street and Gateway Boulevard, Pittsburgh, Pennsylvania

1.3 Summary of Contract Work

   A. Without force and effect on requirements of the Contract Documents, Contract Work can be summarized as follows:

      1. The Project consists of three interconnected buildings, structural steel frame with face brick veneer, aluminum curtainwall, and fixed windows. The Project includes all interior finishes, specialties, equipment, elevators, mechanical, electrical, and plumbing systems. The Project includes all site development and site work as indicated.

1.4 Contracts

   A. The Project will be constructed under Four (4) Contracts.

      1. General Construction, including associated site work
      2. Heating, Ventilating, and Air Conditioning
      3. Plumbing and Fire Protection
      4. Electrical Work

   B. The General Contractor will be assigned all contracts and serve as the Lead Contractor for the Project.

   C. Each trade shall secure and pay for its own permits.

      1. The Owner will obtain the water tap-in permit.

   D. Each trade shall make all submittals through the General Contractor.

   E. Each trade shall:

      1. Perform its own excavation and backfill.
      2. Provide its own concrete, other than equipment pads.

         a. The General Contractor will provide all mechanical and electrical equipment pads.

      3. Perform its own cutting and patching to the extent that it is required.

1.5 Contractors' Use of Site

   A. During the construction period the Contractors shall have full use of the premises for construction operations, including use of the site. The Contractors' use of the premises is limited only by the Owner's right to perform work or to retain other contractors on portions of the Project.

1.6 Erection Sequence

   A. Office Building

   B. School Building

   C. Conference Center

1.7 Owner Occupancy

   A. Partial Owner Occupancy: The Owner reserves the right to occupy and to place and install equipment in completed areas of the building prior to Substantial Completion, provided such occupancy does not interfere with completion of the Work. Such placing of equipment and partial occupancy shall not constitute acceptance of the total Work.

1. The Architect will prepare a Certificate of Substantial Completion for each specific portion of the Work to be occupied prior to Owner occupancy.
2. Obtain a Certificate of Occupancy from local building officials prior to Owner occupancy.
3. Prior to partial Owner occupancy, mechanical and electrical systems shall be fully operational. Required inspections and tests shall have been successfully completed. Upon occupancy, the Owner will operate and maintain mechanical and electrical systems serving occupied portions of the building.
4. Upon occupancy, the Owner will assume responsibility for maintenance and custodial service for occupied portions of the building.

Part 2. Products

Not used.

Part 3. Execution

Not used.

END OF SECTION

SECTION 01200

Project Meetings

Part 1. General

1.1 Related Documents

A. Drawings and general provisions of the Contract, including General and Supplementary Conditions and Division 1—General Require-ments apply to work of this section.

1.2 Application

A. Prior to and during the progress of work, conferences will be held at the Project Site or an agreed location. The purpose of the conferences will be to discuss job problems, coordination, expedite progress, and receive instructions. Project Meetings include the following:
   1. Pre-Construction Conference
   2. Progress Meetings
   3. Special Purpose Meetings

B. Representatives of Contractors, subcontractors, and suppliers attending meeting shall be qualified and authorized to act on the behalf of entity each represents.

C. Unless otherwise noted, Contractor will prepare a report covering each conference.

Part 2. Products

Not used.

Part 3. Execution

3.1 Pre-Construction Conference

A. Schedule within fifteen (15) calendar days after the date of notice to proceed.

B. Attendance:

  1. Owner's Representative
  2. Architect and his Consultants
  3. Contractor's Superintendent
  4. Major Subcontractors
  5. Others as appropriate

C. Suggested Agenda:

  1. Distribution and Discussion: List of major subcontractors and suppliers and Projected Construction Schedule
  2. Critical work sequencing
  3. Major equipment deliveries and priorities
  4. Project coordination
  5. Designation of responsible personnel
  6. Procedures and Processing of field decisions, proposal requests, submittals, change orders, and applications for payment
  7. Adequacy of distribution of Contract Documents
  8. Submittal of shop drawings, project data, and samples
  9. Procedures for maintaining record documents
  10. Use of Premises: Office, work and storage areas; Owner's requirements
  11. Temporary utilities
  12. Safety and first-aid procedures
  13. Security procedures
  14. Housekeeping procedures

3.2 Progress Meetings

  A. Schedule regular periodic meetings at Pre-Construction Conference as agreed to by all parties.
  B. Location of Meetings: As designated by Owner.
  C. Attendance:

  1. Owner's Representative
  2. Architect and his Consultants
  3. Contractor's Superintendent
  4. Subcontractors as pertinent to agenda

  D. Suggested Agenda:

  1. Review of work progress since last meeting.
  2. Note field observations, problems, and conflicts.
  3. Identify problems which impede construction schedule.
  4. Review off-site fabrication, delivery schedules.
  5. Develop corrective measures and procedures to regain projected schedule.
  6. Revise construction schedule indicated.
  7. Review proposed changes as they affect construction schedule and completion date.
  8. Other business.

3.3 Special Purpose Meetings

  A. Special purpose meetings will be held when an occasion arises to resolve a problem or emergency.

1. Arrangements for such meetings will originate from the Owner, Architect, or General Contractor and will be attended by parties affected by meeting subject.
2. If a party other than those parties listed above is in a situation he or she feels requires a meeting, one of the above-mentioned parties should be contacted.

END OF SECTION

DIVISION 16—ELECTRICAL

SECTION 16001

Summary of Work

Part 1. General

1.1 Work Included

A. Refer to the Instructions to Bidders, General Conditions, General Requirements, which form a part of this Contract and have the same force and effect as if printed herein in full.

B. Any reference in the Electrical Specifications to "this Contractor" is intended as indication of the work to be furnished and installed by the Electrical Contractor. Where work is indicated on the electrical drawings or specified in the electrical specification without reference or mention as to the Contractor, it is understood such work is to be provided by the Electrical Contractor.

C. Wherever the word "provide" appears in these specifications or on the accompanying drawings, it means "furnish and install, with all associated wiring, raceway, supports and appurtenances, final connections, and place same in proper electrical operating condition." Where equipment is listed but the words "provide" or "furnish and install" are not mentioned, it is understood that such material and/or work is the responsibility of this Contractor.

D. Wherever the word "equipment" appears in these specifications or on the accompanying drawings, it means "all wiring, raceways, fixtures, supports, panels, boxes, switches, devices, and appurtenances." Where the word "work" appears in these specifications or on the accompanying drawings, it means "to include all equipment, all labor, rigging, scaffolding, tools, and appurtenances of an auxiliary nature."

Part 2. Products

2.1 Performance

A. Perform the work under this Contract in accordance with the Contract Documents. The work includes the furnishing and installing of all materials, equipment, appurtenances, and other necessary work required for proper completion, operation, and use of the electrical facilities, and includes the furnishing of all tools, equipment, labor, transportation, supervision, and other such items incidental to the execution of the work, all as intended or reasonably implied by the Contract Documents.

Part 3. Execution

3.1 Scope of Work

A. Without limiting or restricting the volume of work and solely for the convenience of this Contractor, the work to be performed is, in general, to be comprised of the following:

1. Electrical service facilities from Utility System to the building
2. Electrical service equipment
3. Electrical distribution system for power and lighting
4. Grounding system
5. Branch circuit system including protection and wiring for lighting, devices, and other electrical utilization equipment
6. Wiring devices, disconnect switches, and lighting fixtures complete with lamps
7. Power wiring for motors and other electrical utilization equipment including the furnishing of combination motor starters and pilot devices only where so specifically indicated
8. Temporary light and power
9. Exterior lighting system
10. Emergency power and lighting system
11. Fire detection and alarm system
12. Life safety system
13. Structured cabling system
14. Building management system
15. Clock program sound system
16. Auditorium local sound system
17. Audio/video door intercom system
18. Auditorium dimming system
19. Paging system
20. Door access system
21. Auditorium dimming system
22. Lightning protection system
23. Television distribution system
24. Obtain and pay for all permits and inspections required by any and all lawful authorities
25. Panel directories, identification nameplates, operation and maintenance instructions, record drawings, and cleaning premises of debris resulting from this Contractor's work

3.2 Work Not Included in This Contract

A. In general, all special control equipment required for the heating, ventilating system, plumbing system, air conditioning system, elevator, and kitchen equipment will be furnished by the respective Contractors.

B. All control wiring for the heating and ventilating system and air conditioning system will be furnished and installed by this Contractor as a Subcontractor to the Mechanical Contractor.

C. The metering cabinet will be furnished by the Utility and installed by this Contractor, and the metering leads will be furnished and installed by the Utility.

D. All telephone service entrance cable, demarcation terminal blocks, circuit protectors, and instruments will be furnished and installed by the telephone company.

END OF SECTION

SECTION 16051

Wiring Methods—Exterior

Part 1. General

1.1 Work Included

A. Install underground primary service conductors in concrete encased plastic PVC ducts with a minimum earth cover of 30 inches.

B. Install underground secondary service conductors in concrete encased plastic PVC ducts as indicated on the Drawings with a minimum earth cover of 24 inches.

C. Install exterior lighting conductors utilizing plastic PVC ducts with a minimum earth cover of 18 inches.

Part 2. Products

2.1 Materials

A. Refer to appropriate accompanying sections of the specification.

Part 3. Execution

3.1 Installation

A. Excavation, Installation, and Backfilling: Provide all excavating required for the installation of electrical facilities. This Contractor is responsible for coordinating and protecting all new and/or existing underground service lines such as gas, water, electric, and telephone services, and sanitary and storm water piping.

Perform the excavation work on an "Unclassified Basis"; that is, all earth or rock formations shall be removed regardless of the type or hardness of such formations. Include the cost of such excavation in the Contract Price at the time of bidding. Excavate bottoms of trenches for nonencased plastic ducts and trenchlay cable in areas where rock is encountered an additional 6 inches and refill with a 6-inch layer of wetted-down sand that has been tamped and made smooth before ducts or cables are installed. To the extent practicable, grade trenches so that the raceways have a fall of at least 3 inches per 100 feet away from the building.

Generally, encase all conduits with a concrete envelope having a minimum cover of 3 inches. Reinforce concrete envelopes at all points where they cross fill or loose soil, foreign piping, or under vehicular roadways. Minimum reinforcing shall consist of three longitudinal runs of No. 5 reinforcing bar, plus cross runs of No. 5 bar tied to longitudinal runs on 12-inch centers, reinforcing shall be installed 1 inch above the bottom of the encasement, and shall extend 6 feet beyond each limit of fill, roadway, and/or foreign pipe.

Where concrete encased ducts connect to manholes, the encasement shall be continuous into the window of the manhole and the encasement shall be reinforced from the manhole to a point 6 feet beyond the perimeter of the manhole excavation.

After underground electrical facilities are installed and tested, backfill all excavations with selected earth, placed in layers not exceeding 6 inches to thickness with each layer mechanically compacted to a density not less than the surrounding earth. Backfill all excavations under paved areas the full depth with "run-of-bank" gravel with approximately 50% sand content by volume.

Provide a plastic marker strip along the entire length of the underground duct system, installed 12 inches below grade. Marker strip shall read "CAUTION—BURIED ELECTRIC LINE BELOW" and shall be "Terra-Tape" as manufactured by Griffolyn Company, Houston, Texas, or an approved equal.

All excavations that are improperly backfilled or where settlement occurs shall be reopened to the depth required for proper compaction, then refilled and compacted with the surface restored to the required grade, and compaction mounded and contoured.

Restore all surfaces to their original condition including replacement of topsoil, seeding, and restoration of any paved surfaces damaged by these operations.

B. Restoration and Maintenance of Surfaces: This Contractor shall replace all surface material and shall restore paving, curbing, sidewalks, gutters, shrubbery, fences, sod, and other surfaces disturbed to a condition equal to that before work began, furnishing all labor and materials required.

Restore damaged paved surfaces with new pavement matching that damaged. No permanent paving shall be placed less than thirty (30) days after backfilling, unless approved by the Architect. This Contractor shall construct a concrete base beneath public roads where indicated on the Contract Drawings or where required by Pennsylvania State Department of Transportation, or local municipal subdivisions. All material and labor required for repair of defects in trench backfill and paved surfaces shall be supplied by this Contractor for the period required by the Maintenance Bond.

END OF SECTION

SECTION 16052

Wiring Methods—Interior

Part 1. General

1.1 Work Included

A. In all portions of the building, install all wire and cable in raceways as hereinafter specified, concealing all such wiring insofar as possible.

B. Plastic PVC conduit may be used for conduit runs in or under the ground floor slab.

Part 2. Products

2.1 Materials

A. Utilize rigid steel or intermediate metal conduit for all raceways 2-1/2 inches and larger; for all conduit runs, other than PVC under floor slabs, where slabs are on grade; for all conduit runs in damp locations; for all exposed runs in the Boiler and other unfinished Mechanical Room(s); or where otherwise required by the National Electrical Code (NEC). Electrical metallic tubing may be used in sizes 2 inches and smaller for all raceway runs, except as noted. Rigid aluminum conduit may be used in lieu of rigid steel conduit for exposed work in dry areas or above finished ceilings; it shall not be used in poured concrete or masonry walls.

B. Use liquid-tight flexible steel conduit in lengths not exceeding 18 inches for the connection of motors and other items subject to vibration in all exposed areas and all damp locations.

C. Flexible steel conduit (Greenfield) may be used in lengths not exceeding 48 inches for the connection of recessed lighting fixtures, and for motors or other items subject to vibration, that are installed above finished ceilings.

D. Wireways shall be of the size indicated and shall have hinged covers. Covers shall not have individual lengths greater than 48 inches. Wireways shall be code-gauge steel with baked enamel finish and shall be as made by Keystone-Columbia, Inc.

Part 3. Execution

3.1 Installation

A. The number of branch circuit home runs in each raceway is indicated by arrows on the drawings. Single-phase branch circuits shown in separate raceways may be grouped together in a single raceway, where each "hot" wire is indicated to be supplied by a separate bus bar in the supplying panel, but only to the extent of three current-carrying conductors per raceway. Three-phase home runs shall not be combined.

B. Each branch circuit "home run" circuits shall have a separate neutral conductor. Combining of two or more such circuits with one common neutral will not be permitted.

END OF SECTION

SECTION 16112

Bus Duct

Part 1. General

1.1 Scope

A. Provide a complete system of interconnected bus ducts where indicated on the drawings. The bus duct system shall consist of prefabricated sections, so formed that the complete assembly is rigid in construction and neat and symmetrical in appearance. Interconnected bus duct runs shall be comprised of sections that are listed by the Underwriters' Laboratories (UL) and are so labeled. The bus ducts shall be designed and arranged for 480Y/277 volt, three-phase, four-wire distribution.

Part 2. Products

2.1 Materials

A. Bus duct shall have full capacity neutrals and in addition shall contain a separate ground bus inside the metallic housings.

B. The bus duct construction and bracing shall be such as to withstand without damage for a period of at least three cycles, a short circuit current of up to 50,000 symmetrical amperes. The bus duct shall have a nonventilated metallic housing.

C. Fabricate bus bars from high conductivity aluminum with a hardness rating that will ensure mechanical strength under short-circuit conditions. The entire surface of the aluminum bars shall be zinc-copper plated to prevent formation of aluminum oxides. In addition, the bars shall be tin or silver plated at all joints and plug-in tap-offs. Bus bars shall be insulated their entire length, except at joints and tap-off points with polyvinyl chloride tubing or other approved insulation.

D. The bus duct installation shall include 10 foot straight sections, elbows, offsets, cable tap boxes, expansion joints, special sections, tap-off devices, protective devices, and additional accessories as required, for a complete system.

E. Provide tap-off devices for connections to the bus duct where shown on the drawings. The tap-off devices shall be circuit breakers with NEMA I enclosures and with external operating handles. Devices shall have built-in interlock to prevent connection or disconnection to the bus duct with the device in the "ON" position. The contact fingers shall be silver plated and shall be spring reinforced. The tap-off devices shall be of the voltage rating, pole arrangement, and current ratings indicated on the drawings. Fuses for fused tap-offs shall be of the sizes and types indicated on the drawings.

F. Bus duct and all associated components shall be the product of one manufacturer and shall be General Electric Company "Armor-Clad," Square D "ILine," Cutler-Hammer "Pow-R-Way," or approved equal.

Part 3. Execution

3.1 Installation

A. The manufacturer shall submit for approval certified temperature rise and voltage drop curves for each rating and type of bus duct to be used in the installation, certified factory megger test on each section of the bus duct, and complete layout shop drawing of all components making up each bus duct run.

B. Utilize proper procedures in the handling, storing, aligning, supporting, and cleaning of joints prior to assembly and lubrication of and torqueing all connection bolts. Perform a 1,000-volt DC megger test of each run prior to energizing. Follow the procedures, in general, as published by the supplying manufacturer.

C. Submit a letter from the manufacturer stating that their field engineer has inspected each bus duct run and found them all to be properly installed.

END OF SECTION

SECTION 16120

Wire and Cable (600 Volt)

Part 1. General

1.1 Scope

A. All conductors shall be of soft-drawn copper having a conductivity of at least 97%. Conductors No. 10 and smaller may be solid. Conductors No. 8 and larger shall be stranded. All wire and cable shall be of the single conductor type unless otherwise indicated, and shall be color coded so as to identify each wire in a raceway or cable.

Part 2. Products

2.1 Materials

A. All wire for installation in exterior underground raceways shall have 600-volt Type USE-RHH-RHW, cross-linked thermosetting type polyethylene insulation, temperature rating of 908C. Minimum insulation thickness shall be based on ICEA Standard S-66-524.

B. All wire for installation in interior raceways shall have 600-volt Type THW insulation for sizes No. 8 and larger, and THWN for sizes No. 10 and smaller. Wire shall comply with the National Electrical Code and Federal Specification J-C-30A, and minimum insulation thickness shall comply with ICEA Standard S-61-1402. Lighting branch circuit, power and motor wires shall not be smaller than No. 12 AWG. Control wires shall not be smaller than No. 14.

C. Wires for use in miscellaneous systems shall have a grade of insulation as specified for light and power, except as elsewhere specifically indicated. Voltage ratings shall be consistent with the maximum voltage applied to the wires or to other wires in same conduit. Sizes shall be as elsewhere specified or indicated and as required.

D. The wire installed in fixture channels, where fixture channel is used as a raceway for a continuous row of fluorescent units, shall be Type SA with No. 12 stranded copper conductors.

E. All wire shall be color coded and a given color wire shall be used for a single purpose throughout any one given system. The phase or ungrounded conductors, as well as the neutral or grounded conductor of the $480/277$ volt and $208/120$ systems shall be of different colors and shall be consistent throughout each system.

F. Where colors are established by National Electrical Code, local code, or inspecting agency, these shall be followed; where not so established, the following color code shall be used:

1. $208/120$ volt, three-phase, four-wire—black, white, red, blue.
2. $480/277$ volt, three-phase, four-wire—brown, gray or white striped, yellow and orange.
3. All ground wires—green.
4. Control wires—color coded to identify each conductor in a raceway.
5. Wires for miscellaneous systems—color coded or tagged to identify each wire at each termination and pull box.
6. Every coil or reel of wire shall bear the manufacturer's name, the Underwriters' label, type, voltage, size, length, and manufacturing date, and shall be delivered to the job in original containers for inspection. Wire shall be manufactured in the United States.
7. Avoid installation of wire and cable in areas of high ambient temperatures insofar as possible. Where wires must be installed in areas with ambient temperatures above 30°C, the conductors shall have Type SA silicon rubber insulation with glass cover.
8. Number all control wires at all terminations and junction points with adhesive numbered labels, in accordance with the approved wiring diagrams. Numbered labels shall have vinyl plastic face with adhesive back as made

by T & B, or approved equal. Submit samples of adhesive labels for approval.

Part 3. Execution

3.1 Installation

A. Install all wire and cable in raceways and in a manner so as not to damage the insulation. Use only UL-approved wire-pulling compounds to decrease the friction of pulling in wires.

B. Raceways shall be clean and dry before conductors are installed. Should the drag wire or pulling cable indicate that the raceway is not clean or dry, correct the condition before wires are installed.

C. Locate the cable reel from which wire is being installed so that surrounding dirt and debris is not drawn into the raceways with the conductors.

D. Wires shall not be "skinned" to a point closer than 8 inches to the bushing from which they emerge.

END OF SECTION

SECTION 16121

Splices and Terminations (600 V)

Part 1. General

1.1 Scope

A. Splices and terminations shall be made using splicing devices and lugs that are the most appropriate type and size for the particular splice.

Part 2. Products

2.1 Materials

A. Utilize mechanical splicing devices and lugs for splices and terminations in wires No. 6 and larger. For wires No. 2 and larger, use a type of splicing device in which the contact pressure on the wire is obtained by two or more screws or bolts, and so designed that the failure of any one screw, bolt, or nut will not result in a total loss of contact pressure. Mechanical splicing devices shall have insulating covers whenever obtainable, and shall otherwise be insulated with plastic tape, Scotch No. 88, or approved equal.

B. Splices in conductors No. 8 and smaller shall be made with pressure connectors consisting of cone-shaped coiled springs with insulating covers, "Scotchlok," or approved equal.

C. The terminations of No. 8 and smaller wires shall be made by forming the wires about the terminal screws in the case of solid wires. All stranded wires shall be provided with spade lugs attached to the wires by a crimped sleeve.

Part 3. Execution

3.1 Installation

A. All materials and devices used for making joints and splices shall be approved by sample, catalog designation, or shop drawings.

END OF SECTION

SECTION 16131

Junction and Pull Boxes

Part 1. General

1.1 Scope

    A. Provide junction and pull boxes where shown on the drawings, or as may be required, to facilitate the pulling of wires.

Part 2. Products

2.1 Material

    A. Junction and pull boxes shall be made of code-gauge galvanized steel and shall not be smaller than the minimum code dimensions. In general, they shall be of such size that it is not necessary to bend wires to radii of less than 10 inches diameter, and lengths shall be adequate for staggering joints or splices. Boxes shall be welded construction and shall have plain sheet steel machine screw-attached covers. Flush mounted boxes shall be fitted with a cover that overlaps the box by 1 inch on all sides. Wireway is not an acceptable substitute for junction or pull boxes.

Part 3. Execution

3.1 Installation

    A. Provide holes in pull and junction boxes to receive entering raceways. There shall be no holes except those used by raceways. Provide a separation section for each wiring system of light, power, and telecommunications in each pull box. Boxes shall not be smaller than the size required by the National Electrical Code.

    B. Boxes shall not be placed in locations made inaccessible by piping, ducts, raceways, or other equipment.

END OF SECTION

SECTION 16134

Outlet Boxes

Part 1. General

1.1 Scope

    A. In general, outlet boxes for ceiling outlets and flush wall outlets shall be steel.

    B. Exposed outlet boxes shall be cast aluminum.

Part 2. Products

2.1 Material

    A. Outlet boxes for ceiling outlets and flush wall outlets shall be not less than $1/16$-inch thick pressed sherardized or galvanized steel, and shall be of ample size to properly accommodate the conductors passing through and the splices contained therein, as required by NEC Table 370-6(b). They shall not be less than 4-inch square or octagonal and 1-$1/2$-inch deep. Boxes to which a 1-inch raceway or three or more smaller raceways are connected shall not be smaller than 4-$11/16$ inches square and 2-$1/8$ inches deep. Where two or more systems enter the outlet boxes, they shall be at least 2-1/4 inches deep and shall have separators properly located to divide one system from another.

B. Ceiling and wall outlet boxes, which are to support lighting fixtures, shall have short galvanized or sherardized 3/8-inch fixture studs inserted from back of box.

C. All outlet boxes in finished walls and ceilings, other than brick, masonry, or tile, shall have plaster covers that come flush or slightly under the plaster surface. Ceiling and bracket outlet boxes shall have covers that reduce the surface opening to approximately 2 inches and shall have ears tap-ped for mounting wiring devices where required.

D. At locations where switch or receptacle outlets occur in finished brick, masonry, or glazed tile walls, use special outlet boxes designed for this type of wiring. Outlet boxes shall be at least 3-1/2 inches deep having square corners with support ears mounted on the inside of the box. Knock-outs in boxes shall be centered 2-1/2 inches from face of box to allow a solid piece of 2-inch brick or tile to be placed in front of raceway entering box without cutting. Provide special larger coverplate for this type of outlet. Boxes shall be Midland Ross Model GW-135, or equal. If special boxes are furnished by others, this Contractor shall receive, store, and install such boxes.

E. Where raceways are exposed, outlet boxes for wiring devices on walls, partitions, or steel framework shall be Crouse-Hinds, Type FS or FD condulets as required with covers to fit. All outlet boxes that are exposed to weather shall be cast aluminum with gasketed covers.

Part 3. Execution

3.1 Installation

A. No outlet boxes shall be installed with open holes other than for the entering conduits. Use approved knockout closers where spare or unused openings occur.

B. Mounting bolts of outlet boxes shall not be used for the support of fixtures. Adequately support each outlet box containing a fixture stud so it is capable of sustaining, without damage, a dead load of at least twice the fixture weight. Where the ceiling construction requires it, anchor each fixture outlet box with an approved bar or stud box hanger.

C. Cover an outlet box that serves as a junction box with a blank switch plate if it occurs on the wall, and by a round flat sherardized cover fastened by flat-head countersunk screws if it occurs on the ceiling.

D. Where the location of ceiling outlets indicated on the drawings occurs directly under a ceiling obstruction, move the lo-cation of the boxes to the nearest unobstructed point, as directed by the Architect.

END OF SECTION

SECTION 16155

Motor Controls—Single Phase

Part 1. General

1.1 Scope

A. Provide separately mounted manual motor starters for various items of equipment, as indicated on the drawings that have single-phase motors.

Part 2. Products

2.1 Materials

A. All manual starters shall be equipped with overload heaters, and heaters shall be sized in accordance with the nameplate current ratings of the motors they are to protect. Where indicated, starters shall also be equipped with pilot lights.

B. The manual motor starters shall be Westinghouse Type MS, General Electric, or approved equal.

Part 3. Execution

3.1 Installation

A. Where starters are indicated in finished areas, they shall be flush mounted and equipped with stainless steel coverplates, and where they are located in unfinished areas they shall be surface mounted in die-cast enclosures having provisions for padlocking.

END OF SECTION

SECTION 16157

Pilot Controls—Separately Mounted

Part 1. General

1.1 Scope

A. Provide separately mounted pushbuttons, selector switches, and pilot light stations for various items of equipment as indicated on the drawings.

Part 2. Products

2.1 Materials

A. Pilot controls shall be of the oil-proof type and shall have sheet steel boxes and covers. Pilot controls shall be of the heavy duty oil-tight type and pilot lights shall be of the self-contained transformer type with No. 55 Type, 6- to 8-volt lamps and with push-to-test feature. The pilot controls shall be as manufactured by General Electric, Westinghouse, Cutler-Hammer, or approved equal.

Part 3. Execution

3.1 Installation

A. Where pilot devices occur in finished areas, they shall be flush mounted.

SECTION 16161

Switchboard

Part 1. General

1.1 Work Included

A. Switchboard shall be of the required number of vertical sections bolted together to form one metal enclosed rigid switchboard. Coordinate the dimensions of the switchboard with the installation space available and the dimensions of the shipping sections with the dimensions of the access pathway for moving the switchboard into the indicated lo-cation within the building. The sides, top, and rear shall be covered with removable, screw-on, code-gauge steel plates. Switchboard shall include all protective devices and equipment as listed on drawings with necessary interconnections, instrumentation, and control wiring.

B. Switchboard shall be manufactured to meet both UL. Standard No. UL-891 and NEMA Standard PB2. UL label shall be furnished on each vertical section that contains all UL-listed devices.

C. The main bus throughout the switchboard shall be nontapered and of the ampacity indicated on the drawings, except in no case shall its capacity be less than the ampere rating of the main protective device. Neutral bus rating shall be 100% of the phase bus rating. All busses in the switchboard shall be in accordance with the latest ANSI, UL, and NEMA Standards and may be either copper or aluminum.

D. Copper bussing shall be high-conductivity copper in sizes not to exceed a current density of 1,000 amperes per square inch or a temperature rise of 308C above an ambient temperature of 40°C at full-load current in still but unconfined air. Bus bar interconnections shall be made by means of neatly fitted scarf lap joints. All contact surfaces shall be silvered and all joints shall be made by standard bolted connections.

E. Aluminum bussing shall be NEC Grade or Aluminum Industry Grade 6101 and shall be tin or silver plated. The preparation of aluminum busses for the plating shall be equal in all respects to the Alstan 70 process. Aluminum busses shall have cross-sectional areas large enough to limit the temperature rise when carrying full load to 35°C (795°F) above an ambient inside enclosure of 55°C (131°F) as defined in the IEEE Standard Rules, but with a current density not exceeding 800 amperes per square inch in any case. All aluminum busses shall be tin or silver plated by first removing all aluminum oxide and then depositing an intermediate conductive plating or platings followed by final electroplating of tin or silver. The bus plating and joints shall conform to the following:

  1. Plating thickness shall not be less than .0002-inch or more than .00035-inch per bar, so that plated area will not be porous. All plated areas must be treated by immersion in a 500°F-600°F salt bath and inspected to show there is no corrosion or blistering.

F. Bolted plated joint areas shall be clean. Bolting shall be torqued to NEMA Standards and shall be 25 ft./lb. for $1/2$-inch bolts and 40 ft./lb. for $5/8$-inch bolts. Belleville washers shall be used with all bolts that are not made of aluminum.

G. Each bus shall be designed and supported such that within its rating limits, it will not sag or permanently distort.

H. Bus supports shall be of high-shock, noncarbonizing, and moisture-resistant molded phenolic or glass fiber. Bus supports shall adequately support and brace the bus bars to withstand mechanical forces exerted during short-circuit conditions where the available fault current is 42,000 A rms symmetrical.

I. A-B-C type bus arrangement—left-to-right, top-to-bottom, and front-to-rear shall be used throughout to assure convenient and safe testing and maintenance.

J. A ground bus shall be furnished, secured to each vertical section structure, and shall extend the entire length of the switchboard.

K. Small wiring, necessary fuse blocks, and terminal blocks within the switchboard shall be furnished when required. All groups of control wires leaving the switchboard shall be provided with terminal blocks with suitable numbering strips.

L. All steel surfaces shall be chemically cleaned and treated to provide a bond between paint and metal surfaces to help prevent the entrance of moisture and formation of rust under the paint film. All hardware used on conductors shall have a high tensile strength and a suitable protective finish.

M. Switchboard shall be provided with adequate lifting means and shall be capable of being rolled or moved into installation position and bolted directly to the floor without the use of floor sills.

Part 2. Products

2.1 Materials

A. Switchboard shall be as manufactured by General Electric, Cutler-Hammer, or approved equal.

B. Switchboard service entrance section shall be 90 inches high, self-supporting, and shall be of sufficient depth to accommodate the equipment. All vertical sections shall align in front and rear. The main and branch protective devices shall be individually mounted with front coverplate and necessary bus-connection straps. Where called for on schedule, "space for future" shall mean to include all necessary busses, device supports, and provide a Cutler-Hammer IQ Data Plus, or approved, instrument package.

C. The bus structure shall be arranged to permit future additions.

D. Switchboard distribution sections shall be 90 inches high and self-supported. All vertical sections shall align in front and rear. (See Compartmented Option.) All protective devices shall be individually mounted with hinged front coverplate and necessary bus-connection straps. To ensure maximum safety of personnel, all busses shall be completely isolated from the devices by means of glass polyester insulation arranged so that the only exposed energized parts will be at the point of connection to devices. Further, devices shall be completely isolated between sections by vertical steel barriers. The vertical bus shall be so arranged as to have insulating barriers interposed between phases to inhibit phase-to-phase faults. The device load terminals shall be extended to the rear so that it will be unnecessary to reach across or beyond a live bus to make load connections.

E. Handles for all devices in the distribution sections will be of the same design and method of external operation. Handles will be prominently labeled to indicate device ampere rating, and color coded for device type. On-off indication shall be clearly shown by prominent marking and handle position.

F. Where called for on schedule, "space for future" shall mean to include all necessary busses, device supports, and connections. Arrange the bus structure to permit future additions. The switchboard exterior shall be finished in ANSI-33 dark gray.

2.2 Protective Devices

A. The switchboard shall be equipped with protective devices in the quantities and ratings as indicated on the drawings, and all shall be capable of withstanding, safely and without damage, a short-circuit current of 42,000 A rms symmetrical.

B. The main switch shall be a fused, three-pole, four-wire with solid neutral, 480/277-volt AC switch of the current rating indicated on the drawings. The switch shall be bolted-pressure type as manufactured by the Pringle Manufacturing Company Type CBC, or approved equal. The switch shall be dead-front with external operating handle that shall be interlocked with the door over the main fuses so as to prevent access except when the switch is in the "OFF" position. The switch shall have provisions for NEMA Class L fuses. The switch shall be equipped with blown-fuse protection. Main switch shall be listed under UL Subject 977, Appendix B, and shall be provided with a UL label. Switch and fuse combination shall be capable of withstanding the electromechanical stresses imposed from a 200,000-ampere fault current.

C. Main switch shall be complete with integral ground fault protection including stored energy operator, electrical solenoid trip coil, ground fault current sensor, ground fault relay, and monitor panel. The switch shall have a maximum opening time of eight cycles. A pushbutton shall be provided for manual opening of main switch. Ground fault-sensing system shall utilize a zero sequence current sensor, and a current sensitive relay with calibrated adjustable pickup and time-delay settings.

2.3 Insulated Case Breakers—Static Trip (Note: Trip Ratings 1600 A through 2500 A Only)

A. The feeder circuit breakers shall be of the stationary multipole, insulatedcase type with integral solid state trip devices to provide long-time shorttime instantaneous and ground fault tripping. Breakers shall be rated for 100% application. No external source of power shall be required to trip the circuit breaker under overload or short-circuit conditions. The trip elements shall have the following characteristics:

1. The long-time trip unit shall have a minimum adjustment of 70% to 100% of the breaker trip rating and long-time delay shall be adjustable in three bands: minimum, intermediate, and maximum. The short-time delay pickup shall be adjustable from three to eight times the trip setting. The short-time delay shall be adjustable in three bands: minimum, intermediate, and maximum. The instantaneous trip unit shall have a minimum adjustment of from three to eight times the breaker trip rating.

2. Ground fault tripping shall be provided as an integral part of the breaker static trip unit with neutral current sensor where required. The ground fault tripping shall have an adjustable pickup range of approximately 600 to 1200 amperes, and time delay shall be adjustable in three bands: minimum, intermediate, and maximum from approximately instantaneous to .5 seconds or thirty cycles.

2.4 Molded Case Breakers—Static Trip (Note: Trip Ratings 400 A through 1200 A only)

A. The feeder circuit breakers shall be of the stationary multipole, molded-case type with integral solid state trip devices to provide long-time and instantaneous tripping. No external source of power shall be required to trip the circuit breaker under overload or short-circuit conditions. The trip elements shall have the following characteristics:

1. The long-time trip unit shall have a minimum adjustment of 70% to 100% of the breaker trip rating. The instantaneous trip unit shall have a minimum adjustment of from five to ten times the breaker trip rating.

2.5 Molded Case Breakers—Thermal Magnetic Trip (Note: Trip rating below 400 A)

A. The feeder circuit breakers shall be of the stationary mounted, multipole, molded-case type with integral thermal-magnetic trip elements. Trip elements rated 250 amperes and above shall be of the interchangeable type. The thermal trip elements shall have fixed calibrations, and the magnetic trip unit for trip elements rated 250 amperes and above, shall be adjustable from approximately three to ten times the thermal trip rating.

B. Circuit breakers shall have quick-make, quick-break operating mechanisms; shall have silver alloy contacts; shall be electrically and mechanically trip free; shall have common trip bar for all poles; and the operating handle shall indicate ON, TRIPPED, and OFF positions.

Part 3. Execution

3.1 Installation

A. Furnish shop drawings containing the following information: complete rating, short-circuit withstandability of bus and of lowest rated device, overall outline dimensions including space available for conduits, circuit schedule showing circuit number, device description, device frame ampere rating, trip or fuse clip ampere rating, feeder circuit identification, conductor ratings, and one-line diagram with each circuit device numbered.

END OF SECTION

SECTION 16163

Distribution Panelboards

Part 1. General

1.1 Scope

A. Distribution panelboards shall be of the dead-front type, and shall incorporate switching and protective devices of the number, type, and rating as noted herein or shown on the drawings. Panelboards shall be rated for the intended voltage and shall be constructed in accordance with the Underwriters' Laboratories, Inc. "Standard for Panelboards" and "Standard for Cabinets and Boxes" and shall be so labeled. Where panelboards are to be used as service entrance equipment, they shall be so labeled. Panelboards shall also comply with NEMA Standard for Panelboards, National Electrical Code, and Federal Specification W-P-115a, where applicable.

Part 2. Products

2.1 Materials

A. Refer also to section "Panelboard Cabinets and Trims." Panelboards shall be Cutler-Hammer Type(s) PRL-4B, General Electric Type(s) SCP, or approved equal.

B. Interiors shall be completely factory assembled with switching devices, wire connectors, etc. All wire connectors, except screw terminals, shall be of the anti-turn solderless type. Interiors shall be so designed that switching and protective devices can be replaced without disturbing adjacent units and without removing the main bus connectors and shall be so designed that circuits may be changed without machining, drilling, or tapping.

C. Bus bars for the mains shall be full height without reduction and shall be of copper, tin, or silver-plated aluminum, sized in accordance with UL Standards. Unless otherwise noted, provide a full-size neutral bar. In addition to the solid neutral, include a copper ground bar with terminals in each panel for all the grounding wires. Bus bar taps to protective devices shall be copper, tin, or silver-plated aluminum and shall be arranged for sequence phasing of these devices. Spaces for future switching and protective devices shall be bussed for the maximum current rated device that can be fitted into the space.

D. Bus bar joints shall be tin or silver plated, constant high-pressure type with high strength copper-silicon alloy bolts, nuts, and belleville-type spring-cup washers.

E. Bus bracing shall be as indicated on the drawings or required by the available fault current, but in no case shall be less than the interrupting rating of the smallest overcurrent device in the panel.

F. The switching and protective devices for distribution panels shall be of the stationary mounted, multipole, molded-case type with integral thermal-magnetic trip elements. Trip elements rated 250 amperes and above shall be of the interchangeable type. The thermal trip elements shall have fixed calibrations, and the magnetic trip unit for trip elements rated 250 amperes and above shall be adjustable from approximately three to ten times the thermal trip rating.

G. Circuit breakers shall have quick-make, quick-break operating mechanisms; shall have silver alloy contacts; shall be electrically and mechanically trip free; and shall have common trip bar for all poles. The operating handle shall indicate ON, TRIPPED, and OFF positions. Fault current interrupting ratings shall not be less than the short-circuit current rating indicated on the drawings, nor less than the available short-circuit current where the breaker is being installed.

Part 3. Execution

3.1 Installation

A. Identify all switching and protective devices by engraved nameplate or by neatly typed directory card with nonflammable, transparent cover mounted on the inside of the doors of the cabinets.

B. Coordinate the dimensions of each panelboard and its cabinet with the dimensions of the space designated for its installation. Detailed shop drawings of all panelboards, indicating size, rating, and arrangement of all switching and protective devices, shall be submitted.

END OF SECTION

SECTION 16166

Combination Motor Starters

Part 1. General

1.1 Contract Documents

A. The General Provisions of the Contract, including General and Supplementary Conditions and General Requirements, apply to work specified in this section.

B. This section is a Division 16 Basic Materials and Methods section and is a part of each Division 16 section making reference to motor starters specified herein.

C. Submit complete data on all starters and controls to the Architect for review by the Engineer.

1.2 Description of Work

A. Extent of motor starter work is indicated by drawings.

B. Types of motor starters in this section include the following:
   1. Manual
   2. Full voltage, nonreversing magnetic
   3. Combination full voltage, nonreversing magnetic

1.3 Related Work Specified Elsewhere

A Section 01300—Submittals

B. Division 16—Electrical

1.4 Quality Assurance

A. Manufacturers shall be firms regularly engaged in manufacture of motor starters of types, ratings, and characteristics required, whose products have been in satisfactory use in similar service for not less than five years.

B. Installer shall be qualified with at least three years of successful installation experience on projects with electrical work similar to that required for this project.

C. National Electrical Code Compliance: Comply with NEC as applicable to wiring methods, construction, and installation of motor starters.

D. Underwriters' Laboratory (UL) Compliance and Labeling: Comply with applicable requirements of UL 508, "Electrical Industrial Control Equipment," pertaining to electrical motor starters. Provide units that have been UL listed and labeled.

E. National Electrical Manufacturer's Association (NEMA) Compliance: Comply with applicable portions of NEMA Standards pertaining to motor controllers/starters and enclosures.

Part 2. Products

2.1 General

Except as otherwise indicated, provide motor starters and ancillary components that comply with manufacturers' standard materials, design, and construction in accordance with published product information, and as required for complete installation.

2.2 Full Voltage, Nonreversing Starters

A. Full voltage, nonreversing (FVNR) starters—Provide full voltage alternating current non-reversing starters, consisting of types, ratings, and NEMA sizes indicated. The motor starter shall provide overload protection accurate to 2%, phase loss and unbalance protection, selectable trip class, automatic/manual reset, and class II ground fault protection. The circuit monitoring the overload protection functions of the motor starters shall be pro-vided by three current sensors. The sensor-microprocessor combination yields a protection system that closely regulates power supplied to the operating coil. Motor starters that require heater elements are unacceptable. Equip starters with auxiliary contacts for interlocks where indicated and molded-case circuit breaker that trips all phases of the circuit breaker on specified occurrences of electrical service deviation. Starters to have individually fused control transformers.

B. Combination full voltage, nonreversing starters (FVNR)—Provide full voltage, alternating current combination nonreversing starters, consisting of motor starters and disconnect switches mounted in common enclosures of types, sizes, ratings, and NEMA sizes indicated. The motor starter shall provide overload protection accurate to 2%, phase loss and unbalance protection, selectable trip class, automatic/manual reset, and class II ground fault protection. The circuit monitoring the overload protection functions of the motor starters shall be provided by three current sensors. The sensormicroprocessor combination yields a protection system that closely regulates power supplied to the operating coil. Motor starters that require heater elements are unacceptable. Equip starters with auxiliary contacts for interlocks where indicated and molded-case circuit breaker that trips all phases of the circuit breaker on specified occurrences of electrical service deviation. Starters to have individually fused control transformers. Provide operating handle for motor circuit protection mechanism providing indication and control of switch position with enclosure door open or closed, and capable of being locked in OFF position with three padlocks.

2.4 Acceptable Manufacturers

Subject to compliance with requirements, provide the Advantage motor starters as manufactured by Cutler-Hammer (for each type and rating of motor starters) or approved equivalent.

Part 3. Execution

3.1 Installation of Motor Starters

A. Install motor starters as indicated, in accor-dance with manufacturer's written instruc-tions, following applicable requirements.

END OF SECTION

SECTION 16321

Transformers—Dry Type, K-Rated

Part 1. General

1.1 Scope

A. Transformers of the dry-type shall be enclosed in ventilated sheet steel housings and shall be designed in full accordance with the latest revisions of ANSI/IEEE Standard C57.110, Underwriters' Laboratory Standard 1561 for 600 volt and below, and shall include k-factor listing.

B. Transformers shall have the KVA, voltage ratings, phasing, and connections types as indicated on the drawings; and shall be wall and/or floor mounted as indicated.

Part 2. Products

2.1 Materials

A. Transformers shall be General Electric Type GHT with k-factor of 4, Cutler-Hammer Type DT-3 with k-factor of 4, or Square D Type NL with k-factor of 4.

B. Transformer coils shall be of continuous wound construction.

C. Full-capacity taps shall be provided on the primary side of the transformer having two 2.5% above normal and four 2.5% below normal.

D. Each transformer shall be UL listed and suitable for non-sinusoidal current loads with a k-factor not exceeding 4.

E. Transformers shall incorporate a 220°C insulation system and be designed not to exceed 150°C temperature rise above a 40°C ambient under full load.

F. Cores shall be common core construction having low hysterisis and eddy current losses. The core flux density shall be below the saturation point to prevent overheating caused by harmonic distortion.

G. Each transformer shall incorporate an electrostatic shield grounded to the transformer core for attenuation of spikes, line noise, and transients.

H. Transformers shall operate at audible sound levels substantially below American National Standards Institute (ANSI) Standard C89.2. Sound levels shall not exceed the following:
1. 45 decibels (db) for units 15 to 50 KVA,
2. 50 db for 51 to 150 KVA units,
3. 55 db for 151 to 300 KVA units, and
4. 60 db for 301 to 500 KVA units.

I. The neutral bus of each transformer shall be sized and configured for at least 200% of the secondary full-load current.

J. Impedance of transformer shall be a minimum of 3% and a maximum of 5%.

Part 3. Execution

3.1 Installation

A. Ground secondary neutral of each transformer to the nearest structural member of the building frame, cold water line, or as otherwise shown on the drawings.

B. Ground the enclosure of each transformer either by a bonding jumper to the neutral or to the building steel where there is no neutral. Ground the core of the transformer visible to the enclosure by means of a flexible grounding conductor sized in accordance with applicable NEMA and NEC standards.

END OF SECTION

SECTION 16450

Grounding

Part 1. General

1.1 Scope

A. Provide all electrical system grounds as indicated on the drawings, mentioned herein, and/or as required by the National Electrical Code.

B. All grounding fittings for service and equipment grounds shall be of an approved type.

Part 2. Products

2.1 Materials

A. The following shall be solidly grounded:

1. Neutral leads of secondary service(s)
2. Main service entrance switch
3. Main switchboard
4. Panelboards
5. Transformer(s), enclosures, and neutral leads
6. Generator frames and neutral leads
7. Wireways and pull boxes
8. Raceway system
9. Bus duct(s)
10. Kitchen equipment
11. Shop equipment
12. Motor frames
13. Electric heating equipment
14. Lighting fixtures

Part 3. Execution

3.1 Installation

A. The main service grounding system shall consist of two branches, one being a grounding conductor to the water piping system, which shall be sized in accordance with Table 250-94 of the National Electrical Code; and the second being a grounding conductor to the electrode grounding system, which shall be sized in accordance with Table 250-79(c) of the National Electrical Code. In both instances, the grounding conductor shall be bonded at both ends to the conduit in which it is installed. The main service ground to the water piping system shall be connected on the street side of the water meter, or on a cold water pipe as near as practicable to the water service entrance to the building. The grounding electrode system shall include the metal building frame and/or footer reinforcing steel wherever available, and a ground ring and/or ground rod system in all other cases. Bonding jumpers shall be provided where required by Article 250-112 of the National Electrical Code. Connections to the water line and the grounding electrode system shall comply with Articles 250-112 and 250-115.

B. All conduit that enters equipment through open bottoms, insulated openings, or otherwise without having intimate contact with the enclosures shall be terminated with grounding bushings. Copper conductors from grounding bushings to equipment grounding bus shall be sized in accordance with Table 250-95 of the National Electrical Code.

C. All raceways for lighting and power use, for both feeders and branch circuits, shall contain a separate equipment ground wire with green insulation, the minimum size of which shall be in accordance with Article 250-95 of the National Electrical Code. Ground wire shall be continuous from circuit origination to termination. (Exception: Ground wire shall not be installed in raceways serving isolated wiring systems.)

END OF SECTION

SECTION 16601

Lightning Protection System

Part 1. General

1.1 Section Includes

A. Furnish and install a complete lightning protection system for protection of the entire campus.

B. Test existing site soil conditions to determine the average soil resistivity for use in the grounding calculations. Utilize 4-Point method, in accordance with IEEE-81 standards. Calculate the anticipated resistance to ground of various grounding elements, both ground rods and copper plates.

1.2 Scope

A. The work covered by this section consists of furnishing all labor, materials, and items of service required for the completion of a functional and unobtrusive lightning protection system as approved by the Architect, and in strict accordance with this section of the specifications. Protect entire building.

B. The following specifications and standard of the latest issue form a part of this specification: (a) Lightning Protection Institute Installation Code LPI-175; (b) IEEE Standards 81; and (c) UL-96A.

1.3 Submittals

A. Complete shop drawings showing the type, size, and locations of all equipment, grounds, cable routings, etc., shall be submitted to the Architect for approval prior to the start of work.

B. Submit earth ground resistance test results for approval before proceeding with grounding design.

C. Submit calculations to determine anticipated ground resistance of system electrodes and/or ground plates. Design system to deliver 5.0 ohms maximum resistance.

D. Submit shop drawings indicating lightning protection system layout and ground system components installation.

E. Product data for all roof-mounted, down conductor, and grounding materials.

F. Product data and MSDS for all roof adhesives and ground-enhancement materials.

1.4 Quality Assurance

A. Shop drawings shall bear the seal of a Registered Professional Engineer (PE) in the State of Pennsylvania. Drawings submitted without PE seal will be rejected.

B. The lightning protection system shall conform to the requirements of Underwriters' Laboratories standard UL-96A and NFPA standards for lightning protection systems.

C. The equipment manufacturer shall be a UL Listed and Approved Manufacturer and a fully certified manufacturer member in good standing of the Lightning Protection Institute. All material specified for this work is manufactured by Thompson Lightning Protection, Inc., 901 Sibley Highway, St. Paul, Minnesota, 55118, or approved LPI equal.

Part 2. Products

2.1 Standard

A. All equipment shall be new, and of a design and construction to suit the application where it is used in accordance with accepted industry standards and UL and NFPA Code requirements. All cable and air terminals must bear UL listing label. Fixtures installed must bear the UL stamp.

2.2 Equipment

A. All materials shall be copper or aluminum, in accordance with UL and NFPA Code requirements for Class I structures.

B. All main roof conductors shall be aluminum of 28 strands, 14 gauge minimum, 115 lb./thousand feet, Thompson Catalog No. A28. Copper conductors from structural steel shall be a minimum 4/0, 660 lbs./thousand feet. All buried counterpoise conductors shall be a minimum 4/0 MCM bare copper.

C. Air terminals shall be solid, round aluminum bar of 1/2″ minimum diameter, Cat. No. A55, and shall project 10″ minimum above the object to be protected. Locate and space according to UL and NFPA requirements.

D. Cable fasteners shall be substantial in construction, electrolytically compatible with the conductor and mounting surface, and shall be spaced according to UL and NFPA Code requirements (Thompson Cat. No. A730, or A166, etc.).

E. Bonding devices, cable splicers, and miscellaneous connectors shall be of cast aluminum with bolt pressure connections to cable. Cast or stamped crimp fittings are not acceptable.

F. Ground rods shall be 3/4″ × 10″ copperclad steel and shall be connected to counterpoise conductors exothermically with cadweld mold No. GTC-182G. Provide complete counterpoise.

G. Ground plates shall be 18″ × 18″, 20-gauge copper with mechanical attachments, Thompson Cat. No. 233M. Each plate shall be augmented with Erico

GEM conductive enhancement material for continuous one-cubic-inch coverage on top and bottom of plate.

H. All metal coping attachments shall be with ITW Buildex Traxx screws 1/4-14-7/8, with an 18/8 stainless steel head and neoprene washer. No substitutions will be accepted.

I. All masonry attachments shall be with lead expansion anchors and stainless steel bolts.

J. All miscellaneous bolts, nuts, and screws shall be stainless steel.

Part 3. Execution

3.1 Equipment

A. Equipment shall be as manufactured by Ace Lightning Protection, 4707 Logans Ferry Road, Murrysville, Pennsylvania, or approved equal.

3.2 Installation

A. The installation shall be accomplished by an experienced installer who is a Certi-fied Master Installer of the Lightning Protection Institute. This contractor's LPI Master Installer number shall be listed on the Contractor's bid form. The Contractor must have a minimum of ten years documented experience and submit a list of similar projects completed for the Architect's review. Failure to meet experience standards disqualifies the Contractor from the project.

B. All equipment shall be installed in a neat manner, as inconspicuously as possible. All cable runs shall be chalk-line straight and plumb. All air terminals shall be plumbed. All adhesive fastening shall be accomplished with approved compatible built-up/SBS modified adhesive. (Use same sealer as roof manufacturer.) Air terminal bases and adhesive cable pads shall be installed in straight lines; apply adhesive to the base and fastener and let dry for a minimum of forty-eight (48) hours before inserting cable. Pouring adhesive over cable, base, and fastener is strictly prohibited. Ensure air terminals are plumb (within 10°).

C. On Parapets: A complete network of lightning protection shall be installed utilizing parapet bases fastened to the inside vertical surface of the aluminum coping. The base and clips shall be attached to the coping with stainless steel Scots Traxx screws.

D. On EPDM Roofs: All adhesive fastening shall be accomplished with EPDM pourable sealer. (Use same sealer as roof manufacturer.) Air terminal bases and adhesive cable pads shall be installed in straight lines; apply pourable sealer to the base and fastener and let dry for a minimum of forty-eight (48) hours before inserting cable. Pouring adhesive over cable and base and fastener is strictly prohibited. Utilize cast adhesive base No. A688 with adhesive cable holder No. A730.

E. On Standing Seam Roofs: A complete network of lightning protection shall be installed utilizing through-roof electrodes bonded to the structural steel. The electrodes shall extend through the main ridge and prominent features of all vaulted roofs. The electrodes shall be flashed with Dektite-type flashings. Cable on vaulted roofs shall be run within deck flutes; no exposed cabling on vaulted roofs. The supply of the Dektite flashing is by the Lightning Protection Contractor.

F. The installation of all flashing is by the Division 7 Contractor at no charge to the Lightning Protection Contractor.

G. Structural steel shall be utilized as down conductors. Use 8-square-inch beam clamps, Thompson Catalog No. A701X.

H. The limitations on areas of usage for aluminum cables and for copper and aluminum materials together as outlined in NFPA 78 and LPI 175 shall be observed.

I. Verify completed system resistance to ground is 5.0 ohms or less. Perform "3-Point Fall of Potential" measurement in two opposite directions with a minimum current probe distance of 300'. Collect data at 30%, 40%, 50%, 62%, and 70% intervals. The test shall be accomplished according to IEEE Standards 81 and 142, in the presence of the Architect's representative. If test results exceed the 5.0 ohm requirement, the Lightning Protection Contractor will be required to install additional grounding and/or conductive enhancement material at his or her expense to achieve the desired 5.0 ohm results.

3.3 Completion

A. The trade shall submit copies of as-built shop drawings with the Underwriters' Laboratories Master Label System Certification—UL-96A.

END OF SECTION

SECTION 16720

Fire Alarm System

Part 1. General

1.1 Related Documents

A. Drawings and general provisions of Contract, including General and Supplementary Conditions and Division 1 Specification sections, apply to work of this section.

B. Division 16 Basic Electrical Materials and Methods sections apply to work specified in this section.

1.2 General

A. Furnish and install an electrically supervised Fire Alarm System, as covered by these specifications, to be wired, connected, and left in first-class operating condition. New point addressable devices shall be installed, where indicated on the drawings. All equipment shall be listed by the Underwriters' Laboratories, and shall conform with the requirements of the 1996 BOCA Code and the City of Pittsburgh.

B. The catalogue numbers herein specified are those of the Simplex Time Recorder Company and constitute the type and quality of equipment to be furnished. The Fire Alarm System shall be complete in every respect including all necessary equipment shown or not shown on the drawings to perform the functions relative to the system operation. All published specifications of the aforementioned manufacturer shall be considered as part of this specification even though they may not be shown in complete detail.

C. The equipment manufacturer shall furnish the services of a Factory Trained Technical Representative to supervise the installation of the system and to verify that the system has been installed and is functioning properly. The equipment manufacturer shall furnish installation drawings and technical assistance to the Installing Contractor. At the completion of the installation, the Technical Representative shall completely test the system including each initiating device and signaling device and each circuit of the system shall be tested for trouble reporting. Documentation shall be provided to the Owner and authority having jurisdiction that these tests were completed. The documentation shall list each device of the system, when it was tested, and the name of the Technical Representative.

1.3 Operation

A. The actuation of any addressable initiating device shall cause the appropriate zone description label to be displayed on the 80-character LCD display on the Fire Alarm Control Panel and remote annunciators. Operation shall cause all alarm signal devices (horns and visual lights) to activate. Horns shall sound the Temporal Pattern until silenced, and visual strobe lights shall flash, until the system is reset. All magnetically held doors shall be released.

B. In addition to the aforementioned, activation of any Elevator Lobby smoke detector shall cause elevators to return to the main level of exit. If the activated detector is on the main level, then the elevators shall return to a predetermined alternate floor.

C. Municipal Connection—Any alarm condition shall be transmitted to the local Central Station monitoring service via the digital communicator. The Central Station shall be selected and contracted with by the Owner.

D. Fire Suppression System Supervision—The activation of a Sprinkler System Tamper Switch shall sound the system trouble signal, illuminate the dedicated system supervisory LED, and cause the appropriate zone description label to be displayed on the 80-character LCD display, and shall cause the appropriate 80-character descriptor to be displayed on the remote annunciator panel.

E. System Wiring Supervision—Any abnormal condition, such as power failure, open circuit, or a short circuit, shall cause the system trouble signal to sound and the appropriate trouble LED shall illuminate. The trouble signal may be silenced by momentarily depressing the trouble silence switch, but the trouble LED shall illuminate until the trouble condition has been corrected. If a second trouble condition should arise, the trouble signal shall be reactivated.

F. System Power Sources—The system shall normally operate from a single 120-VAC, 60-Hz Normal/Emergency Source. Standby batteries shall be provided to operate the system during complete power outages. Batteries shall be sized to operate the system in the standby mode for a minimum of four hours and in the alarm mode for a minimum of five minutes at the end of the four hour period.

G. Output to Building Management System—The system shall be provided with a compatible output to the Building Management System (BMS). The output shall be a data stream readable by the BMS system software.

Part 2. Products

2.1 Fire Alarm Control Panel

A. Furnish and install, where shown on the drawings, a Simplex type 4100-8001 Fire Alarm Control Panel. It shall be capable of the following as a minimum:

1. Expansion capacity of 1000 circuits. Circuits may be any one of the following:

    a. Addressable Input Device (Pull Station, Smoke Sensor, Monitor ZAM [Zone Adaptor Module], etc.)
    b. Addressable Output Device (Signal ZAM, Control ZAM)
    c. Conventional Fire Alarm Zone
    d. Conventional Fire Alarm Signal Circuit
    e. Conventional Relay Output

2. Printer and/or CRT/Keyboard Interface (Future)
3. Alarm Verification
4. Walk Test
5. Historical Alarm and Trouble Log
6. True Alarm—Smoke Sensor Environmental Compensation
7. Manual Control of auxiliary functions, such as fans, dampers, etc.
8. Data stream output to the BMS system
9. Digital Communicator (DACT), for alarm transmission to the Owner's Central Station Monitoring service

B. The system must provide addressable communication with initiating and control devices individually. All of these devices will be individually annunciated at the control panel. All addressable devices shall have the capability of being disabled or enabled individually. Up to 127 addressable devices may be multidropped from a single pair of wires. The communication format must be a completely digital poll/response protocol to allow tapping of the circuit wiring. Each addressable device must be uniquely identified by an address code entered on each device at time of installation. **Supply sufficient device capacity to handle the number of addressable devices shown on the drawings, plus 20% spare capacity.**

C. The Fire Alarm Control Panel shall allow for loading and editing of special instructions and operating sequences as required. The system shall be capable of on-site programming to accommodate system expansion and facilitate changes in operation. All software operations shall be stored in a nonvolatile programmable memory within the Fire Alarm Control Panel. Loss of primary and secondary power shall not erase the instructions stored in memory.

D. Control panel construction shall be modular with solid state, microprocessor-based electronics, and shall include 80-character LED, alarm and trouble LEDs, and acknowledge, silence, and reset switches. Secondary control switches and LEDs shall include a City disconnect/switch, Signal Circuit Bypass, and Manual evacuation (drill).

E. True Alarm Detection: The control panel shall be capable of Simplex True Alarm operation. The smoke sensors shall be smoke-density measuring devices having no self-contained alarm set-point. The alarm decision for each sensor shall be determined by the control panel. The control panel shall determine the condition of each sensor by comparing the sensor value to stored values. The control panel shall maintain a moving average of the sensor's smoke chamber value. The system shall automatically indicate when an individual sensor needs cleaning. When a sensor's average value reaches a predetermined value, a "Dirty Sensor" trouble condition shall be audibly and visually indicated at the control panel for the individual sensor. Additionally, the LED on the sensor base shall glow steady, giving a visual indication at the sensor location. If a "Dirty Sensor" is left unattended, and its average value increases to a second predetermined value, an "Excessively Dirty Sensor" trouble condition shall be indicated at the control panel for the individual sensor.

F. The control panel shall automatically perform a daily self-test on each sensor. A sensor which fails the self-test will cause a "Self Test Abnormal" trouble condition at the control panel. An operator at the control panel, having a proper access level, shall have the capability to manually access the following information for each sensor:

1. Primary status
2. Device type
3. Present average value
4. Present sensitivity selected*
5. Peak detection values*
6. Sensor range (normal, dirty, etc.)

*Values shall be in "percent of smoke obscuration" format so that no interpretation is required by the operator.

An operator at the control panel, having a proper access level, shall have the capability to manually control the following for each sensor:

1. Clear peak detection values
2. Enable or disable the point
3. Clear verification tally
4. Control a sensor's relay driver output

It shall be possible to program the control panel to automatically change the sensitivity settings of each sensor based on time-of-day and day-of-week. (For example, to be more sensitive at during unoccupied times and less sensitive during occupied.) There shall be seven sensitivity settings available for each sensor. The control panel shall have the capability of being programmed for a pre-alarm or two-stage function. This function allows an indication to occur when, for example, a 3% sensor reaches a threshold of 1.5% smoke obscuration.

G. For increased smoke detection assurance, all individually addressed smoke sensors shall be provided with alarm verification. Only a verified alarm shall initiate the alarm sequence operation. The alarm verification feature shall be software selectable, ON or OFF, for each detector.

H. Provide cabinets of sufficient size to accommodate the aforementioned equipment. It shall be possible to expand the system to its maximum capacity at any time by the addition of components within its cabinet or into additional cabinets. Cabinet shall be equipped with locks and transparent door panel providing freedom from tampering yet allowing full view of the various lights and controls.

2.2 Remote Annunciator Panel

A. Furnish and install, where shown on the drawings, a Simplex type 4603-9101 Flush Mounted Alphanumeric LCD Remote Annunciator Panel. It shall operate from the Fire Alarm Control Panel using serial transmission techniques and shall require only a four-wire connection regardless of the number of annunciation zones or manual controls. An 80-character (2 lines of 40) LCD display shall be included for annunciation of all system status changes including alarms, troubles, and supervisory service indications. Manual control switches with associated LEDs shall be provided as an integral part of the annunciator for alarm acknowledge, trouble acknowledge, supervisory service acknowledge, and alarm silence. System reset switch, display time switch, local audible alert, and system master control enable switch shall also be provided on the annunciator. To prevent unauthorized operation of the control switches, the enable switch shall be key operated and be used to enable the other control switches. The annunciator shall also be provided with a power-on indicator and four auxiliary control switches with associated status LEDs programmed as follows:

1. City Disconnect
2. Signal Circuit Bypass
3. Control Circuit Bypass
4. Disaster Alarm Signal Activate

All wiring to the remote annunciator shall be supervised by the Fire Alarm Control Panel for opens, shorts, and ground faults.

2.3 Transponder Panels

A. Transponder Panels: Furnish and install, where shown on the drawings, Simplex type 4100-8019 Transponder Panels. The system transponders shall be microprocessor based, housed in all-metal cabinet suitable for surface wall mounting. They shall be connected to the Fire Alarm Control Panel via the communication network in a NFPA 72A Style 6 configuration. They shall provide direct connection of addressable device circuits (NFPA 72A Style 4), indicating appliance circuits (NFPA 72A Style Y), and unsupervised control outputs.

These units shall be polled by the Fire Alarm Control Panel on a continuous basis. Any change in status of a system point shall report to the Central Processing Unit (CPU). If, due to a trouble condition, a transponder fails to respond to a communication poll from the Fire Alarm Control Panel, three more attempts shall be made. Failure of a transponder to respond to the fourth attempt will result in a communications failure, with a resulting LED display, audible signal, and print-out.

The transponders shall be modular in design to allow variables in the number and type of circuits to meet the application requirements. Transponders shall be equipped with control point outputs utilizing relays with two-ampere, 24 VDC contacts. The location and circuit configuration of transponders shall be as indicated on the drawings.

Transponder panels shall contain integral power supplies with built-in charger and standby batteries. The power supplies shall provide outputs of 8 amps of 24 VDC nonregulated power. All outputs shall be individually fused. Each power supply shall have an integral regulated charging circuit, and sealed standby storage batteries to provide four hours standby in the event of AC power failure at the transponder panel.

2.4 Manual Pull Stations

A. Manual Pull Stations: Furnish and install, where shown on the drawings, Simplex type 2099-9795 Single Action Addressable Manual Pull Stations. Stations shall be constructed of high-impact red lexan and shall operate with or without glass rod. Station shall contain a key lock keyed identical to the control panel to be used for resetting the station. Once activated, the pull station handle shall remain at a 90° angle from the front of the station to provide visual indication as to which pull station was activated. Manual stations shall be suitable for semiflush or surface mounting. Where stations are to be surface mounted, the enclosures shall be provided by the manufacturer. Stations shall be semiflush mounted wherever possible.

2.5 Smoke Sensors

A. True Alarm Smoke Sensor: The addressable smoke sensors shall be Simplex type 4098-9714, photoelectric type, with 4098-9792 addressable base, and shall communicate actual smoke chamber values to the system control panel. The sensors shall be listed to UL Standard 268 and shall be documented compatible with the control equipment to which they are connected. The sensors shall be listed for both ceiling- and wall-mount applications. Each sensor base shall contain a LED that will flash each time it is scanned by the control panel, which shall be once every four seconds. When the control panel determines that a sensor is in an alarm or trouble condition, the control panel shall command the LED on that sensor's base to turn on steady, indicating the abnormal condition. Sensors that do not provide a visible indication of an abnormal condition at the sensor location shall not be acceptable.

Each sensor shall contain a magnetically actuated test switch to provide for easy alarm testing at the sensor location. Each sensor shall be scanned by the control panel for its type identification to prevent inadvertent substitution of another sensor type. The control panel shall operate with the installed device but shall initiate a "Wrong Device" trouble condition until the proper type is installed or the programmed sensor type is changed. The sensor's electronics shall be immune from false alarms caused by Electro Motive Inductance (EMI) and Radio Frequency Inductance (RFI).

B. Duct Smoke Sensors: Furnish and install, where indicated on the plans, Simplex 4098-9753 Duct Housings, with type 4098-9714 True Alarm sensors. The sensors shall meet the requirements of UL Standard 268A and shall be documented compatible with the control equipment to which they are connected. The addressable duct smoke sensors shall operate on the lightscattering, photodiode principle, and shall communicate actual smoke chamber values to the system control. The sensors shall not have a self-contained smoke sensitivity setting and shall automatically compensate for environmental changes. The sensor's electronics shall be completely shielded to protect against false alarms from EMI and RFI. The duct housing shall contain a red LED that shall pulse to indicate power-on and glow continuously to indicate an alarm or a sensor trouble condition. It shall also contain a SPDT auxiliary relay rated at 2 amps at 120 VAC or at 28 VDC for unit shutdown. The detectors shall be installed, wired, and connected to the Fire Alarm

Control Panel by the electrical contractor. They shall obtain their operating power from 24 VDC from the Fire Alarm Control Panel. The duct detectors shall utilize cross-sectional sampling principle by which a sampling tube is extended across the duct to continuously sample the air movement through the duct, after which the sampled air is returned to the duct via an exhaust tube. Sampling shall be properly sized for the duct in which it is installed. Provide a Simplex type 2098-9808 Remote Alarm Indicator for each duct smoke sensor installed.

C. Heat Sensors: Furnish and install, where indicated on the drawings, Simplex type 4098-9733 True Alarm Heat Sensors, with 4098-9792 addressable base. They shall be a combination rate-of-rise and rate-compensated fixed temperature sensors of which both operations are self-restoring. The sensor's small thermal mass shall allow an accurate, up-to-date temperature reading of each sensor to be logged at the control panel. The rate-of-rise operation shall be selectable in either a 15° per minute or a 20° per minute rate of temperature rise. The fixed-temperature principle shall operate entirely independent of the rate-of-rise principle and shall be selectable for either 117° or 135°F.

2.6 Notification Appliances

A. Audible/Visible Units: Furnish and install, where shown on drawings, Simplex type 4903 Series Horn/Strobe Units. Horn shall produce a sound output of 87 decibels, with a current consumption of .038 amps at 24 VDC. Strobe light shall produce a 15-, 30, or 110-candela flash. In order to meet the criteria of the American Disabilities Act, the Strobe Unit shall be UL listed to Standard 1971. A compound reflector shall provide light output in Key Axis Directions. The lens shall be tamper resistant, and the lamp assembly shall also be shock resistant. The lens shall be clear with the word "FIRE" printed in red. In order to prevent seizures of persons with photosensitive epilepsy, the strobe lights shall incorporate a synchronized flash feature. Furnish and install a Simplex type 4905-9914 Synchronized Flash Module for each strobe circuit. The horn/strobe units shall be suitable for wall mounting to a standard electrical outlet box.

B. Visible Only Units: Furnish and install, where shown on drawings, Simplex type 4904 Series Strobe Visible Units. Strobe light shall produce a 15-, 30, or 110-candela flash. In order to meet the criteria of the American Disabilities Act, the Strobe Unit shall be UL listed to Standard 1971. The strobe lights shall incorporate a 60-flashes per-minute flash rate, in order to prevent seizures in photosensitive epileptics. A compound reflector shall provide light output in Key Axis Directions. The lens shall be tamper resistant, and the lamp assembly shall also be shock resistant. The lens shall be clear with the word "FIRE" printed in red. Furnish and install a Simplex type 4905-9914 Synchronized Flash Module for each strobe circuit. The horn/strobe units shall be suitable for wall mounting to a standard electrical outlet box.

2.7 Addressable Modules

A. Individual Addressable Module: Furnish and install, where shown on the drawings, Simplex type 2190-9172 Individual Addressable Module (IAM). An IAM shall be provided for interfacing normally open direct-contact devices, such as waterflow switches, tamper switches, temperature sensors, etc. IAMs will be capable of mounting in a standard electric outlet box. IAMs shall communicate with the Fire Alarm Control Panel over, the MAPNET addressable loop.

B. Control Zone Adaptor Module: Furnish and install, where shown on the drawings, Simplex type 2190-9163 Control ZAM to be used for nonsupervised control of door-holder circuits and other control devices not requiring supervised wiring. ZAMs will be capable of mounting in a standard electric outlet box. ZAMs will include coverplates to allow surface or flush mounting. These ZAMs will communicate the supervised wiring status (normal, trouble) to the Fire Alarm Control panel and will receive a command to transfer the relay from the Fire Alarm Control Panel.

2.8 Door Holders

A. Door Holders: Furnish and install, where shown on the drawings, Simplex type 2088 Series Door Holders. They shall be rated for a minimum of 35 lbs. of holding force and shall be wall or floor mounted as indicated on the drawings. They shall be powered from the Fire Alarm Control Panel by a fused 24 VDC circuit and shall release the door on any alarm.

2.9 Sprinkler System

A. Waterflow and OS&Y Monitor Switches: Waterflow and OS&Y Monitor Switches shall be furnished under other sections of these specifications but shall be wired and connected to the Fire Alarm System by the Electrical Contractor.

Part 3. Execution

3.1 Wiring

A. The Electrical Contractor shall furnish and install in accordance with the manufacturers' instructions all wiring, conduit, raceways, outlet boxes, and auxiliary equipment required for the installation of the system.

B. The wiring system shall meet the requirements of all applicable national, state, and local electrical codes and shall conform with the requirements of Standard No. 72 of the National Fire Protection Association. All wiring shall be copper and shall be No. 18 AWG minimum or as recommended by the equipment manufacturer. The number of the conductors shall be as shown or as required by the equipment manufacturer. All wiring shall be color coded. All wires shall be tagged in all junction points and shall test free from grounds or crosses between conductors. The Electrical Contractor shall furnish and install all necessary outlet boxes and mounting boxes required.

C. Final connections between equipment and wiring system shall be made under the direct supervision of a representative of the fire alarm equipment manufacturer.

D. The exact wiring arrangement shall be in accordance with the fire alarm equipment manufacturer's requirements and the exact number of initiating and signaling devices to be furnished and installed shall be as shown on the drawings.

3.2 Guarantee

A. All the aforementioned equipment shall be unconditionally guaranteed against all electrical and mechanical defects for a period of:

1. Eighteen (18) months following shipment of any system or subsystem from Simplex to the customer site, or

2. Twelve (12) months following the final connection, operation, and beneficial use of all or any part of the system.

END OF SECTION

SECTION 16762

Television Distribution System

Part 1. General

1.1 Work Included

A. Furnish and install new TV head-end electronics and a new TV distribution system in the building as shown and herein specified. The owner shall install the directional multiports and TV jacks.

B. It is the intent of this specification to outline the requirements for a complete and satisfactory operating CATV distribution system to be furnished and installed in the building. The system shall be wired, connected, adjusted, and left in first-class operating condition.

C. Numbers in these specifications have been taken from the catalogs of Blonder Tongue Laboratories, Inc. and other manufacturers and are intended to denote a standard of quality and type of equipment.

D. All published specifications of the manufacturers mentioned in these specifications shall be considered as being a part of these specifications even though they have not been shown here in complete detail.

E. The electronic equipment supplier must be a franchised distributor of all equipment utilized in the CATV distribution system. Further, the supplier must have a minimum of ten years of experience in the specific application of the equipment proposed for these systems. At the request of the Architect, the electronic equipment supplier will provide a letter from the manufacturer certifying that they are a franchised dealer for the equipment they intend to supply.

F. The Contractor shall provide with the shop drawings a factory-engineered system design detailing the head-end components along with a building floor plan indicating all field devices and their values.

G. The Contractor shall furnish two sets of instructions (requisite material) for the proper operation of the equipment by the Owner's operating personnel.

H. The manufacturer or his or her agent shall show satisfactory evidence, upon request, that it maintains a fully equipped service organization capable of furnishing adequate inspection and service to the equipment specified including standard replacement parts, and such services shall be within 75 miles of the project site.

1.2 Scope

A. A television distribution system shall be provided exactly as specified by the following conditions and as shown on the plans.

B. The system shall consist of a lockable equipment cabinet that will house the amplifiers and other head-end equipment. Distribution system will consist of .500 aluminum cable feeding future directional multitaps with RG-6 drop cables to feed through tap-offs in the classrooms.

C. The system shall provide for the distribution of all available television channels from a feed by the CATV company. Bandwidth of all passive devices shall be 5 to 550 MHz. Bandwidth of amplifier shall be 54 to 550 MHz. Amplifiers shall be of the push-pull design to minimize harmonics and beats. The system shall be designed for −57 dB cross modulation or better and a carrier to noise ratio of at least 46 dB.

D. Isolation between any two outlets in the system shall be a minimum of 25 dB on any frequency between 5 and 550 MHz.

Part 2. Products

2.1 Materials

A. The channel combiner shall have eight broadband input ports for combining modulators and/or processors in the head end. The combiner shall use radiation-proof passives. The combiner shall have a −20 dB test point for testing of signals without interruption of service. The combiner shall have a frequency response of 1000 MHz with an insertion loss of 14 dB. The combiner shall be Blonder Tongue Model OC-8d.

B. The radiation-proof RF splitters shall comply with FCC specifications concerning radiation shielding. Construction shall be a cast housing with mounting tabs and grounding block. Splitters shall be Blonder Tongue CRS Series with two or three ports.

C. The rack-mounted distribution amplifier shall be a broadband push-pull type with 50 dB of operational gain. The amplifier shall have a bandwidth of 40-550 MHz with front panel gain and slope controls. The amplifier shall have −20 dB input and output test points for testing of signals without interruption of service. The amplifier shall be Blonder Tongue RMDA-550-50.

D. The future directional couplers shall comply with FCC specifications concerning radiation shielding. The couplers shall be available in one-, two-, or four-port models with tap values between 4 dB and 35 dB and for operation in 750 MHz CATV systems. The directional couplers shall be Blonder Tongue CRT series.

E. The future wall taps shall be a radiation-proof design utilizing a steel backplate and a feed-through outlet for use with trunk line tap-offs. The wall tap shall mount in a standard single gang configuration and will require an ivory duplex receptacle plate. The wall tap shall be Blonder Tongue V-1GF-FT.

F. The trunk cable shall be plenum rated .500 aluminum with 100% shield construction. The cable shall be Comscope P375JCASS or approved equal.

G. The future drop cable shall be plenum rated RG-6U with 100% shielding. The cable shall be Comscope 2279K or equal.

H. The headend equipment cabinet shall be a lockable wall mount unit installed in the MDF Room as shown on the drawings. The cabinet shall have 28" of standard 19" rack space and be constructed of 16-gauge CRS, factory welded for stability. The equipment cabinet shall be Soundolier 300-28 or approved equal.

2.2 Equipment

A. (1) Soundolier 300-28 Equipment Cabinet

B. (1) Blonder Tongue OC-8d Combiner

C. (1) Blonder Tongue RMDA-550-50 Broadband Amplifier

D. (*) Blonder Tongue CRS Series Splitters

Part 3. Execution

3.1 Installation

A. Install in accordance with manufacturers' instructions.

B. Coordinate the conduit and wire installation with TV distribution equipment supplier.

C. Install all equipment power wiring and grounding to conform to NEC requirements and applicable local codes.

D. Adequately support cable. Use connectors specifically designed for the type of cable being used.

E. Mount all components except in-line pads in cabinets or other solid support. Equipment suspended by coaxial connectors is unacceptable.

3.2 TV System Test

A. Upon completion of system installation, perform the necessary adjustments and balancing of all signals and amplifier level controls to ensure proper system operation.

B. Before final acceptance, conduct an operating test at all outlet locations for approval. The system shall be demonstrated to operate in accord with requirements of these specifications. The test shall be performed in the presence of an authorized representative or the Owner. Contractor shall furnish all equipment and personnel required for the tests.

C. Should such a demonstration of performance show that the Contractor has not properly balanced the system and that picture degradation is present or that output is not as specified, make all necessary changes or adjustments and a second demonstration will be arranged.

D. After the system is accepted, prepare "as-built" drawings to indicate all field changes. These drawings shall include the recording of all pertinent signal levels through the system as measured at the time of final acceptance.

E. Head-end Tests:

1. System balance test shall employ a Blonder Tongue Model FSM-10 field strength meter or approved equal. Measurements shall be made at the combined output of the head-end system. The level of each channel's picture and sound carrier shall be measured and recorded.

2. All levels shall be within +3 dB from design levels specified. In no case shall the levels measured exceed the maximum output rating for the head end amplifier(s) employed.

3. The level difference between channel picture carriers shall not exceed 2 dB for adjacent channels nor 6 dB between the strongest and weakest channel normally carried.

4. Using a field strength meter, measure the signal level at the last outlet on each feeder line and other randomly selected outlets totaling not less than 5% of the total number of outlets. The signal level on each channel shall not read less than 3 dBMV, nor more than 7 dB, unless specified otherwise.

5. Using a standard TV receiver connected to randomly selected outlets, not less than one (1) per feeder, observe picture quality. No visible components of cross-modulation (windshield wiper effect), ghosting, or beat interference shall appear on the screen of the receiver tuned to any normal signal.

3.3 Warranty

A. Provide a one (1) year warranty on material and labor on the TV distribution system.

END OF SECTION

SECTION 16851

Electrical Equipment and Motor Control

Part 1. General

1.1 Scope

A. The Electrical Contractor is hereinafter referred to as the EC. The HVAC Contractor and the Plumbing Contractor are hereinafter referred to as the Mechanical Contractor (MC).

Part 2. Products

2.1 Performance

A. In general, motors less than $1/2$ horsepower will be wound for single phase, 120 volt, and motors $1/2$ horsepower and over will be wound for three phase, 460 volts. EC shall be responsible for establishing the proper direction of rotation for all three-phase motors.

B. This Contractor shall not install any work pertaining to power and/or control wiring for equipment furnished by other contractors under other contracts until the required wiring diagrams have been approved by the Architect for use as installation drawing by this Contractor.

Part 3. Execution

3.1 Installation

A. Wiring to HVAC Equipment

1. EC shall provide all power wiring and conduit from the electrical source to an equipment junction box, control panel, or disconnect device serving each item of equipment, and between the equipment junction box, control panel, or disconnect device and the motor or other load device.

2. EC shall provide all indicated and/or required disconnecting means at motors and/or controllers.

3. For single-phase equipment furnished by the MC, the control wiring and control equipment shall be as follows:

   a. MC shall furnish and install each control device such as motor sentinel switch, PE switch, thermostat, etc.

    b. EC shall furnish and install as a subcontractor to the MC all control conduit and wiring including that between control devices, and that from each control device to the equipment junction box, control panel, and/or disconnect device, as applicable.

4. For three-phase equipment furnished by the MC, the control wiring and control equipment shall be as follows:

    a. MC shall furnish and install all control devices, such as contactors, thermal overload protection, pushbuttons, PE and EP switches, thermostats, etc.

    b. EC shall furnish all combination motor starters and install the starters. All staters shall be tagged by the EC with an identification label indicating the motor served by a specific starter.

    c. EC shall furnish and install all control conduit and wiring from each control device to motor starter or contactor, and between control devices and/or to control panels, as a Subcontractor to the Mechanical Contractor.

B. Wiring to Equipment Other Than HVAC

1. For equipment furnished by other than the MC, all control wiring for motors and equipment shall be furnished and installed by the EC. All control devices such as motor starters, pushbuttons, limit switches, etc., will be delivered to the EC for installation and final connection.

END OF SECTION

INDUSTRIAL SPECIFICATIONS

NOTE: This section contains a partial set of industrial specifications taken from the original industrial specifications supplied to the NJATC. This reduced industrial specification set is sufficient for the lessons in industrial print and specification reading. It contains all information generally necessary for an electrical worker. The original industrial specification set was over 1,200 pages in length. Much of the information had no direct bearing on electrical installation.

SECTION 00001
TABLE OF CONTENTS

This project manual follows the MasterFormat Identifying System.

Nonapplicable division and section references have been omitted.

Recipients of bidding instruments must consult the Table of Contents to determine the full scope of the work involved and to ensure that all pages of the project manual and drawings have been included.

Neither the Owner nor the Architect/Engineer (A/E) will be responsible for bids submitted that are based on incomplete bidding instruments.

Section	No. of Pages

Division 1: General Requirements

Division 2: Site Work

Division 3: Concrete

Division 4: Mortar

Division 5: Metals

SECTION 16010

Electrical Scope of Work

Part 1. General

1.1 Separation of the specifications into divisions does not establish the limits of the work. See other divisions for related work.

1.2 The work covered by this division of the specifications consists of furnishing all plant, labor, equipment, and materials and performing all operations in connection with the installation of electrical work and doing so completely, in accordance with the drawings and specifications, and subject to the terms of the contract. Bidders shall visit the project site and verify existing conditions prior to the submission of their bids.

1.3 Work Included

    A. A source of construction electrical power

    B. Lighting and power wiring systems

    C. Instrumentation and controls specified herein and under other divisions, calibrated and in complete working order

    D. Connection of all motors, other items requiring electrical service, and installation of starters and control devices where called for under other sections and as indicated on the drawings

    E. Cutting and patching

    F. Counterflashing of conduits where they penetrate roofs and outside walls

    G. Record Drawings: Mark up one copy of drawings to indicate accurately as-built conditions during the progress of the job. Show all changes, additions, and deviations from the original drawings. Do not use these drawings for construction. Mark and dimension locations of capped underground service conduit stubouts.

    H. Conduits for future systems or other contracts as shown on the drawings

    I. Provision for telephone service

    J. Installation of Owner-furnished equipment

    K. Excavation and backfill (all of which is unclassified) for electrical work

    L. Connection of all control devices such as thermostats, firestats, relays, etc., furnished under other divisions of the specifications. Install all control circuitry in accordance with approved diagrams furnished by the manufacturers of the control equipment.

Part 2. Products

Not used

Part 3. Execution

Not used

END OF SECTION

SECTION 16011

Related Work

Part 1. General

1.1 The following work is described in other divisions of these specifications:

A. Flashing of conduits into roofing and outside walls

B. Painting, including outside lighting standards and poles

C. Concrete for lighting standards, foundations, curbs, and pads

D. Concrete pad for transformer installation in accordance with power company requirements (when specifically required)

Part 2. Products

Not used

Part 3. Execution

Not used

END OF SECTION

SECTION 16012

Submittals

Part 1. General

1.1 Shop drawings and submittals shall bear the stamp of approval of the Contractor as evidence that the material has been checked by him or her. Drawings submitted without this stamp of approval will not be considered but will be returned for proper resubmission. If the submittals show variances from the requirements of the contract, the Contractor shall make specific mention of such variation in the letter of transmittal so that, if acceptable, suitable action may be taken for proper adjustment; otherwise, the Contractor shall not be relieved of the responsibility for executing the work in accordance with the contract.

Part 2. Products

Not used

Part 3. Execution

3.1 Submit for the A/E's approval a list of proposed materials and equipment within 15 days after the award of contract.

3.2 Submit for the A/E's approval, prior to fabrication and within 15 days after the award of contract, six sets of shop drawings for the items listed below:

    A. Panelboards

    B. Paging System

    C. Fire Alarm System

    D. Security Alarm System

    E. Closed Circuit TV System

    F. Switchgear

3.3 Submit for the A/E's approval, within 15 days after the award of contract, six sets of specifications or data sheets bound and identified according to the drawings and specifications for the following items:

    A. Wire and Cable

    B. Lighting Fixtures

    C. Emergency Lighting Equipment

    D. Wiring Devices

    E. Safety Switches

    F. Raceways

    G. Transformers

    H. Conduit and Fittings

    I. Panelboards

    J. Electrical Disconnects

    K. Instrumentation Components

    L. Bus Duct

END OF SECTION

SECTION 16023

Codes and Fees

Part 1. General

1.1 The work shall meet the requirements and recommendations of applicable portions of the latest editions of these standards:

    A. National Electric Code (NFPA 70)

    B. Life Safety Code (NFPA 101)

    C. Southern Standard Building Code (SSBC)

    D. National Electrical Safety Code (ANSI C2)

    E. NEMA Standards

    F. Underwriters' Laboratories (UL)

    G. Institute of Electrical and Electronic Engineers (IEEE)

    H. Certified Ballast Manufacturers (CBM)

    I. Lightning Protection Code (NFPA 78 and UL 96A)

    J. Chapter 0780-2-1, 1984 of the Tennessee Department of Insurance 1.2 In the event of a conflict between the drawings, specifications, and codes, request a ruling from the A/E.

Part 2. Products

Not used

Part 3. Execution

3.1 Comply with state and local electrical and building codes and with special codes having jurisdiction over specific portions within the complete project.

3.2 Observe ordinances pertaining to electrical work.

3.3 Complete work so that it will pass the tests required by agencies having authority over this work.

3.4 Obtain and pay for permits and certificates required by local and state laws and ordinances.

END OF SECTION

SECTION 16044

Acceptance Tests

Part 1. General

1.1 Scope

    A. This section provides specifications for material, equipment, labor, and technical supervision to perform and complete the electrical field acceptance tests as required for the Contractor and Owner-furnished electrical equipment.

1.2 Definitions

    A. Acceptance tests as herein specified are defined as those tests and inspections required to determine that the equipment involved may be energized for final operational test.

    B. Final acceptance shall depend upon equipment performance characteristics as determined by the subject tests, in addition to completion of operational tests on all electrical equipment to show that the equipment will perform the functions for which it was designated.

    C. These tests are intended to ensure that the workmanship, methods, inspections, and materials used in erection and installation of the subject equipment shall conform with accepted engineering practices, IEEE and IPCEA standards, the NEC, and more specifically specifications for the electrical work of the project as prepared by the A/E.

Part 2. Products

Not used

Part 3. Execution

3.1 Responsibility

    A. The Contractor shall be responsible for all tests and test records. Testing shall be performed by and under the immediate supervision of the Contractor and shall be made only by qualified personnel fully experienced in this type of testing.

    B. All testing shall be done in the presence of the A/E.

    C. Records of all tests and inspections, with complete data of all readings taken, shall be made and incorporated into a report for each piece of equipment tested. A copy of all test reports shall be delivered to the A/E at the end of each test period.

    D. The Contractor shall provide all necessary test equipment and shall be responsible for setting up all test equipment, wire checks of factory wiring, and any other preliminary work in preparation for the electrical acceptance tests.

    E. All tests shall be scheduled by the Contractor and cleared by the A/E. No testing shall be done without this clearance.

    F. The Contractor shall be responsible for cleanup and visual inspection of the equipment immediately prior to testing.

    G. Manufacturer's representatives shall be advised of all tests on their equipment. A list of manufacturers of the Owner's furnished equipment will be furnished to the Contractor by the Owner. Reasonable cooperation shall be extended to permit witnessing by a representative of the manufacturer of the material under test, should the manufacturer so request.

3.2 Megger Test

A. All tests shall be performed with motor-driven meggers, unless otherwise approved by the Owner.

B. Megger voltages shall be as specified in Table I.

TABLE I.

Equipment Voltage	Megger Voltage
Over 150 to 600	1000
150 and Under	500

C. The tests shall be applied from phase-to-ground with the other phases grounded. Each phase shall be tested in a similar manner.

D. All 1,000 V and 500 V motor-driven megger tests shall be held for a minimum of one minute and until the reading reaches a constant value for 15 seconds, unless specified otherwise.

E. Phase matching and phase identification shall be finally checked immediately prior to energization of equipment covered by the specifications.

3.3 Lighting Transformers

A. All 600 V and lower primary transformers shall be meggered prior to connecting.

 1. The minimum megger reading with 1,000 V megger on 480 V windings shall be 75 Megohms.

 2. The minimum megger reading with 500 V megger on 120 V windings shall be 25 Megohms.

 3. Correctness of phase and tap connections of all windings shall be checked.

3.4 Switchgear and Substations

A. Prior to energizing the switchgear, all control and instrument wiring connections shall be inspected for workmanship and proper identification and a check for installation of proper fuses shall be made.

B. After energizing and before breakers are initially closed, a check for proper phase rotation shall be made and recorded.

C. On switchboards with a tie breaker arrangement, a check of voltage and phase sequence between bus on each end shall be made.

D. All current and potential transformer polarity wiring shall be checked.

E. Switchgear shall be operated through all remote control operations including actuation of all alarms and mechanical and electrical operation from protective relays.

F. After removing all fuses on potential transformers and all breakers open, bus shall be given megger test.

G. Where specified, relays shall be checked for calibration and operation by qualified personnel, properly equipped with test equipment and manufacturer's data.

3.5 Cable—600 V and Lower

A. All new or reconnected power service cables shall be tested for continuity, shall be free of grounds and short circuits, and shall be given a megger test using a 1,000 V motor-driven megger.

B. Each 480 V feeder cable from the substations or switch room shall be meggered with the feed end of the cable connected to racked-out breaker at the substation or switch room and the load end as follows:

1. Cables to busways shall be meggered while connected to the busways and all bus plug switches open and loads disconnected.

2. Power feeders other than busways shall be disconnected from the loads and fanned out before meggering.

C. The minimum acceptable megger reading for unconnected cables shall be 100 Megohms. Any cable having a megger reading markedly lower than average, even though meeting minimum requirements, shall require approval by the A/E.

3.6 Power and Control Equipment

A. Prior to energization of equipment, all compartments and equipment shall receive a visual inspection by representatives of the Contractor and the Owner.

B. All control devices, breakers, switches, and contactors shall be given functional tests to determine satisfactory performance under operating conditions.

3.7 Rotating Equipment

A. General

1. Each electric motor above 600 V or over 100 HP shall be given an insulation resistance test. Insulation resistance tests shall be made between all windings and the ground frame. Megger tests for 3 phase motors shall be applied between all phases tied together and the grounded frame and shall include cable back to the open starter or circuit breaker as applicable.

2. Motors shall be checked for proper ro-tation and alignment. If rotation could damage the equipment, arrangements shall be made to mechanically uncou-ple the motor. Any connections found to be in error shall be corrected by the Contractor and reinsulated as required. No machine shall have excessive vibration.

B. Insulation Resistance Tests

1. All windings shall be at ambient temperature.

2. Any machine not passing this test shall be dried and retested until it either passes the test or is rejected.

3. The minimum acceptable readings for insulation resistance shall be in accordance with Table II.

TABLE II.

Equipment	Megger Voltage	Minimum Megger Reading (Megohms)
Motors over 600 V	2500	100
Motors under 600 V	1000	30

C. Grounding

   1. Ground Resistance Measurements

      a. Ground resistance measurements shall be made on each individual ground rod and on each basic grounding system installed under this contract. Tests shall be performed with a Biddle Ground Ohmer, or by the 3-point method described in EIII Standard 550, Paragraph 3.42, using 2 auxiliary ground rods. The latter method requires the use of AC test current. Auxiliary grounds, where used, shall be located sufficiently far away from the test rod or system and from each other so that the regions in which their resistance is localized do not overlap. Ground resistance measurements shall be made only in normally dry weather and not less than 48 hours after rainfall.

      b. Test Records: The ground resistance measurements required above shall be taken and certified by the Contractor to the Owner. The Contractor shall submit in writing to the Owner upon completion of the contract, the measured ground resistance of each ground rod and each basic grounding system, including all supplementary electrodes required, and the soil conditions at the time measurements were taken. The Contractor shall also furnish complete data on the methods of testing and the instruments used in the tests.

3.8 Records

   A. All test data and results shall be recorded.

   B. All test reports shall be prepared by the Contractor, signed by the authorized witnesses and approved by the A/E.

   C. A minimum of 2 bound copies of the approved test reports shall be furnished to the Owner, unless otherwise specified.

END OF SECTION

SECTION 16050

Basic Materials and Methods

Part 1. General

1.1 The specifications covering this work are open. Reference in the specifications to any article, device, product, material, fixture, form, or type of construction by name, make, or catalog number shall be interpreted as establishing a standard of quality and shall not be construed as limiting competition. The Contractor may use any article, device, product, material, fixture, form, or type of construction that, in the judgment of the A/E, expressed in writing, is equal to that specified.

1.2 Electrical drawings are diagrammatic and shall not be scaled for exact sizes or locations.

1.3 Conduit routed through the building that interferes with other equipment and construction shall not constitute a reason for an extra charge. Equipment, conduit, and fixtures shall fit into available spaces in the building; do not introduce these into the building at such times or in such a manner as to cause damage to the structure. Equipment that requires servicing shall be readily accessible.

1.4 Keep cutting and patching to an absolute minimum. Insofar as possible, determine in advance the proper chases and openings for the work.

1.5 Where cutting and patching are required due to an error of the Contractor, or where the Contractor has not been given enough advance notice of the need for holes, recesses, and chases, patching shall be performed by those trades skilled in the use of the materials involved and shall be done at the Contractor's expense.

1.6 The approximate location of fixtures, receptacles, and wall switches is indicated on the drawings. Exact locations shall be determined by the A/E as building work progresses. The indicated locations of outlets may be changed by 10' in any direction without additional cost before the outlets are installed.

1.7 Maintain one set of electrical white prints on the job site, marked to show as-built conditions and installations. At job completion, give this set of drawings to the Owner's representative and obtain written acknowledgement of receipt.

Part 2. Products

2.1 The materials, appliances, fixtures, and equipment shall be new and of the quality specified and shall bear the UL label when such labels are available.

2.2 Metal parts of conduit, boxes, fittings, enclosures, hangers, straps, screws, etc., shall be made of corrosion-resistant materials or protected by corrosion-resistant materials.

2.3 Identify the electrical equipment listed below with nameplates that correspond to the markings shown on the drawings. Equipment shall be provided with nameplates made of black bakelite engraved with white letters 3/8-inch high and core-screw attached to the equipment:

    A. Panelboards

    B. Electrical Disconnects

    C. Motor Control Centers

D. Switchgear

E. Instrumentation Components

2.4 WIRE MARKERS: Markers for wire number identification at all terminal points shall be white, polyvinyl chloride tubular type with black imprinting. Markers shall be designed to slip on the conductors and shall be equipped with integral wire grippers. Markers shall be Thomas and Betts Company, EZ Code Type SM, or equal.

2.5 CONDUIT MARKERS: All conduits at both ends shall be provided with metal tag identifying conduit numbers as assigned by Contractor.

Part 3. Execution

3.1 Provide the bracing, shoring, rails, guards, and covers necessary to prevent damage or injury. Do not leave energized electrical items unnecessarily exposed or unprotected. Protect personnel from exposure to contact with electricity. Protect work and materials from damage by weather and the entrance of water or dirt. Cap conduit during installation. Avoid damage to materials and equipment in place. Satisfactorily repair or remove and replace damaged work and materials. Deliver equipment and materials to the job site in their original, unopened, labeled containers. Store ferrous materials so as to prevent rusting. Store finished materials and equipment so as to prevent staining and discoloring.

3.2 All sheathing, shoring, and cleaning necessary to keep trenches and their grades in proper condition for the work to be carried on, including the removal of water by mechanical means, shall be the Contractor's responsibility. Excavate trenches 3 inches below the elevation of the bottom of conduit. Then fill the trench to the proper elevation, and tamp in layers 6 inches deep until firm and even. During backfilling, the final layer of fill shall be topsoil. Backfill carefully, and restore the surface to its original condition. The backfill used may be earth up to the final topsoil layer, but in no case shall it contain large rocks, tree roots, trash, or debris. Omit topsoil under paved areas.

3.3 Arrange for the temporary electrical service necessary for the entire project during construction. Provide a minimum of one 100 W lamp and one duplex receptacle for each 500 square feet of floor area. Arrange for permanent electrical service and for orderly transfer between temporary and permanent electrical services.

3.4 Make electrical connections to transformers, mechanical equipment, and controls with approximately 2 feet of flexible conduit. Determine the requirements from the drawings, these specifications, and the approved manufacturer drawings. Where conflicts occur between the bidding instruments and the manufacturer's recommendations, request a ruling from the A/E.

3.5 Provide inserts, hangers, supports, braces, and anchor bolts as necessary for all work called for under this division.

3.6 Install conduit and outlets for telephones as shown on the drawings. Coordinate the work with the telephone company.

3.7 Megger test the feeder circuits and major equipment both phase-to-phase and phase-to-ground at 500 V DC before energizing. Insulation resistance of less than 1 Megohm is not acceptable.

3.8 Perform testing in the presence of the A/E and officials of authorities having jurisdiction. Do not cover or conceal work until testing is completed and the installation has been approved. Furnish the instruments, devices, and equipment necessary for testing. Correct defects discovered during testing.

3.9 Turn equipment over to the Owner in lubricated condition. Include instructions on further lubrication in the operating instructions.

3.10 At the termination of work under this division, furnish the Owner with 3 complete bound sets of operating instructions on equipment furnished under this division.

3.11 As work progresses and once it is completed, remove from the premises all dirt, debris, rubbish, and waste materials resulting from this work.

3.12 Remove all tools, scaffolding, and surplus materials after completion and acceptance of the work.

3.13 Turn the entire job over to the Owner in a complete and satisfactory condition.

3.14 The Contractor shall furnish a written guarantee of his or her work and materials for a period of one year from the date of final payment for the work, and shall repair and make good any defects occurring during this period due to defective materials or workmanship.

3.15 In the event that the project is occupied or systems placed in operation in several phases at the request of the Owner, then the guarantee of each system or piece of equipment so used shall begin on the date each system or piece of equipment was placed in satisfactory operation and accepted as such by the Owner.

3.16 Circuit Identification to be done in accordance with the following guidelines and the as-built drawings marked to acknowledge incorporation by Contractor.

    A. Wire and cable for lighting circuits shall be color coded as follows:

       1. 480/208 Volts Line 1—Black
          Line 2—Red
          Line 3—Blue
          Neutral—White

    277/120 Volts Line (Hot)—Black
          Neutral—White

    B. Wire for fixture drops shall be coded as indicated above.

    C. All control conduit and motor branch circuit wiring shall be color coded in accordance with IPCEA standards.

    D. All feeders entering or leaving distribution equipment, junction and pull boxes, etc., shall have conductors tagged as to phase identification, i.e., "A", "B", and "C," and with circuit designation, including conduit number.

    E. All motor leads, power branch, and control circuit conductors and conduits shall be tagged with identification numbers, as designated by the Contractor at each termination such as motor control centers, starters, operator's stations, control panels, etc.

    F. All control wiring of voltages different from that of power wiring shall be run in separate conduits.

    G. All control panels shall be provided with necessary relays, starters, and controls to provide a complete operating system.

END OF SECTION

SECTION 16110

Raceways

Part 1. General

1.1 Unless otherwise noted on the drawings, install all wiring in conduit.

1.2 Use PVC-coated conduit and fittings in highly corrosive areas noted on the drawings.

Part 2. Products

2.1 Conduit shall be rigid, hot-dip galvanized steel. Rigid aluminum conduit may be used except where run underground or in concrete.

2.2 Install rigid conduit in or under concrete slabs on grade and all exposed areas to 7' 0" above the finished floor.

2.3 Use electrical metallic tubing where the drawings call for conduit to be concealed in walls, run exposed, 7' or higher above the floor, or installed above suspended ceiling. Connectors and couplings shall be of the steel compression type.

2.4 Use Schedule 40 PVC heavy wall conduit unless otherwise noted for underground circuits beginning 5' outside of the building or other structure. Make transitions between rigid galvanized steel conduit and PVC conduit with the manufacturer's standard adaptors designed for this purpose. All joints shall be solvent welded in accordance with the manufacturer's recommendations. Conduit fittings, elbows, and cement shall be produced by the same manufacturer.

2.5 Where conduit enters or leaves cabinets and boxes, use standard locknuts on the outside of the box and a locknut and bushing on the inside. Use OZ insulated bushings, Type B, OZ Steel City No. B1-900, or matching insulating grounding bushings on conduit inside boxes instead of standard pipe bushings.

Part 3. Execution

3.1 All conduits within building shall be routed overhead with proper supports unless shown specifically under slab.

3.2 Install branch circuit and feeder wiring in conduit unless otherwise noted on the drawings. Comply with the requirements of NEC and local authorities having jurisdiction, including requirements concerning grounding and supporting arrangements.

3.3 Install conduit to avoid trapping moisture and with as few bends as practicable. Running thread couplings are not permitted.

3.4 Use expansion fittings, properly bonded, to ensure ground continuity across expansion joints in floors and ceilings. Use double lock nuts and bushings on panel feeders at panel tubs.

3.5 For bends made in the field, use a radius of not less than that allowed in Section 346-10 of NEC. Keep bends free from dents and flattening. Use no more than the equivalent of 4 bends at 90 degrees between any 2 outlets, counting bends at outlets. Do not heat metal conduit.

3.6 Conduit shall be continuous from outlet or cabinet, with no wires spliced in conduit, and shall be secured to the building structure. Stuff boxes and cork fittings to prevent the entrance of water during concrete pouring and at other times during construction prior to completion of conduit installations.

3.7 Support conduit vertically and horizontally in accordance with NEC Article 346-12. Do not exceed these intervals:

$^3/_4$''	5'
1'' through 1-$^1/_2$''	7'
2'' and larger	10'

3.8 Install grounding bushings on all conduit that enters or leaves the main panelboard in the building.

3.9 Where conduit is installed in groups on common supports, angles, or channels, secure each conduit to each support with U or J bolts. Unistrut fittings are acceptable. All hardware shall be galvanized.

3.10 Unless otherwise noted on the drawings, use only rigid galvanized conduit or intermediate metal conduit when conduit is to be run underground, exposed to severe mechanical damage, or used for panel feeders.

3.11 Where galvanized rigid steel conduit is placed directly in the ground or gravel, thoroughly coat the outer surface with asphaltum or PVC plastic before installation. Underground conduit shall be at least 24" below grade.

3.12 In concrete slabs, block conduit up from forms, and securely fasten in place. All conduit in slabs shall have a minimum of 1-$^1/_2$" concrete coverage where concrete is exposed to earth or weather, and $^3/_4$" where not exposed to earth or weather.

3.13 Seal conduit that passes through floor slabs (except ground floor) with concrete grout. The grout shall be 3" on sides of conduit at the bottom surface of the slab and shall taper in conical form to the conduit surface at a point 4" below the bottom of the slab. Seal around conduit or other wiring materials passing through partitions that extend to the underside of the slab above and around those passing through fire-rated walls. Use grout or drywall cement to prevent the passage of smoke or fire.

3.14 Underground conduit lines shall be spaced a minimum of 3" apart and shall have a minimum earth cover of 24".

3.15 Install a plastic, detectable, magnetic, 2" wide tape 8" below grade above all underground conduit. Tape shall be printed continuously with "Electric Line," or equal.

3.16 Run exposed conduit parallel to, or at right angles with, the lines of the structure.When exposed, make right-angle bends with standard conduit ells by bending conduit as specified or by using screw-jointed conduit fittings.

3.17 Exercise particular care in cutting conduit to the proper length in order to ensure that the ends fit exactly in outlet boxes, couplings, and cabinets. All threads shall be clean cut and no longer than necessary. Ream and file conduit ends to remove burrs and fins. Make up threaded joints with aluminum paint applied to the male threads only.

3.18 Support exposed conduit work with hot-dip galvanized steel clamps, straps, or pipe hangers.

3.19 At each motor, transformer, mechanical equipment, water heater, etc., use a short section (approximately 2' to 6') of watertight, PVC-jacketed, galvanized flexible conduit of proper size for connection to the building wiring system. This connection shall be arranged to prevent vibration from being transferred to the structure and to allow for expansion and contraction of the equipment and structure.

3.20 For pulling, use soapstone, Yellow 77, Yer-Eas, or equal.

3.21 Leave No. 12 Type TW or THWN copper pull wire in each conduit when permanent wiring is not installed.

3.22 Follow the manufacturer's recommendations regarding the handling, bending, coupling, and installation of PVC conduit.

END OF SECTION

SECTION 16113

Plug-In Bus Duct

Part 1. General

1.1 Furnish plug-in bus of the capacities and arrangement shown on the drawings.

1.2 Plug-in fusible switch units, complete with fuses, where indicated.

1.3 The plug-in busway system shall consist of straight sections and fittings arranged and rated as shown on the drawings. Establish exact plant arrangement and elevations by site measurements.

1.4 Busway and accessories shall be the products of the same manufacturer. The entire system shall be UL listed.

1.5 Use ITE, Square D, Cutler-Hammer, or A/E-approved equal.

1.6 Busways shall be marked with the voltage and current rating and the manufacturer's name or trademark to be visible.

Part 2. Products

2.1 Plug-in bus shall be 480 V, 3 phase, 3 wire, and 208 V, 3 phase, 4 wire, totally enclosed, low-impedance type. Bus ducts shall include a ground bus in addition to phase and neutral busses.

2.2 Bus bar shall be electrolytically tin-plated aluminum, insulated over entire length. Bus shall carry full rated load when mounted in any position without derating. Temperature rise shall not exceed 55°C above ambient at any point. Make bolted connections with Belleville washers.

2.3 Plug-in bus short-circuit rating shall be 22,000 A, RMS, symmetrical.

2.4 The entire system of plug-in bus shall be UL listed.

2.5 Provide hangers every 5'-0", with every other hanger also serving as a sway brace.

2.6 Plug-in units shall be of the fusible switch type suitable for operation at 480 V or 208 V as required, and of the ampere ratings required per Owner-furnished equipment rating being served. Plug-in units shall be equipped with safety interlocks and insulated stabs. Units shall have an external handle or operating mechanism for operation from the floor with hookstick.

Part 3. Execution

3.1 Make field measurements for preparation of shop drawings. Obtain the A/E's approval of both field measurements and shop drawings before fabrication.

3.2 Connectors and fittings for connections to feeder bus and terminations of plugin bus furnished by the Owner.

END OF SECTION

SECTION 16114

Cable Tray

Part 1. General

1.1 Cable tray system shall consist of straight sections and directional change fittings of the nominal widths shown on the drawings. Exact plan arrangements and elevations shall be established by site measurements by the Contractor and approved by the A/E.

1.2 Trays and accessories shall be the products of the same manufacturer.

1.3 Use Globe, B-Line, Husky, or Mono Systems Inc.

Part 2. Products

2.1 Ladder cable trays and ventilated channels shall be heavy duty, 6-inch nominal depth, and cross rungs attached to a center spine at 3-inch intervals.

2.2 Ladder rungs shall be extra heavy duty, $^5/_8$ inch 1-$^1/_8$ inches, 14-gauge minimum.

2.3 Fittings and accessories shall be of the same material and equivalent strength as the cable tray system.

2.4 All trays, fittings, and accessories shall be aluminum and supported at 12-foot intervals, by $^1/_2$-inch threaded rods.

Part 3. Execution

3.1 Make field measurements for the preparation of shop drawings. Obtain the A/E's approval for the use of vertical and/or horizontal splices, the location of expansion joints, field measurements, and shop drawings before fabrication.

3.2 Provide and install hangers, tray box connectors, dropout plates, cable grip supports, clamps, and fittings as necessary to protect wire insulation and maintain cable bend radii.

3.3 Provide and install all straight sections, bends, risers, fittings, vented covers, splice plates, and installation accessories as necessary to comply with NEC Article 318 and as indicated on the drawings. Obtain a ruling from the A/E if there are conflicts in requirements.

END OF SECTION

SECTION 16120

Conductors

Part 1. General

1.1 Use only copper conductors.

1.2 No conductor for branch circuit wiring shall be smaller than No. 12.

Part 2. Products

2.1 Use solid, 600 V, Type THWN wire for No. 10 AWG and smaller branch circuit wiring, color coded in accordance with NEC 210-5 and as follows: Phase A, black; Phase B, red; and Phase C, blue. Size conduit for THW.

2.2 Use stranded 600 V, Type XHHW with cross-linked polyethylene insulation for No. 8 AWG and larger wire unless otherwise noted on the drawings.

2.3 Use stranded Type AF, No. 14 AWG fixture wire at fixture outlets unless otherwise noted on the drawings. The wire in fluorescent fixtures shall be at least 90°C, stranded, 600 V, and may be thermoplastic.

2.4 Cable 3/c No. 18 shielded shall be General Electric No. S1-58196, Belden 8791, or equal.

2.5 Cable 2/c No. 18 shielded shall be General Electric No. S1-58196, Belden 8760, or equal.

2.6 Use Scotchlok or Buchanan connectors for No. 14 AWG through No. 8 AWG conductors. Use Burndy or T&B compression connectors with crimpit cover, Type CC, for No. 6 AWG through 600 MCM conductors. Use Burndy or T&B compression spade lugs for bus and stud connections on No. 6 AWG through 600 MCM conductors. Use Scotchlok or Buchanan connectors at fixture outlets.

2.7 Use TC-type cable for conductors in cable trays.

2.8 Cables run in air plenums shall be rated for application.

Part 3. Execution

3.1 Complete conduit system before pulling any wire or cable.

3.2 Conductors shall be continuous from outlet to outlet and to branch circuit overcurrent devices. Make splices only in junction boxes. Splices shall not be made in panelboards or underground.

3.3 Make 600 V splices and connections in accessible boxes and cabinets with pressure terminals and connectors. Tape connectors to an insulation value at least 50° greater than the conductor insulation with Scotch No. 70 (rubber) tape covered with Scotch No. 88 (all-weather) tape.

3.4 Unless otherwise noted on the drawings, all branch circuit conductors shall be No. 12 AWG. Any branch circuit run over 50 feet 0 inches in length, measuring one way from the first outlet of the circuit to the panel, shall be No. 10 AWG to the first outlet.

3.5 Where outlets only are indicated, leave 48 inches leads of conductors for connection of equipment. All conductors shall be identified with Brady tape at terminals and junction boxes indicating circuit numbers.

END OF SECTION

SECTION 16134

Outlet Boxes and Fittings

Part 1. General

1.1 All ceiling and wall outlet boxes for lights, switches, wall receptacles, etc. shall be standard galvanized outlet boxes with $^3/_8$" fixture studs where needed. Junction boxes shall have blank metal covers.

Part 2. Products

2.1 Boxes shall be as manufactured by Steel City Electric Company, Appleton Electric Company, Raco, Bowers, or equal.

2.2 Pull boxes shall be a minimum of 16-gauge galvanized steel painted with a prime coat, and provided with a screw cover. Size each pull box according to the intended use, meeting NEC requirements.

2.3 For weatherproof, exposed, or surface-mounted work, use suitable FS or FD cast boxes and covers unless otherwise noted on the drawings.

2.4 All conduit fittings shall be by Crouse-Hinds, Appleton, Thomas & Betts, or Russell & Stoll and have cadmium plating or galvanized finish. All outdoor fittings and boxes shall be gasketed and watertight. Fittings and boxes in hazardous areas shall also be suitable for damp locations.

2.5 Select boxes suitable for the intended use and type of outlet. For ceiling- or wall-mounted lighting fixtures, use boxes with a minimum diameter of 4" and a minimum depth of $1-^1/_2$". Provide plaster or masonry rings for flush mounted outlets to conceal the joint between the box and wall finish materials.

2.6 For flush switch and receptacle outlets, provide boxes 4" square.

2.7 Size each box according to the number of conductors in the box or the type of service to be provided. The minimum size of each junction box shall be 4" square and $2-^1/_8$" deep. Provide screw covers for junction boxes.

2.8 Outlet boxes for telephones, television, or computer shall be 4" square boxes with plaster rings for plastered walls or square cut device rings for block walls. Do not use single-gang device boxes for telephone outlets.

Part 3. Execution

3.1 Furnish and install an outlet box for each wiring device and fixture. For concealed boxes, use galvanized steel. For exposed boxes, use cast iron conduit fittings that are similar to Condulets and Unilets, with threaded hubs.

3.2 Provide a pull box every 100' of conduit run and wherever an excessive number of bends necessitates a pull box for ease of wire installation. The maximum number of cumulative bends shall not exceed 360° in any single run.

3.3 Securely anchor outlet boxes, and set them level and plumb with no part of the box extending beyond the finished wall or ceiling.

3.4 All outlet boxes left for future use shall have single-gang or 2-gang blank device plates installed to match device plates being used on the project.

3.5 Use Condulets with proper configurations for changes of direction of exposed conduit. A conduit ell may be used if it does not cause interference or damage or mar the appearance of the installation.

3.6 Plug open knockouts or holes in boxes with suitable blanking devices.

END OF SECTION

SECTION 16140

Switches and Disconnects

Part 1. General

1.1 Wiring devices shall be by Bryant or Hubbell, as indicated on the drawings, or A/E-approved equal. They shall bear the UL label.

1.2 Coverplates for recessed switches shall be stainless steel. Coverplates for surface-mounted switches shall be cast metal when installed in cast boxes and sheet steel when installed in sheet steel boxes.

Part 2. Products

2.1 20A SP TOGGLE SWITCHES: Hubbell No. 1221-GRY, Bryant No. 4901-GRY; 3-way, 4-way, or 2-pole switches to be the same series.

2.2 20A SP LOCK TYPE SWITCH: Hubbell No. 1221-L, Bryant No. 4901-L, 3-way, 4-way, or 2-pole switches to be the same series.

2.3 MANUAL MOTOR STARTERS: 120 V, single pole with overload protection; Square D Class 2510 or Westinghouse equivalent, with enclosure or stainless steel plate as required by the drawings

2.4 SAFETY SWITCHES: heavy duty, HP rated, quick-make/quick-break, with arc shields, enclosed construction, and cover interlock; fusible or nonfusible as indicated on the drawings. Swit-ches shall be rated for either 250 V AC or 600 V AC service as necessary. Switches shall be capable of interrupting the locked rotor current of each motor for which they are to be used. Enclosures shall be NEMA 1 for the interior, and NEMA 4 stainless steel or 4X for exterior loca-tions, with provisions for watertight pushbutton stations.

2.5 MANUAL DISCONNECT: 20 A, 480 V, 3 phase, without overload protection; Square D No. 2510-KG-2, or approved equal. Mount on each 3 phase unit heater.

2.6 START/STOP/GREEN PILOT PUSHBUTTON STATIONS: surface mounted, NEMA 1, heavy duty, oiltight, 120 V, as manufactured by Square D, Westinghouse, or General Electric.

2.7 WALL PLATES, INDOORS: stainless steel, Type 302 Alloy 18-8, satin finish. Use galvanized pressed steel in unfinished areas.

Part 3. Execution

3.1 Provide and install switches with coverplates as noted on the drawings.

3.2 Provide and install safety switches with the number of poles and fuses noted on the drawings.

3.3 Use Type R dual element fuses and fuse rejection kit in any safety switch serv-ing a motor circuit. Use nonfusible disconnects at remote motor locations.

3.4 Install a manual disconnect switch for each unit heater.

3.5 Mount all switches 4 feet 0 inches above the finished floor unless otherwise noted on the drawings. Where switches are mounted in gang boxes and the voltage exceeds 120 V to ground, install partitions between switches.

3.6 Install all plates in full contact with the wall surface; do not allow them to project out from the wall.

END OF SECTION

SECTION 16141

Receptacles

Part 1. General

1.1 This specification covers convenience and power outlets.

Part 2. Products

2.1 Wiring devices shall be by Hubbell, Bryant, or A/E-approved equal, and shall bear the UL label.

2.2 20 A, 125 V DUPLEX GROUNDING RECEPTACLE: Hubbell No. 5362 GRY or Bryant No. 5362 GRY.

2.3 20 A, 125 V DUPLEX WEATHERPROOF GROUNDING RECEPTACLE: Hubbell No. 5362 Ivory with No. 5205 or No. 5206, Bryant No. 5362 Ivory, Crouse-Hinds equivalent (FS) double flap cover No. WLRD-1.

2.4 20 A, 125 V DUPLEX GROUND FAULT RECEPTACLE: Leviton No. 6395 or Bryant No. GFR53FT-GRY. Use single horizontal flap cover for weatherproof locations.

2.5 20 A, 125 V SINGLE GROUNDING RECEPTACLE: Hubbell No. 5361-GRY or Bryant No. 5361-GRY.

2.6 20 A, 250 V SINGLE GROUNDING RECEPTACLE: Hubbell No. 5461-I or Bryant 54-61-I.

2.7 30 A, 208 V, 2 POLE, 3-WIRE GROUNDING RECEPTACLE: Bryant No. 9630-FR or Hubbell No. 9330.

2.8 WALL PLATES, INDOORS: Use same type plates for receptacles and telephone outlets as specified for switches.

2.9 EXPLOSION-PROOF RECEPTACLE: 20 A, 125 V Duplex Grounding type, Appleton Co. No. EFS110-2023M.

2.10 ISOLATED GROUND RECEPTACLE: 20 A, 125 V Duplex Grounding type, Hubbell Co. No. IG 5262 or equal.

Part 3. Execution

3.1 Provide and install receptacles as noted on the drawings.

3.2 Mount all outlets located in manufacturing areas 3 feet 6 inches above the finished floor unless otherwise noted on the drawings. Mount all outlets located in lean-to's 1 feet 6 inches above the finished floor unless otherwise noted on the drawings. Where outlets are mounted in gang boxes and the voltage exceeds 120 V to ground, install partitions between outlets. All outlets shall be of the grounding type.

3.3 Provide and install one 20 A, 125 V isolated grounding duplex receptacle in each floor box receptacle.

END OF SECTION

SECTION 16150

Motors

Part 1. General

1.1 Motors furnished under this and other sections of these specifications shall be of sufficient size for the duty to be performed and shall not exceed the motor's full rated load when the driven equipment is operating at specified capacity under the most severe conditions likely to be encountered. The horsepower ratings indicated on the electrical drawings are for guidance only and do not limit the equipment size. Insulation shall be Class F with Class B rise and moisture, fungus, and oil resistant treatment and shall be of a type designed and constructed to withstand the severe moisture conditions and the wide range of ambient temperature to which the motors will be subjected. Unless otherwise specified, all motors shall have open dripproof frames and shall be rated for continuous full load operation without exceeding the standard temperature rise permitted for the frame construction and class of insulation used.

1.2 For standby generator loading purposes, limit motors to NEMA KVA Codes D, E, or F.

Part 2. Products

2.1 High-efficiency motors shall be in accordance with NEMA MG-1-12.53b and IEEE 112A, Test Method B, with guaranteed minimum efficiency of 88.5%, an across-the-line minimum power factor of 85% (for synchronous speeds of 1,800 and 3,600 rpm), and a service factor of 1.15.

Part 3. Execution

3.1 When electrically driven equipment furnished under other sections of these specifications materially differs from the contemplated design, make the necessary adjustments to the wiring, disconnect devices, running, and branch circuit protection to accommodate the equipment actually installed.

END OF SECTION

SECTION 16155

Motor Starters

Part 1. General

1.1 Drawings and general provisions of Contract, including General and Supplementary Conditions and Division 1 sections, apply to work of this section.

1.2 Extent of motor starter work is indicated by drawings and schedules.

1.3 Types of combination motor starters in this section include the following:
    A. Full Voltage Nonreversing (FVNR)
    B. Reduced Voltage Nonreversing (RVNR)
    C. Variable Frequency Drive (VFD)

1.4 Firms regularly engaged in manufacture of motor starters, of types, ratings, and characteristics required, whose products have been in satisfactory use in similar service for not less than 5 years.

1.5 Qualified with at least 3 years of successful installation experience on projects with electrical work similar to that required for this project.

1.6 Comply with NEC as applicable to wiring methods, construction, and installation of motor starters.

1.7 Comply with applicable portions of NEMA standards pertaining to motor controllers/starters and enclosures.

1.8 Comply with applicable requirements of UL 508, "Electric Industrial Control Equipment," pertaining to electrical motor starters. Provide units that have been UL listed and labeled.

1.9 Product Data: Submit manufacturer's data on motor starters.

1.10 For types and ratings required, furnish one additional set of fuses for every installed fused unit.

Part 2. Products

2.1 Subject to compliance with requirements, provide products of one of the following (for each type and rating of motor starter):
    A. Allen Bradley Co.
    B. Cutler Hammer Products, Eaton Corp.
    C. General Electric Co.
    D. Gould, Inc.
    E. Square D Co.
    F. Westinghouse Corp.

2.2 Except as otherwise indicated, provide motor starters and ancillary components that comply with manufacturer's standard materials, design, and construction in accordance with published product information, and as required for complete installation. Where more than one type of equipment meets indicated requirements, selection is installer's option.

2.3 Provide full-voltage alternating-current combination starters, consisting of starters and disconnect switches mounted in common enclosures; of types, sizes, ratings, and NEMA sizes indicated. Equip starters with block type manual reset overload relays and with fusible disconnect switch mechanism providing indication and control as switch position, and capable of being locked in an OFF position. Construct and mount starters and disconnect switches in single NEMA Type 12 enclosure indoors and NEMA 3R outdoors.

2.4 Provide reduced voltage autotransformer starters for motors above 30 HP, consisting of starters, contactors, transformers, and disconnect switches mounted in common enclosures; of type, sizes, ratings, and NEMA sizes indicated. Equip starters with blocktype manual reset overload relays and with fusible disconnect switch mechanism providing indication and control as switch position and capable of being locked in an OFF position. Construct and mount starters and disconnect switches in single NEMA Type 12 enclosure indoors and NEMA 3R outdoors.

2.5 The variable frequency AC drives shall operate directly from 480-volt, 3 phase, 60-Hz plant power and produce an adjustable frequency, adjustable voltage, 3-phase output for control of a variable torque pump and fan loads. Units to be provided with as a minimum standard of ground fault protection, instantaneous overcurrent trip, overtemperature shutdown, frequency (speed) stability of 1% of set/8 hours, speed regulation of 3% of no load to full load, and up to 7.5% slip compensation. Speed adjustment shall be via a potentiometer input or a 4-20 ma DC input. Provide unit with a percent speed analog type indicating meter. Variable frequency drives shall be derated and/or provided with forced air ventilation, stripheaters, etc., as required to meet the range of external environmental conditions that may be encountered by its physical location. Provide NEMA Type 12 enclosures indoors and NEMA 3R outdoors.

   A. Provide isolation transformers, filters, capacitors, etc., as required by manufacturer to limit harmonics, transients, and noise feedback by VFDs into plant's AC system.

   B. Provide surge arrestors on VFDs as required by manufacturer for protection against transients and lighting damage.

2.6 Provide the control accessories scheduled on the drawings.

2.7 Provide four auxiliary contacts mounted on each starter.

Part 3. Execution

3.1 Install motor starters as indicated, in accordance with manufacturer's written instructions, applicable requirements of NEC, NEMA Standards, and NECA's "Standard of Installation," and in compliance with recognized industry practices to ensure that products fulfill requirements.

3.2 Coordinate with other work including motor and electrical wiring/cabling work, as necessary to interface installation of motor starters with other work.

3.3 Install dual-element fuses in fusible disconnects.

3.4 Inspect operating mechanisms for malfunctioning and, where necessary, adjust units for free mechanical movement.

3.5 Touch-up scratched or marred surfaces to match original finish.

3.6 Subsequent to wire/cable hook-up, energize motor starters and demonstrate functioning of equipment in accordance with requirements, and under full loads. Where necessary, correct malfunctioning units.

3.7 Each motor $^1/_2$ HP and above, unless otherwise noted, is provided with a combination fused switch and magnetic motor starter, 3-pole, complete with overload devices and a 120-volt coil. Control of starters shall be indicated on the drawings.

3.8 All motor starters shall be housed on NEMA Type 12 enclosures indoors, and NEMA 3R outdoors. Include other NEMA Standards where applicable.

3.9 All motor starters shall be mounted and connected as shown on the drawings.

3.10 Overload relay heaters of the proper size shall be supplied and installed by the Contractor to provide protection for the motors.

3.11 Each motor starter and disconnect switch shall be identified by name and number of drive. Use stencil or engraved legend plate. Do not use embossed self-adhesive labels.

3.12 All starters except those designed for floor mounting shall be mounted on starter racks made of channel or angle steel. The channels will be mounted horizontally and fastened to the wall with iron expansion bolts. The racks shall be fastened to the floor or wall and a minimum of 1 inch from any wall to provide permanently firm support for starters.

3.13 All "wall mounted" starters shall be offset about 3 inches to 5 inches from the wall and bolted to a starter rack. The rack will be made of 1-$^1/_4$-inch $^1/_4$-inch flat steel welded to 1-inch steel pipe with floor flange mounts.

3.14 All pushbutton stations, including lock-out stops, not located on control panels shall be installed and wired by the Contractor.

END OF SECTION

SECTION 16160

Panelboards

Part 1. General

1.1 All panelboards shall be of the dead front safety type, bear a UL-approved device label, and be sequence bussed for required service.

Part 2. Products

2.1 Power distribution panelboards shall be rated 120/208 V, 3 phase, 4 wire, or 277/480 V, 3-phase, 4-wire, as indicated on the drawings, and shall be Square-D I-Line, ITE CDP-6 or approved equal.

2.2 Power or distribution panelboards shall have a main circuit breaker and circuit breaker branch circuits, as shown on the drawings.

2.3 Standard-width lighting and receptacle panelboards shall be Square-D NQOD, ITE Co., CDP-7, or approved equal. Sequence bus for 120/208 V, 3-phase, four-wire, or 277/480 V, 3-phase, 4-wire as shown on the Drawings. Twin and tandem breakers will not be allowed.

2.4 Provide 4-handle locking devices for each lighting and receptacle panelboard.

2.5 Circuit breakers shall be of the indicated type, providing "on," "off," and "tripped" positions on the operating handle, and shall have an interrupting capacity as scheduled on the drawings. All multipole breakers shall be common trip type. All breakers used to switch lighting shall be rated SWD. Circuit-breaker ratings shall be at least 10,000 AIC for 120/208 V panelboards and 14,000 AIC for 277/480 V panelboards.

2.6 For circuit breaker panelboards, use bolt-in thermal-magnetic molded case circuit breakers of the frame and trip ratings shown on the drawings. Number breakers consecutively down the left row and continue similarly down the right row.

2.7 Gutter space shall be a minimum of 4″ or larger as required by code. Where noted on the Panelboard Schedule, panels shall have an extra wide gutter for feed-through feeders and taps.

2.8 Cabinets for panelboards shall be galvanized sheet steel of NEC thickness, properly reinforced. Cabinets shall be for surface or flush mounting as indicated on the drawings. The door and trim shall be steel (not galvanized) with adjustable trim clamps, semiflush hinges, and inside rabbet.

2.9 Furnish doors with a lock and two keys. Key all doors alike.

2.10 Provide space with provisions in panelboards for devices noted as "(space only)" on the drawings.

Part 3. Execution

3.1 Submit shop drawings on all panelboards and cabinets for the A/E's approval before fabrication.

3.2 Provide and install panelboards as noted on the drawings. Equip panelboards with the devices noted in the schedules.

3.3 The holes for entrance of conduit shall be knockouts or be drilled in the field. Burned holes will not be permitted.

3.4 Cards listing the locations of the circuits controlled shall be typewritten and inserted with a plastic cover into the directory frames on the door of each panelboard. List numbers to coincide with the position on the panel.

3.5 Unless otherwise indicated on the drawings, mount protective devices with top of cabinet or enclosure 6'-6" above the finished floor, properly align, and adequately support independently of the connecting raceways. Furnish and install all steel shapes, etc., necessary for the support of equipment where the building structure is not suitable for mounting the equipment directly thereon.

END OF SECTION

SECTION 16165

Automatic Transfer Switches

Part 1. General

1.1 The automatic transfer switch shall be installed in a NEMA 1 enclosure, as indicated on the drawings.

1.2 A test report shall be submitted from an independent testing laboratory verifying that identical switches have met the requirements of UL Standard 1008. The temperature test shall be performed with the switch in a nonventilated enclosure after completion of the short circuit, overload, and endurance tests. The dielectric withstand test shall be repeated after the tests have been completed. During withstand and closing tests, there shall be no welding of contacts and no separation of the contacts, as verified by oscillograph traces.

1.3 The automatic transfer switches shall be the product of Onan Co., Kohler Co., Automatic Switch Company, Russelectric, Inc., or equal.

Part 2. Products

2.1 The automatic transfer switches shall be of the air break, double throw, rated load interrupter type, electrically operated but mechanically held in both the normal and standby positions. The switch operators shall be single solenoid or motor operated devices momentarily energized by the source to which the load is transferred. The transfer time in either direction shall not exceed $1/2$ second.

2.2 The main contacts and current carrying parts of the switches shall be insulated for 600 V. The voltage rating, current rating, and number of poles shall be as indicated on the drawings and shall have 24-hour continuous rating for the switch in a nonventilated enclosure for all classes of loads, including resistance, tungsten lamp, ballast, and inductive. Temperature rise shall conform to NEMA standards.

2.3 The main contacts, normal and standby, shall be surfaced with a silver alloy and protected by arcing contacts and magnetic blowcuts for each pole. The thermal capacity of the main contacts shall be no less than 20 times the continuous duty. In either position, the normal and standby main contacts shall be mechanically locked in position by the operating linkage. Auxiliary contacts shall be attached to and actuated by the same shaft as the main contacts.

2.4 The failure of any device or disarrangement of any part shall not result in a neutral position where both main and standby contacts are open or closed.

2.5 The transfer switches shall be furnished without integral overcurrent or shortcircuit protection.

2.6 The main and auxiliary contacts, operators, coils, springs, and control elements shall be removable from the front without removing the switch from the enclosure or disconnecting the main power cables.

2.7 The switches shall be provided with plastic encapsulated instructions and operating procedures, complete with schematic wiring diagrams, permanently attached in a conspicuous place on the automatic transfer switch enclosures.

2.8 The switch operation shall be controlled by voltage-sensing relays in each phase of the normal power supply, and in 2 phases plus a voltage-frequency sensing relay in the third phase of the standby power supply.

A. When the voltage of one or more phases of the normal source decreases to 70% of rated voltage, the load shall be transferred to the standby source. When the normal source has been restored to 90% of rated voltage on all phases, the load shall be retransferred to the normal source after an adjustable time delay of 5-25 minutes. If the standby source fails at any time while carrying the load, retransfer to the normal source shall be instantaneous upon restoration of the normal source on all phases.

B. Low voltage on the normal source shall close a normally open pilot contact after an adjustable time delay period of one to three seconds. The pilot contact shall actuate the starting system of the engine-generator set.

C. After retransfer to the normal source, the emergency generator plant shall run for five minutes unloaded and then automatically shut down and be ready to start when the next failure of the normal source occurs.

2.9 Time delay relays shall reset automatically. All solid state controls shall have surge protection.

2.10 The automatic transfer switches shall be provided with the following accessories:

A. A two-minute maximum adjustable time delay on transfer to standby (air diaphragm type)

B. An adjustable time delay on retransfer to normal with 5-minute unloaded running time of standby plant (motor driven type—minimum setting for retransfer 2 minutes, maximum 25 minutes, unloaded running time fixed at 5 minutes)

C. A test switch mounted on the enclosure door to simulate failure of the normal power source and to test operation of the transfer switch

D. An auxiliary contact to close when normal source is below 70% of rated voltage

E. An auxiliary contact to open when normal source is below 70% of rated voltage

F. A clear indicating light mounted on the enclosure door to indicate switch in normal position, including nameplate, fuse, and auxiliary contact

G. An amber indicating light mounted on the enclosure door to indicate switch in standby position, including nameplate, fuse, and auxiliary contact

H. An auxiliary contact closed when normal main contacts are closed

I. An auxiliary contact closed when standby main contacts are closed

J. An override switch mounted on the enclosure door to hold the transferred switch indefinitely connected to the standby power source regardless of the condition of the normal power source

K. Programmed transition transfer of contacts or a phase monitor relay shall be employed to prevent phase imbalance and flash over during switching.

L. Provide an exciser clock for automatic cycling of generator.

Part 3. Execution

Not used

END OF SECTION

SECTION 16170

Standby Generator

Part 1. General

1.1 This section defines the requirements for emergency standby engine/generator systems to be installed in accordance with the drawings and specifications. The systems shall provide for completely automatic unattended operation for the duration of any loss of normal utility power. The engine/generator system for the treatment plant shall be rated 70 KW continuous standby, 277/480 V, 3-phase, 4-wire, 60 Hz, 0.8 power factor at 1,800 rpm.

1.2 Each engine/generator shall be furnished with all necessary features and options to comprise a complete operable system when installed according to the manufacturer's recommendations. As a part of submittal data, provide complete composite electrical and mechanical drawings indicating control wiring, routing quantity and size, as well as sizes and suggested routing of coolant, fuel lines, and exhaust lines. If options or special equipment other than that specified below are required, they shall be included in the bill of material. In addition to complete installation drawings, provide a factory-trained engineer for periodic job site visits during installation to ensure that the system is being installed in accordance with the manufacturer's recommendations. After installation, a 4-hour load test shall be conducted by the manufacturer's engineer. Test data shall be recorded and become a part of the three Owner's manuals to be supplied. After the Owner's acceptance, the manufacturer shall conduct a minimum two-hour training session in operation and maintenance for the Owner's personnel. Parts and service shall be available on a 24-hour basis.

1.3 Acceptable manufacturers are Onan, Kohler, Caterpillar, Detroit Diesel, or Cummins Co.

Part 2. Products

2.1 Engine

   A. The engine shall be diesel-powered generator domestically manufactured, and capable of producing KW as specified under job site conditions. The engine shall be equipped with an electric starting, battery charging generator, electronic or mechanical governor, fuel filters, oil filters, air cleaners, blower system, and other equipment to provide a complete operable system.

   B. The engine shall be cooled by factory mounted radiator with engine-driven fan, jacket water pump, and fan guard. It shall be equipped with a thermostatically controlled water jacket heater to maintain a minimum water jacket temperature of 80° under ambient conditions. The complete cooling system shall be filled with antifreeze coolant (0°F).

   C. The exhaust silencers shall be installed on housing. These silencers shall be of the industrial type. Inlet and outlet shall be flanged with gaskets. Silencer arrangements shall be as necessary, and the engine/generator distributor shall provide the exhaust line required for acceptable back pressure.

2.2 Exhaust Line

   A. The exhaust line shall be of Schedule 40 steel pipe and shall be connected to engine manifold(s) through seamless, flexible, carbon steel bellows connection. Exhaust systems installed inside enclosures shall be fully insulated.

2.3 Starting Batteries

A. Provide starting batteries capable of starting the unit in the specified time under installed conditions and also capable of 60 seconds of engine cranking. Install batteries in insulated steel racks. Furnish battery cables.

B. Batteries shall be of the heavy duty, lead acid type Group 8, shipped dry with electrolyte installed at start-up. Furnish battery heater pads.

2.4 Automatic Battery Charger

A. Furnish and install an automatic battery charger in enclosure or in transfer switch. The charger shall operate on 115 V input and shall be AC line compensated. Input and output shall be fused. Charger shall provide continuous taper charging and provide float and equalize function. It shall be in a NEMA 1 cabinet and equipped with DC ammeter, DC voltmeter, high/float switch, and low-DC voltage alarm relay. Maximum continuous DC output shall be 10 A.

2.5 Fuel Tank

A. Furnish and install a base mounted 200-gallon fuel tank. Provide fuel line between fuel source and tank as well as between tank and engine. The tank shall be of heavy duty steel construction, self-supporting. The fuel transfer pump shall be electric motor driven, float controlled, and capable of delivering fuel from source at a rate three times the full load fuel consumption. Provide an interlocking fuel solenoid valve with fuel transfer pump and a check valve shall be provided on pump inlet to prevent loss of prime. Provide a manual hand pump large enough to supply adequate fuel under manual operation. In addition to standard fittings, the tank shall also include manual fill cap and weatherproof cover, fuel level gauge, pump test switch, and drain petcock.

2.6 Generator

A. The alternator shall be of the 4-pole revolving field, 10- or 12-lead, reconnectable, single bearing, brushless type. The entire insulation system shall be Class F or better, and the temperature rise shall be within NEMA MH-1-22.40 requirements for all nameplate voltages at full rated load. All load connections shall be made up in rear or side mount junction box. The generator construction shall allow load connections to be made on top, bottom, or either side of the junction box. The rotating rectifier shall employ three-phase sensing.

B. Voltage regulation shall be plus or minus 1% from no load to full load with steady state modulation of 0.5%. The voltage regulator shall be static type, and a voltage adjusting rheostat shall be furnished in the control panel.

C. Voltage change shall not exceed 15% upon application of the full rated load with recovery to steady state conditions within $2\text{-}1/2$ seconds. Frequency regulation by engine governor shall be with 1% steady state with a maximum 2% drop, no load to full load.

D. The generator shall be capable of supporting 300% rated current for 10 seconds.

2.7 Mounting

A. The complete engine/generator, base fuel tank, and all mounted accessories shall be assembled on a common channel steel base. Fuel oil lines and lube oil drain shall terminate in base. Install flexible fuel line sections (18 inches long) between the base and fuel lines. Install heavy duty, steel spring vibration isolators between the base and mounting pad. Size and locate these isolators as recommended by the engine/generator manufacturer.

2.8 Control Panel

A. The generator control panel shall be NEMA 1, dead front construction, and shall include the following: 2% AC ammeter, 2% AC voltmeter, meter phase selector switch, dial type frequency meter, elapsed time meter, panel illumination lamps, voltage adjusting rheostat, governor speed control switch, engine oil pressure gauge, engine water temperature gauge, battery charging ammeter, and engine emergency stop control. The engine shall be automatic start using a N/O contact that will close to signal engine start. Provide shutdown lamps for "high water temperature," "low oil pressure," "overcrank," and "overspeed," and provide auxiliary contacts for "low water temperature," "low level in fuel tank," "low voltage on battery charge," "Engine running," and "Generator on line."

B. Provide prealarm annunciator box in panel for low fuel level, low water temperature, low oil pressure, high water temperature, and low battery. Provide common alarm contact for remote indication and adjustable timer to provide up to 5 minutes unloaded running of engine after automatic transfer switch returns to normal.

C. Shock mount the generator control panel on the unit.

2.9 Equipment Mounting

A. Batteries, day tank, control gear, and water jacket heater shall be completely installed, wired, and piped prior to delivery to the job site. All devices requiring AC service (fuel tank pump, battery charger, water jacket heater, battery heaters, etc.) shall be connected to an AC service panel. This panel shall be equipped with a main breaker and branch breakers as required. AC wiring inside the enclosure shall be neatly routed and enclosed in the same type of conduit used in the building.

B. Submittal drawings for the engine/generator shall include a plan view of the enclosure indicating entrance areas for control, power, and service wiring conduit, as well as fuel line entrance locations.

2.10 Frame Mounted Fuel Storage Tank

A. Tank: 200-gallon steel construction with a polyurethane coating inside of tank.

B. Construct tank with tappings for installation of accessories.

C. Provide fill and vent lines.

D. Provide antisiphon supply connection to bottom of tank.

E. Provide low-fuel-level alarm.

F. Governing standards:

1. ASTM Standard Draft No. 5
2. UL File MH7991 for storage of flammable liquids
3. NFPA 30, "Flammable and Combustible Liquids Code," and NFPA 31, "Standard for Installation of Oil Burning Equipment"
4. EPA Regulations
5. ASME Code for unfired pressure vessels
6. Federal Register, Appendix E, Friday, April 17, 1987 Part II Environmental Protection Agency pp. 256–290, Vol. 52, No. 74.

G. Submittals:

1. Shop drawings shall include, but not be limited to, all critical dimensions, and show the locations and types of all fittings and accessories.
2. Catalog Data: Submit manufacturer's literature.
3. Submit the manufacturer's latest installation instructions.
4. Submit manufacturer's latest calibration charts.

H. Install a pickup line to within 8 inches of tank bottom, and equip with a check valve.

I. Provide fittings for supply line, return line, fill line, and vent line as well as at least two spare fittings with plugs. Install fill line with two lockable caps. Install tank according to the tank manufacturer's recommendations and as detailed herein.

Part 3. Execution

3.1 The engine/generator set shall be mounted on a common channel iron base rigid enough to maintain alignment of the generator set and minimize stress.

3.2 The channel iron base shall be mounted on the pad by means of heavy duty steel spring vibration isolators. These isolators shall be of a size recommended by the engine/generator manufacturer.

3.3 Before final approval and acceptance, the generator set and all associated equipment shall be operated for a 6-hour period to show that it will operate according to specifications. Test load shall be resistive load bank. The engine/generator supplier shall furnish load bank and conduct 6-hour full load test. The builder shall connect and disconnect load bank for testing.

END OF SECTION

SECTION 16180

Overcurrent Protective Devices

Part 1. General

Not used

Part 2. Products

2.1 INDIVIDUALLY MOUNTED CIRCUIT BREAKERS: Molded case type, with the frame and ampacities noted on the drawings; mounted in NEMA 1 or flush-mounted enclosures as noted on the drawings

2.2 FUSES FOR POWER DISTRIBUTION PANELBOARDS AND SAFETY SWITCHES: Bussman or A/E-approved equal with the following capabilities:

    A. HI-CAP Type KRP-C: rated 601 A and larger; time delay type holding 500% of the rated current for a minimum of 4 seconds and clearing 20 times the rated current in 0.01 second or less; 200,000 A, RMS, symmetrical minimum interrupting rating; UL listed as Class RK1

    B. Low-Peak Type LPS-RK: rated 0 to 600 A, 600 V; dual-element time delay type with separate overload and short-circuit elements, holding 500% of the rated current for a minimum of 10 seconds; 200,000A, RMS, symmetrical minimum interrupting rating; UL listed as Class RK1

Part 3. Execution

3.1 Provide and install fuses and circuit breakers as noted on the drawings.

3.2 Do not install fuses until equipment is ready to be energized.

3.3 Provide 2 spare sets of fuses for each size used.

3.4 Provide Bussman Catalog No. SFC, or equal, spare fuse cabinet. Install near the main switchboard/panelboard as shown on the drawings or as directed by the A/E.

END OF SECTION

SECTION 16400

Service and Distribution

Part 1. General

1.1 Service shall be $^{277}/_{480}$ V, 3-phase, 4-wire, from pad-mounted transformer transclosures as indicated on the drawings.

1.2 Metering will be installed on the pad-mounted transformer transclosures by the electric utility company.

1.3 Power shall be distributed throughout the facility at $^{120}/_{208}$ V, 3-phase, 4-wire, and $^{277}/_{480}$ V, 3-phase, 4-wire.

Part 2. Products

Not used

Part 3. Execution

3.1 Meet with a representative of the electric utility company prior to starting work to determine the exact requirements for service installation. Carefully coordinate all service work with the electric utility company.

3.2 Furnish and install any equipment in connection with metering that is required by the power supplier.

3.3 Furnish all necessary riser conduit, riser cable, meter base service entrance conduit and cable, cable racks, and miscellaneous hardware.

3.4 Furnish and install secondary service entrance conduit and cable. The exact arrangement of conduit and cables at the pad mounted transformers shall be in accordance with the requirements of the utility company.

3.5 Open and close all trenches for service. Furnish sand backfill for service cable wherever designated by the electric utility company. Wherever services pass under walks or paving, install rigid galvanized conduit of the proper size and length.

END OF SECTION

SECTION 16410

Power Factor Correction

Part 1. General

1.1 Section Includes

  A. Capacitors

  B. Automatic power factor controllers

1.2 References

  A. ANSI/NEMA ICS 2—Industrial Control Devices, Controllers and Assemblies

  B. ANSI/NEMA 250—Enclosures for Electrical Equipment (1000 Volts Maximum)

  C. ANSI/NFPA 30—Flammable and Combustible Liquids Code

1.3 Submittals

  A. Submit product data under provisions of Section 01300.

  B. Submit product data showing outline and mounting dimensions, weights, voltage and capacity ratings, fusing, and accessories.

  C. Submit manufacturer's installation instructions under provisions of Section 01300.

1.4 Project Record Documents

  A. Submit record documents under provisions of Section 01700.

  B. Accurately record location, rating, and connection to distribution system of each capacitor.

1.5 Operation and Maintenance Data

  A. Submit maintenance data under provisions of Section 01700.

  B. Include cell and fuse replacement instructions.

  C. Include maintenance and trouble shooting instructions for electronic components.

1.6 Qualifications

  A. Manufacturer: Company specializing in power factor correction capacitors with minimum five years of experience.

Part 2. Products

2.1 Unit Capacitors

  A. Capacity: 5 kVAR increments at rated voltage and frequency.

  B. Voltage: 480 volts, three-phase, 60 Hz.

  C. Operating Temperature Limits: Designed to operate at 110% rated voltage in ambient air temperature between − 40°F and 104°F.

  D. Construction: Enclosed capacitor unit with replaceable current-limiting fuse. Include internal discharge resistor.

  E. Dielectric Impregnant: Non-PCB, noncombustible liquid.

  F. Enclosure: ANSI/NEMA 250; Type 4.

  G. Finish: Manufacturer's standard gray enamel.

H. Mounting: Include brackets and accessories for mounting capacitors in groups on wall, or suspended from structure.

2.2 Automatic Power Factor Controllers

A. Controllers: Factory-assembled and pre-wired equipment consisting of unit capacitors, power factor sensing and control equipment, and switching contactors to provide variable correction within discrete steps to maintain preset value of power factor.

B. Step Capacity: 5 kVAR at rated voltage and frequency.

C. Number of Steps: 20.

D. Voltage: 480 volts, three-phase, 60 Hz.

E. Power Factor Range: 80% to 90%.

F. Indicators: Include indicating light for each step of capacity.

G. Basic Impulse Level: 30 kilovolts.

H. Integrated Equipment Short Circuit Rating: 42,000 RMS amperes symmetrical.

I. Operating Temperature Limits: Designed to operate at 110% rated voltage in ambient air temperature between 32°F (0°C) and 104°F (40°C).

J. Capacitor Construction: Capacitor unit with internal discharge resistor, in ANSI/NEMA 250, Type 5 enclosure.

K. Capacitor Dielectric Impregnant: Non-PCB, noncombustible liquid.

L. Power Factor Sensing and Control: Utilize reactive current sensing and solid-state electronic controller to connect appropriate correction capacitors to line through contactors. Include time delay to accommodate capacitor resistor discharge and prevent hunting.

M. Contactors; Electrically-held general purpose magnetic contactors, sized in accordance with ANSI/NEMA ICS 2, Part 210.

N. System Enclosure: ANSI/NEMA 250; Type 1, floor mounted.

O. Finish: Manufacturer's standard gray enamel.

Part 3. Execution

3.1 Installation

A. Install in accordance with manufacturer's instructions.

B. Locate capacitor systems to allow adequate ventilation around enclosures.

C. Provide a 100 kVAR capacity switched bank of 5 kVAR increments on each main entering the facility (two total).

END OF SECTION

SECTION 16440

Main Switchboard

Part 1. General

1.1 The switchboard shall meet UL enclosure requirements and be furnished with a UL label.

1.2 Each switchboard, as a complete unit, shall be given a single withstand shortcircuit current rating by the manufacturer. The withstand short-circuit current rating shall certify that all equipment is capable of withstanding the stresses of a fault equal to the interrupting rating of the least overcurrent protective device contained therein. Such rating shall be established by actual tests by the manufacturer on equipment constructed similarly to the subject switchboard. These test data shall be available and shall be furnished to the A/E, if requested, with or before the submittal of approval drawings.

1.3 Submit shop drawings for the A/E's approval before fabrication.

Part 2 Products

2.1 General

A. The switchboard shall have the approxi-mate dimensions shown on the drawings and be rated for indoor location in a NEMA 1 enclosure.

B. The entire switchboard shall be dead front, totally enclosed, free standing, and front accessible.

C. The switchboard framework shall consist of steel channels welded or bolted to the frame to support the entire shipping section rigidly for moving on rollers and floor mounting. The framework is to be formed code-gauge steel, rigidly welded and bolted together to support all cover plates, bussing, and component devices during shipment and installation.

D. Each switchboard section shall have an open bottom and an individual removable top plate for installation and termination of conduit. Top and bottom conduit areas are to be clearly shown and dimensioned on the shop drawings. The wireway front covers shall be hinged to permit access to the branch switch load side terminals without removing the covers. All front plates used for mounting meters, selector switches, or other front-mounted devices shall be hinged with all wiring installed and laced with flexibility at the hinged side. All closure plates shall be screw removable and shall be small enough for easy handling by one person. The paint finish shall be gray enamel over a rust-inhibiting phosphate primer.

E. Every switch or other item of the switchboard shall be identified with a lamicoid nameplate (black background with white lettered engraving) stating the panelboard or motor, etc., to be served, voltage, and feeder size to be supplied by each switch. The nameplates shall be large enough to be clearly read at a distance of 4 feet. Switches that are left for spares or spaces shall have no engraving on the plate but shall be left blank for future engraving.

F. The switchboard bussing shall be plated and have enough cross-sectional area to conduct the rated full load current continuously with a maximum average temperature rise of 50°C above an ambient temperature of 25°C. The main horizontal or through bus shall consist of one single continuous conductor per phase and neutral.

G. The bus bars shall be rigidly braced to comply with the withstand rating of the switchboard.

H. The main horizontal bus bars between sections shall be located at the back of the switchboard to permit a maximum of available conduit areas. The end section shall have bus bar provisions for the addition of a future section. The provisions shall include the bus bars installed and extended to the extreme side of the section and fabricated in such a fashion that the addition of a future section would require only the installation of a single splice bus connection per phase (and neutral). The horizontal main bus bar supports, connections, and joints are to be bolted with Grade 5 carriage bolts and Belleville washers to minimize maintenance requirements.

2.2 Incoming Line Sections

A. Service shall be 277/480 V, 3-phase, 4-wire, as shown on the drawings with main switch ground fault protection.

B. The incoming line sections shall include a panel-mounted metering section, to include but not be limited to voltmeter, ammeter, frequency megawatt, demand, and megawatt hours. Use Westinghouse Co., IQ Data Plus, or approved equal.

C. The main and tie switches shall be a solid-state type circuit breaker with ratings as shown on the associated drawings. Breakers shall have adjustable settings for long time and short time amperes, instantaneous and ground fault settings.

D. Provide a keyed sequence of operating main and tie switches for present lineup and to include future lineup. Keys shall be arranged for energizing tie breakers after associated mains have been opened.

2.3 Distribution Section

A. Group mounted molded-case circuit breakers are to be totally front accessible. The circuit breakers are to be mounted in the switchboard sections to permit installation, maintenance, and testing without interference with line side bussing. The circuit breakers are to be removable by the disconnection of only the load side cable terminations and all line and load side connections are to be individual to each circuit breaker. No common mounting brackets or electrical bus connectors will be acceptable.

B. Each circuit breaker is to be furnished with an externally operable mechanical means to trip the circuit breaker.

Part 3. Execution

3.1 Provide and install main switchboard as noted on the drawings and specified herein.

3.2 Provide and install circuit breakers as noted on the drawings. Provide spares as specified herein.

3.3 Provide a rubber mat measuring $1/8$ inch 3 36 inch wide by a length equal to the motor control center's, plus the future section plus 1 foot. Install in front of the motor control center.

END OF SECTION

SECTION 16450

Grounding

Part 1. General

1.1 Ground all electrical equipment as shown on the drawings and in accordance with Article 250 of NEC and the requirements of the local authorities having jurisdiction and as specified herein or detailed on the drawings.

1.2 Only a direct connection with copper wire to either or both of the following will be considered as a direct ground:

   A. A 1″ or larger mechanically and electrically continuous, underground, iron or steel cold water line.

   B. A ³/₄″ diameter, copper weld rod 15′ long, driven vertically into the ground.

1.3 All ground connections, where buried or otherwise inaccessible, shall be brazed or welded.

1.4 Provide bonding bushings and jumpers to grounded bus for all metallic conduit entering switchboard and motor control centers from below. Jumper size shall conform to NEC requirements.

1.5 Do not use flexible metal conduit and fittings as a means of grounding. Pull a green ground wire with lugs on both ends in or around each piece of flexible conduit, and screw it to the conduit system.

1.6 Install green bonding jumpers in flush mounted receptacle boxes, and attach them to the receptacle grounding terminal. Install green bonding jumpers with lugs between outlet box and wall bracket mounted lighting fixtures. Screw jumpers to the fixture chassis (NEC 250-74).

1.7 Install a green bonding jumper with lugs in each hand hole, and attach to the ground bus.

1.8 Install a ground network as indicated on the drawings, around the structure, to switchgear, and other electrical installations. The ground network shall consist of a main cable loop, ground rods, branch cables from loop to individual grounds, above-ground connecting points, and necessary inspection points on ground rods. The cable for the main loop shall be not less than No. ⁴/₀ AWG stranded bare copper, and branch runs shall be not less than No. 2 stranded bare copper.

1.9 Install code-size green ground conductors in all branch circuits feeding receptacles, permanently wired fixed equipment, and lighting fixtures mounted 7′ or less above finished floor. Bond conductors to chassis of fixed equipment and light fixtures and to ground lug on receptacles. All ground conductors shall be bonded to multiterminal ground bus in panelboard. Grouping of ground conductors under a single lug is not acceptable.

1.10 Insofar as possible, lay conductors directly in the ground without breaks or joints and provide a minimum cover of 24″ below grade or 30″ under roads. Where underground joints are unavoidable, connect them with Cadweld SS or AMP Ampact connectors that are suitably covered and protected.

1.11 Attach ground rods to the main cable loop at the intervals necessary to obtain resistance to earth not exceeding 5 ohms at any point in the ground network, provided the intervals between rods are not greater than 75". Cable between ground rods shall be slack. Ground rods shall be copperweld $3/4" \times 8'$ minimum length.

1.12 Measure the resistance of each ground rod or ground field before connecting to the ground network. Make measurements with ground network tied together, and add additional ground rods as necessary to obtain resistance to earth not exceeding 5 ohms at any point in the ground network system.

Part 2. Products

Not used

Part 3. Execution

Not used

END OF SECTION

SECTION 16460

Dry Type Transformers

Part 1. General

1.1 Transformers shall be the dry type, rated as shown on the drawings for 480 V, 3-phase, primary to 120/208 V, 4-wire, 3-phase or 480 V 1 phase to 120/208 V 1-phase, 3-wire.

Part 2. Products

2.1 All single-phase units and 3-phase units up to 30 KVA shall be of the indoor/outdoor type with epoxy filled or encapsulated coil and core.

2.2 Transformers 30 KVA and larger shall be in a heavy gauge, sheet steel, ventilated enclosure. The ventilating openings shall be designed to prevent accidental access to live parts in accordance with UL, NEMA, and ANSI standards for ventilated enclosures.

2.3 Transformers shall have a 115°C temperature rise above 40° ambient and shall be capable of carrying a 15% continuous overload without exceeding a 150°C rise in a 40°C ambient. All insulating materials shall be in accordance with NEMA Standard 20 requirements for a 200°C UL component recognized insulation system.

2.4 Transformers shall be UL listed with four or more 2-$\frac{1}{2}$% taps, two above and two below the rated primary volts. Meet ANSI C89 for maximum sound levels. Insulation system shall be 185°C or 220°C, with 115°C rise.

2.5 The entire transformer enclosure shall be degreased, cleaned, phosphatized, primed, and finished with baked enamel.

2.6 The maximum temperature of the top of the enclosure shall not exceed a rise of 35°C above a 40°C ambient.

2.7 The core of the transformer shall be grounded to the enclosure with a flexible grounding conductor sized in accordance with applicable NEMA and NEC standards.

2.8 Shop drawings for all transformers shall indicate the equipment size, rating, system voltage, sound levels, impedence, insulation class, and other related data specified above.

2.9 Transformers shall be by Heavy Duty, General Electric, Westinghouse, Sorgel, or approved equal.

Part 3. Execution

3.1 Provide and install transformers with all necessary installation accessories. Use flexible conduit connections and resilient mounting in office areas to keep the sound level at a minimum.

END OF SECTION

SECTION 16510

Lighting Fixtures

Part 1. General

1.1 Furnish and install all lighting fixtures, complete with lamps, hangers, etc., as indicated and scheduled on the drawings.

1.2 Equip all fluorescent lighting fixtures with CBM, ETL-approved, Class P, rapid start ballast. Use high efficiency ballast: Advance Mark III, unless otherwise noted on the drawings.

1.3 All linear fluorescent strip light fixtures installed on the ceiling shall be provided with a $1\text{-}1/2''$ spacer between the ballast housing and ceiling.

Part 2. Products

2.1 Lamps shall be of the wattage and type listed and by Phillips Lighting Co. Use high efficiency, low wattage indoor fluorescent lamps unless otherwise noted on the drawings.

2.2 Incandescent lamps shall be of the wattage and type indicated on the drawings and shall be rated at 130 V.

2.3 Determine the type of ceiling to be installed in each space, and furnish fixtures suitable for the exact type.

Part 3. Execution

3.1 Receive, store, uncrate, install, and lamp fixtures that the schedule indicates are to be furnished by others.

3.2 All lighting fixtures shall be structurally supported. Fluorescent fixtures mounted in suspended ceilings shall be supported by the ceiling suspension system. Incandescent fixtures mounted in suspended ceilings shall be supported by fixture channels furnished by the Contractor; incandescent fixture supports shall be laid across ceiling support channels. Each surface mounted fixture shall be supported from the building structure by rods or rods and clamps, or by its outlet box, which in turn shall be supported. Provide fixture studs as necessary. Secure wall mounted fixtures to masonry walls with bolts and lead anchors, and to metal stud dry wall partitions by sheet metal screws driven into the metal studs.

3.3 Connect recessed lighting fixtures to outlet boxes in accordance with NEC 410-67(c).

3.4 Furnish a sample lighting fixture for mock-up sample of integrated ceiling. Refer to Section 13025.

END OF SECTION

SECTION 16530

Outdoor Lighting Fixtures

Part 1. General

Not used

Part 2. Products

2.1 Metal Halide and high-pressure sodium luminaries shall be equipped with 277 V regulated HPF ballast.

2.2 Metal light poles shall be as indicated and scheduled on the drawings. They shall be complete with anchor bolts, hand hole, ground lug and bolt covers, and floodlighting brackets.

2.3 Metal halide ground mounted floodlight luminaries shall be equipped with 208 V regulated HPF ballast.

Part 3. Execution

3.1 Furnish and install lighting fixtures, complete with Phillips Lighting Co. lamps, Advance Co. ballasts, hangers, etc., as indicated and scheduled on the drawings.

END OF SECTION

SECTION 16601

Lightning Protection

Part 1. General

1.1 Provide all labor, materials, and items of service required for the completion of a functional and unobtrusive concealed system of grounds, conductors, and semiconcealed air terminals for protection against damage by lightning striking the flammable storage building, as approved by the A/E.

1.2 The complete system shall conform to NFPA 78 and UL 96A requirements for materials and installation methods.

Part 2. Products

2.1 All materials used in the installation shall be of a type approved for lightning protection systems and approved by the A/E. The manufacturer shall be a member in good standing of the Lightning Protection Institute.

2.2 Specifications as to points, conductors, fittings, fasteners, connectors, etc., shall be in strict accordance with the two codes listed above; these items shall be of copper, bronze, or stainless steel.

2.3 Air terminals shall be made of solid copper, chrome plated. The minimum diameter shall be $1/2''$, and lengths shall be as required to suit the application. Air terminals shall be attached to stud connector inserts by copper couplings with milled hexagonal wrench projections to permit easy attachment and removal.

Part 3. Execution

3.1 The Contractor shall have a minimum of 3 years of experience in installing lightning protection systems. The manufacturer shall furnish eight copies of the proposed system layout for the A/E's approval prior to the start of any work on this system.

3.2 Furnish to the A/E a complete system layout drawing for approval before any material is manufactured for the proposed system.

3.3 Bond all down conductors to the building steel columns, which are to be bonded to a grounding network under Section 16450, Grounding.

3.4 Upon completion of installation, furnish the Owner with certification from the manufacturer that the system, as installed, meets all the requirements mentioned in UL 96A and NFPA 78. Provide a UL Master Label.

END OF SECTION

SECTION 16720

Fire Alarm and Detection System

Part 1. General

1.1 Related Documents

A. Drawings and general provision of the contract, including General and Supplementary conditions and Division 1 General Requirements, apply to work of this section.

1.2 Description of Work

A. Extent of fire alarm and detection system work is indicated by these specifications, drawings, and schedules. The system described herein and shown on the drawings shall be the standard products of a single manufacturer and shall be those as manufactured by ADT Co.

B. Furnish and install a complete fire alarm system as described herein and as shown on the plans; to be wired, connected, and left in first-class operating condition. The system shall use closed-loop initiating device circuits with individual zone supervision, incoming and standby power supervision.

C. The ability for selective input/output control functions based on "ANDing", "ORing", "NOTing", timing and special coded operations shall also be included in the resident software programming of the system. To accommodate and facilitate job site changes, the Control Panel Switches (city disconnect, manual EVAC, auxiliary bypass switches) may be programmed on site to provide control of other output circuits. Initiation circuits shall be individually configurable on site to provide either alarm/trouble operation, alarm only, trouble only, current limited alarm, no alarm, normally closed device monitoring, a nonlatching circuit or alarm verification. Through the use of a program mode dip-switch the fire alarm control panel shall allow for loading or editing any special instructions or operating sequences as required. All instructions shall be stored in a resident nonvolatile programmable memory. Fire alarm panel shall be microprocessor-based software programmable.

D. The programming language shall have the ability to perform the following:

1. ANDing—The ability to require 2 or more points in an alarm state before control point(s) activation.
2. ORing—The ability to allow any number of points to cause control point activation.
3. NOTing—The ability to prevent control point activation if a monitor or control is not in its normal state.
4. Delay—The ability to delay control point activation.
5. Count—The ability to require more than one point in alarm before control point activation.
6. Timing—The ability to cause control point activation for a specified period.
7. Cycle—The ability to cause control point activation on a cycle basis as required for air handling unit restart.

E. All panels and peripheral devices shall be the standard product of a single manufacturer and shall display the manufacturer's name on each component.

F. The work covered by this section of the specifications includes the furnishing of all labor, equipment, materials, and performing all operations in connection with the installation of the Fire Alarm System as shown on the drawing, as hereinafter specified, and as directed by the Contracting Representative.

G. The complete installation shall conform to the applicable sections of NFPA-72 A, B, C, D, E, or F and Local Code Requirements and National Electrical Code Article 760.

H. The work covered by this section of the specifications shall be coordinated with the related work as specified elsewhere under the project specifications.

I. Each item of the Fire Alarm System shall be listed as a product of a single fire alarm system manufacturer under the appropriate category by the UL, and shall bear the UL label.

J. The complete installation shall conform to the applicable sections of NFPA-72, National Electrical Code, Standard Building Code, Life Safety Code 101, UL wiring criteria, and directions of the fire marshal.

K. Types of fire alarm and detection systems in this section include the following:

  1. Manual, noncoded
  2. Automatic, noncoded

1.3 Quality Assurance

  A. Manufacturers: Firms regularly engaged in manufacture of fire alarm and detection systems of types, sizes, and electrical characteristics required, whose products have been in satisfactory use in similar service for not less than 10 years.

  B. Installer: Qualified with at least 10 years of successful installation experience on projects with fire alarm and detection systems similar to that required for this project.

  C. NEC Compliance: Comply with NEC, as applicable to construction and installation of fire alarm and detection systems components and accessories.

  D. UL Compliance: Provide products that have been UL listed and labeled.

  E. F M Compliance: Provide systems and accessories that are Factory Mutual approved.

1.4 Submittals

  A. Product Data: Submit manufacturer's data on fire alarm and detection systems, including but not limited to roughing-in diagrams and instructions for installation, operation and maintenance, suitable for inclusion in maintenance manuals. Also include standard or typical riser and wiring diagrams. The fire alarm system vendor shall be responsible for the actual wire count and circuit arrangement to accomplish the zone and control system required by the contract documents and code directives.

  B. Equipment submissions must include a minimum of the following:

  C. Complete descriptive data indicating UL listing for all system components.

  D. Complete sequence of operations of the system.

1.5 Products

A. Manufacturer: Alternate systems to this specification shall provide similar performance. Exact compliance to construction style and basic components is not necessary. The owner's representative shall have sole authority to determine specification compliance.

1.6 System Operation—Fire Detection and Control— Equipment

A. The system alarm operation subsequent to the alarm activation of any manual station, automatic device, or sprinkler flow switch shall be as follows:

1. The appropriate zone Red LED shall flash on the control panel until the alarm has been silenced at the control panel or the remote annunciator. Once silenced, this same LED shall latch on. A subsequent alarm received after silencing shall flash the subsequent zone alarm LED on the control panel.

B. A pulsing alarm tone shall occur within the control panel until silenced.

1. All alarm-indicating appliances shall sound until silenced by the Alarm Silence Switch at the control panel.
2. All visual alarm lamps shall operate in a flashing mode until system is reset to normal. (All zones are cleared.)
3. Alarm horns and lights shall operate selectively by areas if required.
4. Alarm and/or control functions may be programmed to not operate until two, three, four, etc., zones have been initiated, thus providing "crosszoning" of two or more zones.
5. All doors normally held open by door control devices shall close after the signals have sounded for five (5) sounds.
6. A supervised signal to notify the local area UL-approved central station shall be activated.
7. The mechanical controls shall activate the air handling systems as required.
8. The audible alarms shall automatically time out and switch off after ten (10) minutes of alarm operation.

C. The alarm indicating appliances may be programmed to be silenced, after one (1) minute, after three (3) minutes, or after five (5) minutes as desired by owner, by authorized personnel upon entering the locked control cabinet and operating the Alarm Silence Switch or by use of the key-operated switch at the remote annunciators. A subsequent zone alarm shall reactivate the signals.

D. The activation of any system smoke detec-tor shall initiate an Alarm Verification op-eration whereby the panel shall reset the activated detector and wait for a second activation. If, after reset, a second alarm is reported from the same or any other smoke detector within one (1) minute the system shall process the alarm per the operations listed above. If no second alarm occurs within one minute the system shall resume normal operations. The alarm verification shall operate only for smoke detectors. Other activated initiating devices shall be processed immediately. Smoke detector verification may be programmed for any zone, a combination of zones, or all zones.

E. The activation of any standpipe or sprinkler tamper switch shall activate a distinctive system supervisory audible signal and illuminate a "Sprinkler Supervisory Tamper" LED at the system controls and the remote annunciator. There shall be no confusion between valve tamper activation and opens and/or grounds on fire alarms initiation circuit wiring.

F. Activating the Trouble Silence Switch will silence the supervisory audible signal while maintaining the "Sprinkler Supervisory Tamper" LED indicating the tamper contact is still activated.

G. Restoring the valve to the normal position shall cause the audible signal and LED signal and LED to pulse at a March Time Rate.

   1. Activating the Trouble Silence Switch will silence the supervisory audible signal and restore the system to normal.

H. The alarm activation by manually operated stations shall pulse the signals to follow the code as previously specified.

I. The activation of the program "Test Set-up" switch at the control panel shall activate the "Walk Test or System Test" mode of the system that shall cause the following to occur:

   1. The Remote Central Station circuit shall be disconnected.
   2. Control relay functions shall be bypassed.
   3. The control panel shall show a trouble condition.

J. The alarm activation of any initiation device shall cause the audible signals to pulse one round of code identifying the initiation circuit (e.g., an activated smoke detector connected to Zone 4 shall pulse the audible signals four times in rapid succession).

   1. The panel shall automatically reset itself.
   2. Any momentary opening of an initiating or indicating appliance circuit shall cause the audible signals to sound for 4 seconds to indicate the trouble condition.
   3. A manual evacuation switch shall be provided to operate the systems indicating appliances or initiate "Drill" procedures.

K. Activation of an auxiliary bypass switch shall override the automatic functions either selectively or throughout the system and initiate a trouble condition at the control panel.

L. The system shall have output capabilities and interface with the public address system.

1.7 Supervision

A. All supervised initiation circuits shall be such that a fault in any one zone shall not affect any other zone. The alarm activation of any initiation circuit shall not prevent the subsequent alarm operation of any other initiation circuit.

B. There shall be one supervisory initiation circuit for connection of all sprinkler valve tamper switches to perform the operation specified. Wiring methods which affect any fire alarm initiation circuits to perform this function shall be deemed unacceptable, i.e., sprinkler and standpipe circuits with fire alarm initiation devices (N/O contacts). This independent initiation circuit shall be labeled "Sprinkler Supervisory Tamper" and shall differentiate between tamper switch activation and wiring faults.

C. Furnish the required number of independently supervised and independently fused indicating appliances circuits for alarm horns, bells, chimes, and/or flashing alarm lamps.

D. All auxiliary manual controls shall be supervised so that all switches must be returned to the normal automatic position to clear system trouble.

E. Each independently supervised circuit shall include a discrete amber "Trouble" LED to the normal automatic position per circuit.

F. The incoming power to the system shall be supervised so that any power failure shall be audibly and visually indicated at the control panel and/or annunciator.

G. The system batteries shall be supervised so that disconnection of a battery shall be audibly and visually indicated at the control panel and/or annunciator.

H. The system expansion modules connected by ribbon cables shall be supervised for module placement.

1.8 Power Requirements

A. The control panel shall receive 120 VAC power (as noted on the plans) via a dedicated fused disconnect circuit.

B. The system shall be provided with sufficient battery capacity to operate the entire system upon loss of normal 120 VAC power in a normal supervisory mode for a period of 24 hours with five (5) minutes of alarm indication at the end of this period. The system shall automatically transfer to the standby batteries upon power failure. All battery charging and recharging operations shall be automatic. Batteries, once discharged, shall recharge at a rate to provide a minimum of 70% capacity in 12 hours.

C. All circuits requiring system operating power shall be 24 VDC and shall be individually fused at the control panel.

Part 2. Products

2.1 Fire Alarm Control Panel

A. Where shown on the plans, provide and install an 8-Zone Fire Alarm Control Panel. Construction shall be modular with solid state, microprocessor-based electronics. All visual indicators shall be in high contrast, LED type.

B. The control panel shall contain the following features with quantities as shown on the drawings:

1. Initiation Device Zone as indicated on the drawings
2. Alarm Indicating Appliance Circuit
3. Supervised Annunciator Circuits
4. (selectable) Local Energy, Shunt Master Box, or Reverse
5. Polarity Remote Station Connection
6. Form C Alarm Contacts (2.0 Amps ea.)
7. Form C Trouble Contact (2.0 Amps ea.)
8. Earth Ground Supervision Circuit
9. Automatic Battery Charger
10. Standby Batteries
11. Resident nonvolatile programmable operating system memory for all operating requirements
12. Supervised Manual Evacuation Switch

2.2 External Devices

A. The system shall utilize UL-listed fire alarm initiating devices. To ensure compatibility, these devices shall be supplied by the manufacturer of the fire alarm control panel.

1. Photoelectric Detector: Furnish and install, where indicated on the plans, photoelectric detectors. The detectors shall be interchangeable with ionization detectors using the same bases. The detector shall operate on a two-wire zone with end of line resistor. The (photoelectric/ionization) detector head and twist-lock base shall be UL listed to Standard 268.

2. Manual Fire Alarm Station: Stations shall be noncoded, Break Glass Rod Type. Stations shall be single-action and when operated shall remain mechanically "locked" until reset.

3. Horns: Furnish Xenon Audio Visual units with horns. Horns shall be 24-volt DC horns of metal construction with a minimum sound pressure level output of 87 dB at 10 feet. Provide a PVC chlorine resistant horn in etch room and chlorine room.

4. Visual Flashing Lamps (Xenon Strobe): Furnish Visual only Xenon flash-tubes where indicated on the drawings.

5. Tamper and Waterflow Switches: These devices shall be UL listed. Furnish series as required. Coordinate with sprinkler contractor. Furnish all waterflow, PIV, and Tamper switches.

6. Duct Smoke Detectors: Furnish Photoelectric duct detectors with the appropriate tubes. Coordinate with mechanical contractor. Provide remote indicators and test switches where detectors are not directly visible.

Part 3. Execution

3.1 Installation of Fire Alarm and Detection Systems

A. General: Install work of this section where indicated, complying with manufacturers' written instructions, applicable requirements of NEC and NECA's "Standard of Installation," and in compliance with recognized industry practices to ensure that produces fulfill requirements.

B. Installation of Basic Identification: Install electrical identification in accordance with "Identification" section of Division 16.

C. Installation of Basic Wiring System Materials: Install wiring, faceways, and electrical boxes and fittings in accordance with applicable sections of Division 16.

D. System Wiring: The system wiring and installation shall be in compliance with applicable codes, project drawings, and as required by the manufacturer. Wiring size shall be as recommended by the manufacturer with a minimum of No.14 for appliance signal circuits.

3.2 Wiring and Guarantee

A. The system wiring and installation shall be in compliance with the applicable codes, architectural drawings, and recommendations of the manufacturer. All wiring shall be color coded, tagged, and checked to assure that is free from shorts and grounds.

B. The contractor shall provide and install the system in accordance with the plans and specifications, all national and local applicable codes, UL wiring criteria, and the manufacturer's recommendations. All wiring shall be in a completely separate conduit system. All junction boxes shall be sprayed red and labeled "Fire Alarm." Wiring color code shall be maintained throughout the scope of the work.

C. Installation of equipment and devices that pertain to other work in the contract shall be closely coordinated with the appropriate subcontractors.

D. The manufacturer's representative shall provide all on-site supervision of installation of the complete fire alarm system installation, perform a complete functional test of the system, and submit a written report to the contractor attesting to the proper operation of the completed system.

E. The fire detection and control system shall comply with applicable codes and inspecting authority rules. The fire alarm system Installer shall furnish all necessary components, materials, and labor to result in an approved and acceptable system whether specifically shown on the drawings or implied herein, to result in a complying system.

F. The fire alarm system Installer and the Electrical Contractor shall carefully review the Division 15 drawings and specifications for fire/smoke detection devices and control points required for detection and control of air handling equipment. Provide code-required connections to all HVAC and sprinkler systems.

3.3 Testing

A. The completed fire alarm system shall be fully tested by the Contractor in the presence of the Owner's representative, the Architect, the Consulting Engineer, authority having jurisdiction, and the manufacturer's technical representative. Upon completion of a successful test, the Contractor shall so verify in writing to the Owner, Architect, and General Contractor.

B. The following test shall be performed by the fire alarm manufacturer's authorized representative. Each and every alarm-initiating device shall be tested for alarm and trouble initiation by zone. Verify that each device is located in its appropriate one. Written verification of this test shall be provided to the Owner, Architect, and Electrical Engineer. This test shall be performed in accordance with NFPA Standard 7211.

3.4 Warranty

A. The Contractor shall warrant the completed fire alarm system wiring and equipment to be free from inherent mechanical and electrical defect for a period of one (1) year from the date of the building occupancy and issue of the completed and certified test. A gratis maintenance contract shall be issued on the system for one year for on-site warranty. In addition, the contract may be continued after the first year for a yearly charge, and will include two inspections per year and certification that each device is operational.

B. The system's vendor must employ factory-trained technicians and maintain a service organization within 50 miles of the job site. This organization must have a minimum of 10 years of experience servicing fire alarm systems.

3.5 Training

A. Personalized instructions to the building owner and occupant will be provided by a factory-trained representative of the equipment supplier.

3.6 Instruction Manuals

A. The Contractor shall provide, in addition to one approved copy of the fire alarm system submittal, complete operating instructions; pertinent system orientation documents; and system service, testing, and alarm documentation.

3.7 Field Quality Control

A. Inspect relays and signals for malfunctioning, and where necessary, adjust units for proper operation to fulfill project requirements.

B. Final adjustment shall be performed by specially trained personnel in direct employ of fire alarm and detection system manufacturer.

END OF SECTION

SECTION 16729

Access Control System

Part 1. General

1.1 The access control system specified is that of Simplex Time Recorder Company, ADT, Autocall or Gamewell. Catalog and model numbers are intended to establish the type and quality of equipment and system design as well as exact operating features required.

1.2 Substitutions of products proposed to be equal to those specified herein will be considered only when the following requirements have been met:

1.3 A complete list of such substituted products, with drawings and data sheets, shall be submitted to and approved by the architect and/or consulting engineer, not less than 10 calendar days prior to scheduled date of opening of bids.

1.4 Substitute equipment and its capabilities must be a standard part of that system's current product line and must meet or exceed the capabilities of the equipment specified. Contractors are cautioned to conform to this specification so that the system provided will ensure future options and priorities of the owner with regard to the system's use.

1.5 A bidder intending to use acceptable substitute products shall submit two price quotations: one based on the use of products specified herein; the other shall define the use of substitute products and list cost differentials as related to the base quotation.

1.6 Response to Specification

A. The Contractor shall submit a point-by-point statement of compliance with Sections 3 (EQUIPMENT) and 4 (SOFTWARE).

B. The statement of compliance shall consist of a list of all numbered paragraphs within these sections.

C. Where the proposed system complies fully with the numbered paragraphs as written, such shall be indicated by placing the word "comply" opposite the paragraph number. Where the proposed system does not comply with the paragraph as written, but the Contractor feels it will accomplish the intent of the paragraph in a manner different from that described, a full description of the intent perceived by the Contractor shall be provided as well as a full description of how its proposal will meet its perceived intent.

D. Where a full description is not provided, it shall be assumed that the proposed system does not comply with the paragraph in question. Any submission that does not include a point-by-point statement of compliance as described herein shall be disqualified.

1.7 Support and Training

A. The access control equipment supplied shall be a standard labeled product of the equipment supplier, bearing the company's name and having its exclusive model numbers. This company must be of established reputation and experience, regularly engaged in the access control/alarm business for a period of at least five consecutive years under its current company name. This company shall have a fully staffed office of sales and technical support representatives within 50 miles of this project.

B. Within the first 30 days from system start-up the equipment supplier shall provide eight hours of system operations training at the project site to the Owner's representatives. This training shall be conducted during normal business hours of the equipment supplier at a date and time of mutual convenience.

1.8 Scope

A. Furnish and install per plans and specifications a complete Simplex 3204 access control system that is capable of controlling a minimum of 64 doors; 4,000 cardholders with names assigned; monitoring up to 512 alarm points; and activating up to 512 control points. The system shall provide multiple operator entry devices with an easy-to-use format. The keyboard supplied shall be IBM standard allowing function key control. The system shall also include a "serial" two-button Microsoft mouse to work in conjunction with the keyboard for data entry and system configuration.

B. The day to day operation of the system shall be mostly accomplished by use of the mouse. There can be no substitutions, although the system can function without the mouse. The system shall have a PC (personal computer) with floppy and hard drives, keyboard, mouse controller, and printer. The software shall provide 32 user-defined templates for report generation. It shall provide an easy to use windowed software format with full on-line English-text assistance for every operation. This format shall be enhanced by the use of a mouse to enable quick access to any point in the system by use of the windowed software. The system shall have anti-passback and remote capabilities. Sixty-four access levels shall be available along with 32 time zones containing eight unique intervals per zone.

C. The system software shall be flexible to allow easy understanding of any event transaction in the system. The system shall provide a minimum of 450 help screen windows to ensure the user's ability to receive on-line informational text pertinent to the task being performed. This shall be controlled with a function key.

Part 2. Products

2.1 Human Interface: The system provided shall have as its human interface, a standard off-the-shelf and unmodified PC. Its software package shall interface with the system's 64 maximum card readers and 4,000 maximum card users. The interface shall be provided with a user-defined auto-logoff routine selectable from 1–240. The software package shall provide the ability to archive 400,000 events, generate a minimum of 40 configurable user and archive report templates. It shall also provide the ability to backup and restore system configuration and user's data. The CPU shall be a PS/2 Model 50Z as manufactured by IBM. The configuration shall be:

A. 1.0 MB of RAM

B. 30 MB fixed disk

C. 1.44 MB 3$^{1}/_{2}''$ floppy drive

D. 1 VGA video graphics adaptor

E. 1 keyboard

F. PC Dos 3.3

G. IBM VGA color monitor

2.2 Network Components: The distributed architecture requires the interface of various intelligent controllers. These controllers are to be Simplex Main Controllers and Terminal Controllers, no other substitutes are acceptable. These controllers shall have operating environment to allow complete functionality at a temperature range of 0° to 50°C and a relative humidity of 90% (noncondensing).

2.3 Main Controllers: The controller shall be microprocessor based with on-board time and date generation and battery to allow a minimum of 48 hours data integrity. This device shall be responsible for maintaining the data communications between the terminal controllers attached and the CPU. The system shall have a minimum capacity for 16 terminal controllers attached to any one main controller. The system shall be capable of accepting 32 main controllers. These main controllers shall be modem compatible for communications over standard 3002 voice-grade telephone lines.

2.4 Terminal Controllers

A. The controller shall be microprocessor based with on-board time date generation and battery to allow a minimum of 48 hours of data integrity. This device shall be responsible for all access control decisions and alarm monitoring detection in the system.

B. It shall also be responsible for all output responses to alarm detection and create the appropriate response through its processor logic. In the event that communications are lost to this controller, all card/door-related activity shall continue based upon full card code, access level, and time code. This shall be done without reverting to a degraded mode using a facility code or other lower security techniques. During this off-line condition the controller shall be capable of maintaining a buffer of a minimum of 2,800 event transactions. Events up-loaded upon communications restoral will be tagged with the actual time each event happened and not the uploading time. There shall be a minimum capacity of 32 terminal controllers available in the system.

C. Each terminal controller shall be capable of interfacing the following cardreader technologies, without the necessity of special interfacing. Additional logic panels or personality modules are not acceptable. All reader technologies must be plug-for-plug compatible. The acceptable technologies are Simplex Proximity, Wiegand, scrambling and nonscrambling Wiegand emulation keypads, and magnetic stripe. Each terminal controller shall be capable of accepting two readers.

D. The terminal controller shall have all necessary provisions to implement access control for two doors. The capacity shall be present for two door contacts, two request to exit devices and two door strike outputs. These outputs shall control lock power and are to be rated at a minimum of 1 amp at 24 VDC or .5 amp at 120 VAC. The strike outputs that supply locking power are unacceptable. These devices shall be connected to the terminal controllers through snap-in Buchanan type plugs. These controllers shall be labeled to ensure the correct wire coding is followed and the appropriate devices are wired in the correct locations.

E. The terminal controllers shall be supplied with additional operator-definable alarm and output points. The contacts shall be voltage-free form "C" while the monitor points shall provide four conditions of status: normal, abnormal, cut and short.

F. There shall be two of each of these type of points available per controller with expansion capacity of up to 16 alarm points and 16 outputs total through use of alarm interface module.

G. The system shall provide for the direct interfacing of the I/O board to any and all terminal controllers for a total of 512 inputs and 512 outputs in the system. This interface is accomplished through a Simplex supplied cable allowing the I/O board to share the terminal controller logic and system address. Equipment supplied that requires alarm-monitoring devices that require their own address and take away from the system's total reader capacity are not acceptable.

H. The I/O board shall be logic driven by the connected terminal controller and allow the alarm contact and output control specified. These abilities shall exist whether the terminal controller is on-line or off-line from the main controller.

I. All inputs, including Door Switch and Request to Exit (REX) contacts, shall have supervision capabilities and shall allow a 500-foot, 22-gauge wire run to each alarm device.

J. All outputs, including door relays, shall be single pole, double throw (SPDT) Form C relays. These devices shall be rated for 1 amp at 24 VDC or 0.5 amp at 120 VAC. All outputs shall be made to allow for any of the following states to be programmed. Outputs shall be configured to latch, close momentarily, or set to follow the state of the corresponding inputs. These states shall work in conjunction with outputs being managed under time zones as well.

K. With the software supplied, it shall be possible to initialize a program from the command center defining which of a terminal controller's outputs may be activated by any of the systems 64 maximum access levels when a card is read at that unit. Systems requiring ROM burns, jumpers, or dip switch settings in the network controllers to provide this function will not be accepted.

2.5 Network Wiring

A. The communications network shall be made up of a primary and secondary bus. The communications protocol used shall be RS232 or RS485, networks using 20 mil current loop will not be acceptable. The primary and secondary bus shall use a dual twisted pair, Beldon #8723 (22 AWG). Both bus networks shall each be capable of a distance of 4,000 ft., total cable. This network shall be wired in any configuration desired (star, loop, daisy chain).

B. All reader cabling shall use a 5 conductor West Penn #3280 (18 AWG) with an overall shield. Readers requiring coaxial cable will not be acceptable. A reader shall be wired a minimum of 500 ft. without wire size change. Readers requiring a wire size change or special adapters to drive a signal this distance are not acceptable.

C. The system communications shall be supervised for integrity. If communications is detected as failed, the system shall report the loss and automatically enable the terminal controller buffers in the network. A system that does not buffer event information when communications are lost will not be acceptable.

2.6 Card Reader: Unit shall be acceptable for indoor or outdoor mounting, be vandal resistant, and have an LED indicator.

2.7 System Software

A. Time Zones

1. The system shall have a minimum of 32 time zones that can be assignable to doors, card holders, operators, outputs, and inputs. Each of these time zones shall have a minimum of eight definable intervals and shall allow a minimum 23 character description. The intervals shall be programmable for starting and stopping times and days for the week. The intervals shall also define holiday usage for a minimum of 32 holidays.

2. The system supplied shall be equipped with automatic operator log-off that, when enabled, will log the command center operator out if the system is left unattended for a definable amount of time. This feature shall be selectable to "disable" or "enable" automatic log-off with time selectable from 1 to 240 minutes from last keyboard or mouse activity.

3. The system shall have automatic screen saver to help prolong the useful life of the monitor. This feature shall be selectable to "disable" or "enable" automatic screen saver with time selectable from 1 to 240 minutes from last keyboard or mouse activity. The screen will return from a blank display with any keyboard, mouse or system activity.

B. Access Levels

1. The system shall have the capabilities to relate card holders to doors for access by time. It shall be possible to restrict any single card or group of cards through the use of these access levels. The system shall have a minimum of 64 access levels and shall allow a minimum 31 character descriptions of each. It shall be possible to assign access levels on a per-door basis. All cards in the system must be assigned to an access level. The system shall have the ability to create a "privileged user group" of card numbers.

2. This group shall be assignable to any combination of doors in the system and those cards assigned shall have 24-hour-a-day access every day of the year. There shall be a minimum of eight cards assignable to this group per door in the system. These cards shall not diminish the 64 access level minimum in the system.

C. Operator Levels

1. The system shall have a user defined (up to 4,000) number of operators. The system shall allow the user to distinguish between operator abilities by defining a minimum of 64 levels. The system will provide the definitions yes/no on each individual command within a menu to individually assign to each operator level. Once defined, these levels shall allow an operator to have restrictions placed on them down to the command level. Each of these levels shall be named for their group of users and shall have assignable restrictions down to the system command level. These levels shall be definable with a minimum of 19 characters. Operator restrictions shall have several categories, the first being full usage. An operator assigned this level shall be allowed to make additions or modifications to the database for that particular menu selection. An operator that has been assigned full usage shall still be restricted from performing a selected task or tasks on any command in the system.

2. The second category shall allow an operator to be assigned an access level of view and thereby totally restricting the operator from modifying database entries. The system shall alert any operator trying to access a restricted menu selection with an on screen message. The message shall state "Access Denied—See Your System Administrator."

3. The third category shall allow an operator to be "denied" access and thereby totally restricting the operator from viewing or modifying database entries. The system shall alert any operator trying to access a restricted menu selection with an on screen message. The message shall state "Access Denied—See Your System Administrator."

D. Door Configurations

1. The system shall allow up to 64 doors to be configured in their own unique requirements. The door shall be defined in text form of 30 characters. The door strike time and door open times are to be assignable independently from 1 to 120 seconds through the system's software. Systems requiring parameter adjustments for these functions to be set at their network controllers or data-gathering panels and not through the system software will not be accepted.

2. All door unlocks shall be time-zone configurable and once unlocked shall be configured to report state change at operator discretion.

3. All doors shall be unlockable by a minimum of two input-configured overrides. These inputs shall be described with a minimum of 30 characters and reportable based on time.

4. Any access-controlled doors in the system shall have the ability to generate a local alarm output in the event of the door being forced or left open. The report of these events shall be time-zone definable. This output shall be configured to any of the following states: latched, timed, or reflective. In the event of the output being latched it shall only be reset through operator intervention or by passing a card valid for that door and time through its reader.

5. Request to exit devices shall be assignable for the ability to unlock the door. If the device is not programmed to unlock the door the door contact shall still be shunted by use of the request to exit device.

6. The door status shall be configurable to report the physical state of the door based on time and condition. The system shall provide a minimum of four door status reports: Door Left Open, Door Forced, Door Opened and Closed Under Valid Card, and Door Entries. The system shall be capable of reporting by time valid or invalid code presentations on a per-door basis.

7. The system shall allow a user-programmed message unique to every door in the system. This message shall appear under alarm condition for the door being forced or left open. This message shall be a minimum of 230 characters.

8. The system shall be supplied with the ability to enable a copy command. This function shall permit similar data to be copied when setting up door, input, and output parameters.

E. Alarm Priorities

1. The system shall provide the means to prioritize alarm events. These events shall be generated from any of the seven event origins. These origins shall be as follows: System Events, Main Controller Events, Terminal Controller Events, User Events, Door Events, Input Events, and Output Events. These events shall be configurable to any one of five priority levels and assigned

a corresponding color. It shall be possible to route events for display or printing based on event type. Each alarm event shall have the ability to be assigned different priority levels in accordance with the devices to which they are attached. Events assigned to Alarm Priorities 1–4 will require operator acknowledgment.

2. The system shall have the ability to support 4,000 user files. These files shall have a minimum of four fixed fields. The system shall support user-configurable fields. These user-defined fields shall have two types: index fields or text fields. An index field shall be configurable to allow user-defined classifications.

3. There shall be a minimum of 30 configurable classifications for each index field in the system. The text field capability shall provide any single field to be configured with nine lines, each having 32 characters. It shall allow a minimum of 500 characters for text field use. The system shall provide user-configurable report capabilities that allow the combination of all fields to be sorted to create desired reports. These reports, once generated, shall be displayed and/or printed at the user's discretion.

4. Each user assigned in the system shall be identified as either a "visitor" or "standard" user. The system shall also have the ability to assign a validity period to each user having a "to" and "from" date-month-year.

F. Archive: The system shall allow event history to be written to the fixed disk and accumulated as archives. The system shall have the capacity to store minimum of 400,000 transactions. As the archive files reach approximately 60,000 events, the system shall generate a "backup archive" message. The system does not require that back-up archiving be accomplished when the message appears and will continue to archive additional events. The backup message will reappear each time that the events are enough to fill an additional diskette. The system shall have the capacity to off-load the archive files onto a 3.5″ diskette for off-line storage and later retrieval.

G. Reports Templates

1. The system shall allow reports to be generated from the history accumulated on the system's fixed disk. These reports shall be created on an as-needed basis by selecting the report parameters necessary based on time and date. The system shall allow the creation of preformatted parameters to be stored as archive report templates. These templates shall have a minimum 16-character title to easily identify the format to the operator. The system shall allow any report template to be cleared or modified. Archive templates shall be created through a selection process of event classifications available. This selection allows individual event types to be selected by an all or none choice.

2. An operator choosing an all category shall still be able to exclude any subcategory of an event type. The selection of none excludes the entire event type and all corresponding subcategories. There shall be a minimum of eight archive report templates in the system any of which may be linked to any of the user templates. Once the template is set as desired, the report will be created, and then viewed or printed as desired.

3. The system shall provide a minimum of 32 user-definable report templates, each having a unique 16-character definable name. Systems using numbered reports only will not be acceptable. The operators shall have at their command the ability to create each of the 32 reports from the system's hard disk or from floppy back up. Each report shall be selectable as to its start and stop time. Each report shall be selectable so that it may be displayed and/or printed at the operator's command.

H. Operator Tutorial Screens: The system shall provide help screens that are specific to the area of the system being used. There shall be a minimum of 450 operator help screens. These help screens shall be selected by a single function key.

I. System Status: It shall be possible to query the status of any of the system controllers, access control doors, and input or output devices. This status shall display the current state of the device in question at the time of query. The system shall have the ability to group devices into a minimum of three classes. Each of these classes shall have a minimum of 64 groups. Each group shall be described by a minimum of 16 characters. The three classes shall be doors, inputs, and outputs.

J. System Maintenance

1. The system shall provide the capacity to create and restore backups of the user configuration and network configurations. It shall be possible to create backups of archive files on demand. The system shall allow a diskette to be formatted to ready its ability to accept data from the fixed disk. The system shall provide the ability to select desired mouse sensitivity. There shall be a minimum of five sensitivity settings.

2. Upon operator request, the system shall generate a "system version" report listing the current system software version and serial number as well as the firmware version for all main and terminal controllers active on the system. This shall be accomplished without leaving the access control software and must be part of its standard software.

K. Alarm Response: The system shall have the ability to access the alarm response screen by selecting a single function key or by menu selection through use of the mouse. This function key shall be selectable from any location in the system. Once selected, the alarms shall be displayed based on priority, it shall be impossible to acknowledge any alarm of a lower priority than those displayed on the screen. The system shall be able to maintain a minimum of 256 unacknowledged alarms in the priority queue. All alarms displayed shall have the ability to be acknowledged singularly or as a group based on priority. The system shall provide the ability to enter an operator response documenting action taken. This response shall be a minimum of 30 characters. The system shall provide annunciation of alarm events at the command center based on priority and provide the ability to silence this annunciation in two ways. The annunciation shall be silenced by acknowledgment of the pending alarm or be disabled totally by the command of a privileged operator.

L. On-Line Maintenance: The system shall provide on-line diagnostics and communications maintenance for adjustment to the operating environment. These diagnostics shall allow for the modification of baud rate, system packet information, and network polling. It shall be possible for the system to adjust the data handshake ability through channel commands and channel response. On-line maintenance providing real-time communications conditions of main controllers and terminal controllers is required.

Part 3. Execution

3.1 Warranties: The system shall be warranted for a period of one year from date of acceptance or first beneficial use. The equipment manufacturer shall make available to the owner a maintenance contract proposal to provide a minimum of two inspections and tests per year.

END OF SECTION

SECTION 16731

Security Alarm System

Part 1. General

1.1 Furnish and install a complete intrusion detection monitoring and control system with the performance criteria detailed in this specification. The system shall be inclusive of all necessary functionality, monitoring, and control capability as detailed herein and on accompanying drawings.

1.2 The system shall be completely programmable from any keypad with programming access determined by level of personal identification number (PIN) code. There shall be no need for a removable programming module or PROM burn to accomplish user programming changes.

1.3 The system shall be listed as a Power Limited Device and be listed under the following performance standards.

    A. UL 1610, Central Station Burglar Alarm Units.

    B. UL 1635, Digital Burglar Alarm Communicator System Units.

    C. UL 365, Police Station Connected Burglar Alarm Units and Systems.

    D. UL 864, Control Units for Fire Protective Signaling Systems.

1.4 Each system shall be supplied with complete details on all installation criteria necessary to meet all of the above listings.

1.5 The system supplier shall be a company specializing in the manufacture and supply of security, fire, and access control systems with at least 10 years of experience and shall have local employees available for support during installation and for final hookup and acceptance testing and shall be located within 35 miles of the installation. The local manufacturer's office shall produce system-specific layout and wiring shop drawings for use by the installing contractor.

1.6 System Description

    A. Input/Output Capacity: The system shall be capable of monitoring a minimum of 48 individual loops or zones and controlling a minimum of 2, 8, 16 output relays.

    B. User/Authorization Level Capacity: The sys-tem shall be capable of operation by 100 unique PIN codes with each code being assigned one of nine User Authorization Levels.

    C. The user of the system shall be capable of selectively arming and disarming any one or more of eight areas within the system based on the user PIN code used. Each of the 48 loops or zones shall be assignable to any one of the eight available areas.

    D. Keypads: The system shall support a minimum of eight keypads with alphanumeric display. Each keypad shall be capable of arming and disarming any portion of the intrusion detection system based on PIN authorization. The keypad's alphanumeric display shall provide complete prompt messages during all stages of operation and programming of the system and display all relevant operating and test data.

    E. Loop Configuration: The system shall have a minimum of 16 Class B loops available in the Command Processor control cabinet and a minimum of 4 Class B loops available at each keypad or loop expander on the system. The system shall have the capacity for a minimum of eight keypads or loop expanders total but at least one must be a keypad. All Class B loops shall be two wire, 22 AWG minimum, supervised by an End-of-Line (EOL) device and shall be able to detect open, normal, or short conditions in excess of 200 milliseconds duration.

F. Keypad Communication: Communication be-tween the Command Processor control panel and all keypads and loop expanders shall be multiplexed over a four conductor nonshielded cable. This cable shall also provide power to all keypads, loop expanders, and other power-consuming detection devices.

G. Output Relays: The Command Processor control cabinet shall have, as an integral part of the assembly, 2, 8, 16 output control SPDT form C relays rated 3 amps at 30 VDC or 120 VAC. Each of these relays shall be capable of activation as outlined in this specification.

H. Primary Power: The Command Processor primary power supply shall be a 16 VAC 40 VA Class 2 Wire-in transformer.

I. Secondary Power: The Command Processor secondary power supply shall be two 12 VDC 6 AH sealed lead-acid rechargeable batteries. Each battery shall be protected by an automatic circuit breaker. When initially connected to battery power alone, the Command Processor control panel shall be protected by a cutoff relay until manually started or primary power is applied. The secondary power shall be float charged at 13.8 VDC at a maximum of 1.2 amps.

J. Battery Supervision: The Command Processor control panel shall supervise the secondary power source by placing a load across the batteries once every hour while the primary source is available. If the voltage falls below 11.9 VDC a low battery fault shall be detected. If the primary power source is not available, a low battery fault shall be detected any time the voltage falls below 11.9 VDC. The secondary power supply shall be automatically disconnected from the system when the primary power supply is not available and the secondary power supply drops to 10.0 VDC.

K. Ground Supervision: The Command Processor control panel shall supervise the earth ground connection and annunciate an open circuit condition.

L. Signal Output: The Command Processor control panel shall be capable of supplying a minimum of 2 amps continuous at 10.5–15.0 VDC to power local sounding devices.

M. Auxiliary Output: The Command Processor control panel shall be capable of supplying a minimum of 1 amp continuous at 10.5–15VDC to power keypads, loop expanders, smoke detectors and other power-consuming detection devices such as motion detectors.

N. Keypad Trouble: If at any time a keypad does not detect polling, the alphanumeric display shall indicate "Service Required." If at any time a keypad detects polling but not its particular address, the alphanumeric display shall indicate "Non-Polled Address."

O. Remote Communication Capability

1. Central Station Capability: The system shall be monitored at a Central Monitoring Station using a Digital Alarm Communicating Transmitter.
2. Communicator Program: Single Line Digital Alarm Communicating Transmitter. The system shall be capable of dialing two telephone numbers, of 15 digits each, using the switched telephone network such that if two unsuccessful

attempts are made to the first number the system shall automatically switch to the second number and make two attempts. If these two attempts are unsuccessful the system shall switch between numbers, after two attempts each, until a successful connection is made or a maximum of 10 tries are attempted. Once 10 unsuccessful attempts are made the system shall stop dialing. Should another event occur that requires a message to be transmitted, the dialing process shall be repeated.

1.7 System Capability

A. Arm Display: The system shall display the identity of all armed security areas on keypad alphanumeric displays.

B. Opening Code: The system shall require a valid user code to disarm security areas.

C. Closing Code: The system shall require a valid user code to arm security areas.

D. Any Bypass: The system shall require a valid user code to bypass any security loops during arming.

E. Entry Delay: The system shall permit an entry delay time of 0 to 250 seconds on any loops assigned as exit type loops. When an armed exit type loop is activated, a prewarn tone shall sound and the entry keypad shall display "Enter Code." If a valid user code is not entered prior to the expiration of the entry delay, an alarm will be transmitted.

F. Exit Delay: The system shall permit an exit delay time of 0 to 250 seconds on any loops assigned as exit-type loops. This exit delay shall be displayed and counted down on the exit keypad's alphanumeric display. If any loop is in an alarm condition at the expiration of the exit delay, the entry delay sequence will commence immediately.

G. Loop Retard Delay: The system shall allow a loop retard delay of 0 to 250 seconds to be applied to any loop designated as Fire, Supervisory, Auxiliary 1, or Auxiliary 2. This retard delay shall only function in the short condition.

H. Swinger Bypass: The system shall be able to automatically bypass any loop that trips more than 0 to 7 times within one hour commencing with the first trip. The system shall also transmit a report of automatic bypass to the Central Monitoring Station if Bypass Reports are included as monitored events.

1.8 Output Control Capability

A. Signal Cutoff: The system shall automatically reset the Signal Output 0 to 99 minutes after the Bell Output has been activated.

1.9 User Capability

A. Arm/Disarm: The system shall allow authorized users to arm and disarm the burglary system and display such on the alphanumeric display.

B. Alarm Silence: The system shall allow authorized users to silence the bell output and display such on the alphanumeric display.

C. Armed Areas: The system shall allow authorized users to display a list of armed areas by number and name on the alphanumeric display.

D. User Codes: The system shall allow authorized users to add and delete user codes in the system memory.

1.10 Display Capability

A. System Monitor Trouble: The system shall annunciate and display trouble conditions from the following functions on any or all of the alphanumeric keypads in the system. The functions to be displayed shall be bell circuit, AC power, battery power, bell power fuse, panel tamper auxiliary power fuse, and ground circuit.

B. Area Name: Each of the 8 areas within the system shall be identified by a name consisting of up to 10 alphanumeric characters. This name shall be used to identify an area when displayed on the alphanumeric keypad.

1.11 Loop or Zone Capability

A. Loop Name: Each of the loops within the system shall be identified by a name consisting of up to 10 alphanumeric characters. This name shall be used to identify a loop when displayed on the alphanumeric keypad or card access.

B. Loop Type: The system shall be able to identify each loop as one of 10 different loop types. Each loop type shall have up to eight specifiable characteristics with a default configuration based on loop type.

  1. The system shall allow each individual loop to be configured independently for each of the characteristics related to its selected loop type.

  2. The system shall allow each individual loop to activate one selectable output relay per loop status change and the loop changes that can activate a separate relay each are disarmed opened, disarmed shorted, armed open and armed shorted if available in the selected loop type for that loop. Further, the selected relay shall be either latched activated, pulsed on and off, momentarily activated, or follow the status of the loop.

C. Loop Message: The system shall allow the selection of either an alarm, trouble, or no message to be displayed on the alphanumeric keypad (and transmitted to the Central Monitoring Station) when a loop condition changes. The selection shall be made separately for disarmed open circuit, disarmed short circuit, armed open circuit, and armed short circuit when available for the selected loop type. Each loop shall have a default selection based on the loop type selected.

D. Prewarn Addresses: The system shall allow the selection of any or all keypads to sound a prewarning when entry is made through an exit type loop. The system shall at this time also display "Enter Code" on the alphanumeric display on selected keypads or card access.

1.12 System Operation

A. User Codes: The system shall allow an authorized user to add a minimum of 100 individual PIN codes to the system and each PIN code shall be assigned to one of nine authorization levels.

  1. Each PIN code shall be from two to five digits in length (and allow the assigned user to activate and deactivate a predefined combination of the burglary areas within the system).

B. Output Schedules: The system shall allow an authorized user to establish and change a single on-off permanent schedule for each of the output relays for each of the seven days of the week.

C. Burglary System Schedules: The system shall allow an authorized user to establish a permanent opening and closing schedule for each day for the burglar alarm system such that certain users shall not be able to deactivate the alarm system outside of the established schedule. The system shall also allow an authorized user to establish a temporary opening and closing schedule for each day for the burglar alarm system to operate as a permanent schedule except that this schedule shall be automatically canceled after a single use.

1. The system shall also use the established schedules as a reference for the Closing Check function when openings and closings are reported to a central monitoring station.

Part 2. Products

2.1 The items listed below shall be included as directed by ADT Company.

A. Command Processor Control Panel: The Command Processor control panel shall be supplied with eight output relays.

B. Security Command Keypads: The Security Command Keypads shall be flush mount.

C. (Option) Communication Modules

D. Polarized Signal Module

E. Detection Devices: The various detection devices connected into the protection loops shall be standard Simplex products to ensure compatibility of performance and power consumption with the specified system.

1. Provide, where shown on plans, magnetic flush-mounted door contacts and magnets.
2. Outside High Powered,Weather Resistant, Siren Speaker: Provide 30 watt Siren Speaker. Unit shall feature 125 DB at 10′ sound output and shall be provided with N/C tamper loop to supervise against removal.
3. Outside Tamper Resistant Weatherproof Enclosure: Provide, as shown on plans, tamper resistant enclosure provided with two tamper switches to supervise against front cover and wall mounting.
4. Motion Detectors (Passive Infrared): Provide as shown on plans. Units shall operate from 6-12 volts DC and draw a maximum of 25 Ma. Sensor shall offer wall, corner, ceiling, or flush mounting and offer a tamper switch to supervise cover removal. A gimbal mounting-bracket shall provide mounting to a standard single-gang electrical box and permit the detector pattern to be adjusted as indicated on the drawings.
5. Duress Switch: Provide, as shown on drawings, Holdup/Panic Alarm Switch. Switch actuation shall be recessed to prevent accidental activation. A reset tool shall be provided to restore to normal condition. Contact arrangement shall be either N/O or N/C.

F. Output Devices: The various control devices connected to the system output relays shall be standard Simplex products to ensure compatibility of performance and power consumption with the specified system.

Part 3. Execution

3.1 Codes: The contractor shall provide and install the system in accordance with the plans and specification, all national and local codes, and the manufacturer's installation instructions.

3.2 Other Work: Installation of equipment and devices that pertain to other work in the contract shall be closely coordinated with the appropriate subcontractor.

3.3 Supervision: The manufacturer shall provide all on-site supervision of the installation, perform a complete functional test of the system, and submit a written report to the contractor attesting to the proper operation of the complete system.

3.4 Testing: The complete system shall be fully tested by the contractor in the presence of the Owner's representative, the Architect, the Consulting Engineer, the authority having jurisdiction, and the manufacturer. Upon completion of a successful test, the contractor shall so certify in writing to the Owner, Architect, manufacturer, and General Contractor.

3.5 Warranty

   A. The Contractor shall warrant the complete system wiring and equipment to be free from inherent mechanical and electrical defects for a period of one year from the completed and certified test.

   B. The equipment manufacturer shall make available to the owner a maintenance contract proposal to provide a minimum of two inspections and tests per year.

3.6 Training: The equipment manufacturer shall provide, as part of this contract, a minimum of two (2) hours of system programming and operation training to the building owner and consulting engineer.

END OF SECTION

SECTION 16740

Telephones

Part 1. General

1.1 Coordinate this work with that of other trades a for timely execution. All work shall meet the requirements of the telephone company.

Part 2. Products

2.1 Outlet and Pull Boxes: Standard galvanized, as specified elsewhere.

2.2 Conduit: $3/4''$ diameter, or as specified elsewhere, with pull wire.

Part 3. Execution

3.1 Install a complete raceway system and mounting boards/cabinets for telephone system wiring as indicated on the drawings. The raceway system shall include the conduit, fittings, weatherheads, pull boxes, terminal cabinets, and outlets required for a complete installation.

3.2 Mount telephone outlets 1'6'' above the fin-ished floor unless otherwise noted on the drawings.

END OF SECTION

SECTION 16776

Paging System

Part 1. General

1.1 Furnish and install a complete audio paging system to cover office and plant areas as shown on the drawings and as specified herein.

1.2 It is understood that the Contractor shall be familiar with the on-site conditions and constraints imposed on the installation of all materials resulting from such conditions including the coordination of this work with other trades.

1.3 Time is of the essence in this work. It is essential that this work be 100% complete before plant opening. Nonessential work and final adjustments may be completed after production begins.

1.4 The Contractor shall supply and install complete working sound amplification systems including all apparatus and equipment wiring, labor, and services required to provide systems of professional quality in excellent working order as specified herein and as shown on the included drawings.

1.5 Responsibilities include, but are not limited to, the supplying of total working systems and suitability of the system and its components to meet the functional and performance requirements.

1.6 See the electrical drawings for available conduits, cable trays, location of speaker racks, and related facilities.

1.7 Any incidental equipment needed in order to meet the requirements stated above, even if not specifically mentioned herein or on the drawings, shall be supplied and installed by the Contractor without claim for additional payment.

1.8 Submittals:

    A. Block diagrams indicating proposed connections of all equipment and indicating equipment types, wire sizes, model number, etc.

    B. Layouts of control panels, equipment racks, and cabinets.

    C. Remote loudspeaker mounting arrangements.

    D. Catalog sheets may be used where relevant.

    E. Design information shall be included showing all audio voltage and power levels as well as how acoustical output and sound pressure levels were calculated for all listening areas.

1.9 Retain the services of a full-time, qualified senior technician to oversee the installation. Periodic inspections and supervision shall be carried out by a senior engineer to ensure that the system installation is in accordance with the specifications. Maintain adequate staff on the job to ensure that the scheduling of the system installation coincides with the overall scheduling.

1.10 All reference herein to model numbers and other pertinent information is intended to establish the standards of performance, quality, and appearance based upon equipment designed and manufactured by Dukane Corporation, St. Charles, Illinois.

Part 2. Products

2.1 Paging Rack: Rauland RP-1103 Upright Rack. Proved 61-$\frac{1}{4}$ inches total panel-mounting space. The rack shall be ebony black baked enamel constructed of 16-gauge steel. The rack shall be UL listed. Provide at least 3-inch vent panels between all power amps and blank panels in all unused spaces.

2.2 Monitor Panel: Rauland MPX-100 Rack Mount with channel selector, monitor volume, VU meter. Unit shall be capable of high level aural and visual monitoring of five separate amplifiers.

2.3 Pre-Amp: Rauland AP-4601 8-input mixer pre-amp with automatic muting over one input, 2 adjustable notch filters, built-in limiter, input attenuation pad, and selectable tone controls. Inputs shall accept telephone page input, mic input, or auxiliary input from tone generators and music sources.

2.4 Power Amplifiers: Rauland FAX-250 UL-listed 1480 for use in fire-protective signaling systems. Rated output 250-watts RMS harmonic distortion 2% at rated output. Power amp shall have output transformer for 25-volt or 70-volt outputs at full power. Unit shall be capable of being operated from 24-28 VDC 17-amp backup system. Unit shall incorporate an automatic protective circuit and thermal compensation circuit to protect against shorts in the output. Provide 3 each power amps.

2.5 Power Amp—Office Area: Rauland 1410 rack mounted, 100 watts RMS. 4-inputs with voice op-erated muting 25- or 70-volt output. Unit shall have slide control base, treble, and master volume.

2.6 Tone Generator: Rauland MTG-100 Tone Generator with four separate outputs. Unit shall generate chime, siren, tone, and whooping sounds. This unit shall generate sound to all amplifier circuits upon closer of contacts from fire alarm system.

2.7 Digital Feedback Eliminator: Viking FB1-1 shall store pages up to 16 seconds and automatically adjust to actual page length. Provide one for each zone.

2.8 Speakers—Combinations: Ceiling recessed with white grill similar to Quam C10B70Q/VK 8-inch speaker combination with 70-volt transformer. Unit shall be installed with ERD-8 metal enclosure. All speaker taps will be set at 1-watt and volume control adjusted for best coverage. Volume control shall be on front of speaker.

2.9 Production Area Speakers: Fourjay Model IC8CG suspended speaker housing constructed of polythylene plastic.

2.10 Horn Speakers: Fourjay 216-TD 16-watt, 70 volt paging horn with seven-stop tap selection.

2.11 Wiring: Speaker Wire 16/2 stranded with overall jacket similar to West Penn 225. Cable connections to speaker will be made with crimp type solderless connections.

2.12 The sound system control panel shall be interfaced with the fire alarm system utilizing appropriate and approved interfacing methods. Initiation of any manual or automatic detection device shall cause the fire alarm tone signal to automatically sound over all loudspeakers and override any other audio signals.

2.13 All telephone page amplification circuits shall incorporate an adjustable seize timer (5–50 seconds) and electronic tone circuitry that generates three adjustable tones:

A. An annunciator tone that indicates the system has been accessed by a telephone for an announcement. This tone sounds to all speakers connected to that circuit and back into the telephone earpiece, thereby providing assurance that the amplifier was accessed and is working properly.

B. An access tone that sounds at the telephone earpiece to indicate that the system is ready for a zone number to be dialed.

C. An end of message tone that sounds at the telephone earpiece when the seize timer finishes to warn the pager that the paging system will disconnect (hang up).

2.14 All equipment shall be supplied new and unused.

2.15 All equipment shall be manufactured to both UL and EIA standards.

Part 3. Execution

3.1 General

A. The Contractor shall provide, at his or her own expense, maintenance service for a period of one year after acceptance of installation.

B. Coordinate all related work and interfacing of fire alarm audibles with fire alarm equipment supplier.

C. During this period, the Contractor shall answer all service calls within four normal working hours and carry an ample stock of spare parts for servicing this equipment.

D. The system shall be designed so that once it is completed and sound levels have been properly adjusted, all system amplifiers will not be loaded at more than 80% of their rated output power.

E. The total sound reinforcement system shall be designed so as to provide amplified sound to all listeners in all areas of both office and plant at sound pressure levels sufficiently above average background noises to be heard clearly (a minimum of 5 db above ambient).

3.2 Demonstration and Acceptance Testing

A. Demonstrate operation of each major component, using each microphone and/or telephone paging circuit at all loudspeaker positions, and all input, control, and amplification equipment. After demonstration, the Contractor shall assist as required in the following acceptance tests by representatives of the Owner.

B. Listening Tests: These tests may include speech intelligibility surveys and subjective evaluations by observers listening at various positions under operating conditions, using speech, music, and live or recorded material.

C. Equipment Tests:

1. Any measurements of frequency response, distortion, noise, or other characteristics and any operational tests deemed necessary may be performed on any item or group of items to determine conformity with specifications.

2. If need for adjustment becomes evident during demonstration and testing, the Contractor's work shall be continued until the installation operates properly and to the satisfaction of officials.

3.3 Training

A. Provide on-premise training for personnel in the proper service procedures to be followed in maintaining the complete sound reinforcement system.

END OF SECTION

SECTION 16782

Closed Circuit Television System

Part 1. General

1.1 Furnish and install a system of closed-circuit TV cameras as indicated on the drawings and directed by ADT Co.

Part 2. Products

2.1 Exterior Cameras: High-resolution 580-TVL environmentalized CCD cameras. The housings shall be environmentally sealed and pressurized with dry nitrogen. Ambient temperature −40°F to −13°F. CCD image sensor shall be 2/3 format with 1R filter, 818H × 513V pixels. Lens shall be 8mm F/1.4 auto iris. Finish of housing shall be white. Unit shall be mounted with heavy-duty mount of aluminum with adjustable head bracket. Mount shall have a load rating of 100 pounds with a white epoxy finish. Unit shall be mounted for maximum coverage in reference to scene with minimum glare from sunlight and bright objects.

2.2 Inside Cameras: 1/2-inch CCD image sensor. 510H × 492V pixels. Lens shall be 12mm auto iris F/1.4. Horizontal resolution 380TVL. Furnish and install light weight camera mount of aluminum with adjustable head bracket.

2.3 TV Monitors: Furnish each camera with a 9-inch monochrome video monitor. Resolution of 700 lines. Unit shall be housed in a metal case with front panel controls. Monitor shall accept 0.5 to 2.0 Vp-p composite video, sync negative. Two BNC connectors and HiZ/75 ohm termination switch.

    A. Furnish and install videolink system consisting of transmitter and receiver. This unit shall permit real-time video transmission over twisted pair cable. Furnish repeaters for each unit.

    B. Unit shall accept input impedence of 75 ohm with frequency response of 5 Hz to 5 MHz.

2.4 Cable: All cable shall be CL2 rated 95% braid shield copper of RG-6 type. Where lengths exceed 1,000 feet RG-11 type cable shall be used. All cables shall be labeled at both ends.

Part 3. Execution

3.1 All work shall be performed by persons regularly engaged in CCTV installation with a minimum of five years experience. All material shall be warranteed for one year from date of installation.

END OF SECTION

ELECTRICAL DEVICES AND MATERIALS

NOTE: This section identifies various devices and materials commonly used in electrical construction, including name, photograph, picture, and application.

CONTENTS

Meter Bases
Panel Boxes
Breakers
Disconnects
Fuses
Grounding Devices
Receptacles
Switches
Wire and Cable
Nonmetallic Cable (NMC; Romex)
Armored Cable (AC)
Range Cable
Service Cable
Weather Head
Wire Connectors
Conduit and Fittings
Flexible Conduit
Conduit Bodies
Conduit Straps
Trim Plates
Incandescent Fixtures
Fluorescent Fixtures and Ballasts
Emergency Lighting
Switchgear
Boxes, Metal and Plastic
Relays
Timers
Transformers
Explosion-Proof Devices
Surface Mount Raceway
Floor Duct
Busway
Cable Tray

Meter Bases

The electrical service utility connection point and meter socket location.

FIGURE C-1 Meter base and disconnect.

Panel Boxes

The customer's electrical service point; contains the breaker-mounting framework.

FIGURE C-2 Residential panel and subpanel.

FIGURE C-3 Commercial panel.

FIGURE C-4 Commercial subpanel.

Breakers

Thermomagnetic devices using a bimetallic element to open circuits during overload conditions. They are circuit-protection devices and may be reset if the overload condition is eliminated.

FIGURE C-5 Single-pole breaker.

FIGURE C-6 Two-pole breaker.

FIGURE C-7 Three-pole breaker.

FIGURE C-8 GFCI breaker.

FIGURE C-9 Arc-fault breaker.

Disconnects

Used to disconnect power from individual pieces of equipment (e.g., motors, heating units, and other electrical installations); fused and nonfused.

FIGURE C-10 AC disconnect.

FIGURE C-11 120/240 disconnect.

FIGURE C-12 Disconnect with circuit breaker.

Fuses

Circuit overload protection devices containing short-circuit elements. Excessive current will melt the overload element and open the circuit. Fuses must be replaced after they are opened during overload.

FIGURE C-13 Type W edison-base fuse.

FIGURE C-14 Class G cartridge fuse.

FIGURE C-15 Blade type cartridge fuse.

Grounding Devices

Used to bond all possible current-carrying devices in a building and direct them to the best available ground path.

FIGURE C-16 Acorn clamp.

FIGURE C-17 Ground rod and clamps.

Receptacles

Plug-in contact devices used as an outlet for plug-in electrical devices (lamps, radios, and small appliances). There are many configurations and voltages of receptacles (e.g., grounded, ground fault, range, and others).

FIGURE C-18 Duplex receptacle.

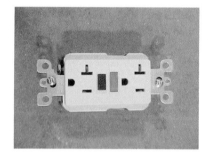

FIGURE C-19 GFI receptacle.

FIGURE C-20 Wire range receptacle.

Switches

Commonly used to switch lights, fans, and other devices off and on. There are various types and voltages of switches (e.g., single-pole, double-pole, three-way, four-way, and more).

FIGURE C-21 Single-pole switch.   **FIGURE C-22** Double-pole switch.

FIGURE C-23 Three-way switch.   **FIGURE C-24** Four-way switch.

Wire and Cable

Current-carrying material, which comes in various sizes, materials, insulation types, and configurations and may be solid and stranded. Solid wire usually comes in No. 6 and smaller, whereas strand-ed wire is usually No. 14 and larger. Common sizes range from 1000 kcmil to No. 18 and smaller in some applications. Wire is usually copper and aluminum. Common insulation types are plastic and rubber. Insulation applications may vary as per NEC Article 310.

FIGURE C-25 Solid wire.   **FIGURE C-26** Stranded wire.

Nonmetallic Cable (NMC; Romex)

Nonmetallic cable (romex) comes in various sizes and configurations (e.g., No. 12/2/G, which means two insulated No. 12 conductors, and one No. 12 bare ground). These are covered with a plastic outer wrap. Romex ranges in size from No. 14 to No. 6 and comes in configurations such as 12/2/G, 12/3/G, and 12/3/ with no ground.

FIGURE C-27 Romex cable.

FIGURE C-28 Underground feeder cable.

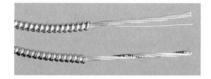

Armored Cable (AC)

Contains two or more insulated conductors covered by a flexible metallic shield. It is used in residential, commercial, and industrial applications as required by code.

FIGURE C-29 Armor-clad cable.

Range Cable

Carries current to ranges and other large appliances. May be in NMC or molded whip form. Size is generally 6/6/8 in copper.

FIGURE C-30 Range cable.

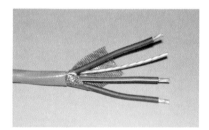

Service Cable

Used from the meter to the utility connection. The size depends on the main service current requirements (e.g., a 200-amp residence uses 2/0–2/0–1/0 copper wire).

FIGURE C-31 Service entrance cable.

Weather Head

Device used on service entrance conductors at exterior conduit entry point to keep moisture out of the meter base.

FIGURE C-32 Weather head.

FIGURE C-33 Weather head on house.

Wire Connectors

Devices used to mechanically connect two or more wires. Common types are twist, compression, and split bolt.

FIGURE C-34 Split bolt connectors.

FIGURE C-35 Crimp connector.

FIGURE C-36 Twist connectors.

Conduit and Fittings

A tubing system used to protect wires from physical damage and environmental elements. Common types include rigid steel, electrical metallic tubing, and plastic (PVC).

Standard conduit sizes are 10′ lengths with inside diameters of ½ to 6 inches. Coupling and connector types are compression, set screw, and threaded. Couplings are used to join two or more pieces of conduit; connectors are used to connect conduit to a box.

FIGURE C–37 EMT conduit.

FIGURE C–38 LB and fittings.

Flexible Conduit

Circular flexible conduit in metal or plastic. Also available in liquid-tight coating, which is steel flex conduit coated with a plastic exterior jacket. Liquid tight is used in wet and hazardous materials locations.

FIGURE C–39 Flexible metal conduit.

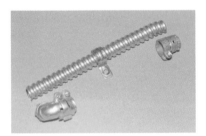

FIGURE C–40 Liquid-tight flexible conduit.

Conduit Bodies

The part of a conduit system that allows access to wires. Common configurations are LB, LR, LL, C, and Ts. Materials and sizes are the same as noted in the conduit section.

FIGURE C–41 Conduit bodies.

FIGURE C–42 Plastic LB.

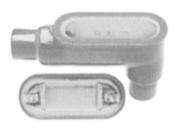

Conduit Straps

Used to mount or support conduit. Common types are nailer, hanger, one- and two-hole straps, and beam clamp type.

FIGURE C–43 Two-hole strap.

FIGURE C–44 Conduit strapped to strut rod.

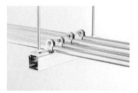

Trim Plates

Used to cover switches and receptacles while preventing physical contact with wires or connections. Trim plates come in a variety of colors and materials.

FIGURE C–45 Plastic trim plates.

FIGURE C–46 Metal coverplate.

FIGURE C–47 Outdoor coverplate.

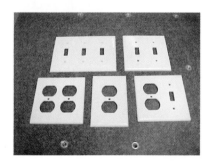

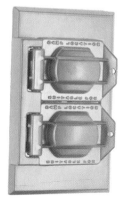

Incandescent Fixtures

This resistance-type lighting passes current through a filament that changes heat to light.

FIGURE C–48 Ceiling fixture.

FIGURE C–49 Ceiling fixture.

FIGURE C–50 Outdoor fixtures.

(A)

(B)

(C)

(D)

(E)

Fluorescent Fixtures and Ballasts

Tubular discharge lamp using vapor ionization to emit light. Produces far less heat than standard incandescent lamps. Lamps are usually straight or round in shape. Use of fluorescent lamps saves energy.

FIGURE C-51 Surface mount fluorescent fixture.

FIGURE C-52 Class P ballast.

Emergency Lighting

Lighting used to indicate emergency exits or to illuminate areas when normal power is interrupted. Emergency lighting usually has a generator or battery backup system.

FIGURE C-53 Emergency light.

FIGURE C-54 Emergency light.

FIGURE C-55 Exit light.

Switch-gear

The electrical distribution point used on commercial and industrial projects. Usually employs a large main breaker (up to several thousand amps) and various distribution breakers. It uses a copper or aluminum bus system to distribute the power to the breakers.

FIGURE C-56 Power-style switchboard.

FIGURE C-57 Metal-enclosed switchgear.

Boxes, Metal and Plastic

Used as a location to join two or more wires, or as pulling points. May also serve as junction points for conduit, switches, receptacles, and other electrical devices. Boxes come in various sizes, shapes, and materials (i.e., single-gang, two-gang, three-gang, aluminum, steel, plastic, special application, and more).

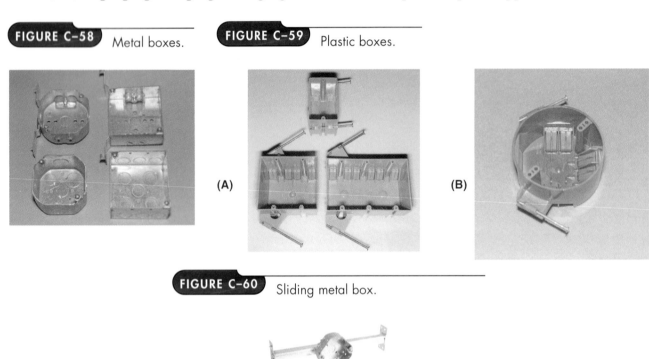

FIGURE C-58 Metal boxes.

FIGURE C-59 Plastic boxes.

(A)

(B)

FIGURE C-60 Sliding metal box.

Relays

Electrical switching device used in various applications of motor controls, lighting, energy management systems, and alarm systems. May be solid state or electromechanical with coil voltage ranging from 24 volts upward.

FIGURE C-61 Plug-in relays.

FIGURE C-62 Overload relay.

FIGURE C-63 Programmable relay.

(A)

(B)

Timers

Time-operated switching devices used to actuate circuits and control the amount of time on and time off. Used to control lights, machinery, and process control.

FIGURE C-64 Timer relay.

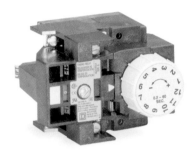

FIGURE C-65 Water heater timer.

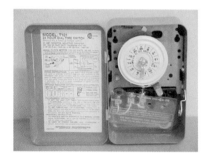

Transformers

Two or more coils coupled by electromagnetic induction used to transfer electrical energy. Frequency remains constant while voltage or current are increased or decreased. Applications range from a 10 VA doorbell transformer to a 1000 kVA transformer used on industrial projects. Transformers are either single- or three-phase. May be either air or liquid cooled.

FIGURE C-66 Transformer.

FIGURE C-67 LV transformers.

FIGURE C-68 Pad mount transformer.

Explosion-Proof Devices

Used to house wires, connections, and switches in flammable, explosive, dust, or vapor areas. Capable of containing sparks and flashes internally. Common in spray booths, grain elevators, and other hazardous locations.

FIGURE C-69 EYS.

FIGURE C-70 Explosion-proof box.

Surface Mount Raceway

Installed as an extension to existing raceways when concealing of the raceway is difficult. Available in various sizes and in plastic or steel construction.

FIGURE C-71 Surface mount raceway.

FIGURE C-72 Wall mount raceway.

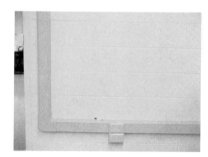

Floor Duct

Metal raceway used under concrete slabs for distribution of power, telephone, computer, or other specialty systems. Monuments are tapped into the duct and used as termination points.

FIGURE C-73 Subway floor duct.

Busway

Power transfer system using large, metal, bar-shaped conductors that act as carriers and connection points. These are called bus bars. The bus system is surrounded by a metal case that protects it and acts as a mounting point for disconnect boxes. These disconnect boxes supply machines, heaters, and other devices.

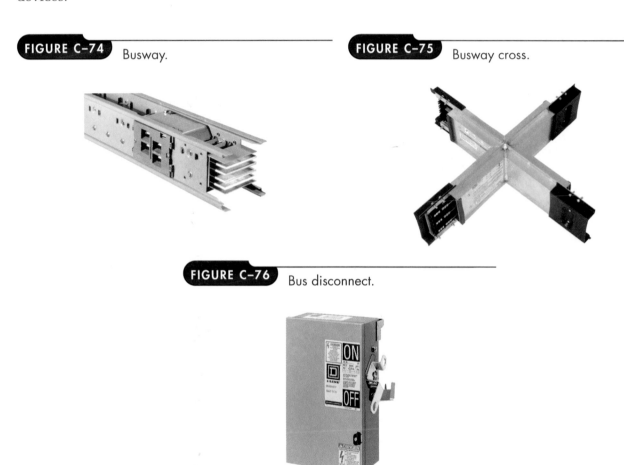

FIGURE C-74 Busway.

FIGURE C-75 Busway cross.

FIGURE C-76 Bus disconnect.

Cable Tray

An external transport system used to carry and support power, control, and communication cables. System components consist of straight sections, curved sections, crossovers, up sections, down sections, and box connectors.

FIGURE C-77 Cable tray.

ABBREVIATIONS

NOTE: This section contains a list of common abbreviations used by electrical workers and in the construction trades. There are specific abbreviations for switches, outlets, wire types, and square and round box notes. These abbreviations are found in this text and in electrical handbooks.

A or amp	amperes
AB	anchor bolt
AC	above counter
AC	air conditioning
AC	alternating current
AFF	above finished floor
AGGR	aggregate
AHU	air-handling unit
AL	aluminum
ALT	alternate
ALUM	aluminum
ANSI	American National Standards Institute
ANT	antenna
AP	access panel
APPROX	approximate
ASB	asbestos
ASPH	asphalt
ASTM	American Society for Testing and Materials
AUTO	automatic
AWG	American wire gauge
A/C	air conditioner
B or BA	bathroom
BALC	balcony
BASM	basement
BATT	batten
BC	broom closet
BD	board
BG	below grade
BL	building
BLDG	building
BLK	black
BLK	block
BLKG	blocking
BLT-IN	built in
BLU	blue
BLO	blower
BM	beam
BM	benchmark

BP	blueprint
BR	bedroom
BRG	bearing
BRK	brick
BRKR	breaker
BRN	brown
BSMT	basement
BTU	British thermal unit
BTM	bottom
BX	flexible armored cable
C	common
C to C	center-to-center
CA	cold air
CAB	cabinet
CALK	caulking
CB	catch basin
CB	circuit breaker
CEIL JST	ceiling joist
CEM	cement
CER	ceramic
CF	cement floor
CI	circuit interrupter
CJ	control joint
CKT	circuit
CL	ceiling joist
CL	center line
CLG	ceiling
CLO	closet
CLR	clear
CMU	concrete masonry unit
CND	conduit
COL	column
COM	common
COMB	combination
CONC	concrete
COND.	conductor
CONST	construction
CONTR	contractor
COP	copper
CORR	corridor
COV	cutoff valve
CSI	Construction Standards Institute
CSK	countersink
CSP	central switch panel
CT	current transformer
CTL	central
CTR	center
CU	copper
CU	cubic
CV	check valve

CW	cold water
D	disposal
D	drain
D	dryer
DB	decibel
DC	direct current
DCP	dimmer control panel
DEG	degree
DET	detail
DF	drinking fountain
DIA	diameter
DIM	dimension
DIM	dimmer
DIF	diffuser
DISP	disposal
DK	decking
DMR	dimmer
DN	down
DNG RM	dining room
DP	double pole
DPR	damper
DPDT	double pole, double throw
DPST	double pole, single throw
DR	dining room
DS	downspout
DSC	disconnect
DT	double throw
DW	dishwasher
DW	drywall
E	east
E	voltage
EG	earth ground
ELEC	electric or electrical
ELEV	elevation
ELV	elevation
EMT	electrical metallic tubing
ENT	entrance
ENT	electrical nonmetallic tubing
EP	electrical panel
EP	explosion proof
EST	estimate
EWC	electric water cooler
EXC	excavate
EXH	exhaust
EXHV	exhaust vent
EXP JT	expansion joint
EXPO	exposed
EXT	exterior
EXT GR	exterior grade
EXTN	extension

F	Fahrenheit
F	fluorescent
FAB	fabricate
FAM RM	family room
FB	fuse block
FBRK	firebrick
FC	foot candle
FC	furred ceiling
FD	floor drain
FIN	finished
FIN FLR	finished floor
FIN GR	finished grade
FIX	fixture
FL	flashing
FL JST	floor joist
FLEX	flexible
FLAM	flammable
FLR	floor
FLUR	fluorescent
FLUOR	fluorescent
FMC	flexible metal conduit
FND	foundation
FO	finished opening
FOS	face of studs
FOUND	foundation
FP	fireplace
FP	fireproof
FPRF	fireproof
FS	flow switch
FS	full size
FSBL	fusible
FSC	full scale
FT	feet
FU	fuse
FUBX	fuse box
FUR	furnace
FURN	furnace
FXTY	fixture
G	gas
GA	gauge
GALV	galvanized
GAR	garage
GD	ground
GFCI	ground fault circuit interrupter
GFI	ground fault interrupter
GIRD	girder
GL	glass
GL	grade line
GND	ground
GR	grade

GRD	grade
GRN	green
GRY	gray
GWB	gypsum wallboard
GVL	gravel
GYP	gypsum
GYP BD	gypsum board
HA	hot air
HB	hose bib
HDR	header
HGT	height
HORIZ	horizontal
HP	horsepower
HT	heater
HT	height
HTG	heating
HTR	heater
HVAC	heating ventilating and air conditioning
HW	hot water
HWH	hot water heater
Hz or cps	hertz (cycles)
I	current
ID	inside diameter
ILLUM	illuminate
IMC	intermediate metal conduit
IN	inches
INCAND	incandescent
INS	insulation
INST	install
INSUL	insulation
INT	interior
J-BOX	junction box
JCT	junction
JST	joist
JT	joint
KIT	kitchen
KO	knockout
KVA	kilovolt ampere
KW	kilowatt
L	inductor
L CL	linen closet
LAU	laundry room
LAV	lavatory
LAU	laundry
LB	pound
LDG	landing
LED	light-emitting diode
LEV	level
LFMC	liquidtight flexible metal conduit
LFNM	liquidtight flexible nonmetallic conduit

LH	left hand
LIB	library
LIN	linen
LIV	living room
LR	living room
LT	light
LTS	lights
LV	louver
LUM	lumber
M	meter
M	motor
M-G	motor generator
MANUF	manufacturer
MAS	masonry
MAT	material
MAX	maximum
MC	medicine cabinet
MCM	thousand (milli) circular mils
MD	medium
MDP	main distribution panel
MECH	mechanical
MEMB	membrane
MH	manhole
MIN	minimum
MIRR	mirror
MN	main
MOT	motor
MRB	marble
MTL	metal
N	north
NA	not applicable
NAT	natural
NAT GR	natural grade
NEC	National Electrical Code
NEG	negative
NELA	National Electric Light Association
NFPA	National Fire Prevention Association
NG	natural grade
NIC	not in contract
NO	number
NOM	nominal
NTS	not to scale
N/C	normally closed
N/O	normally open
O	overload contactor
OC	on center
OC	over current
OD	outside diameter
OH	overhead
OPG	opening

OPP	opposite
OSB	oriented strand board
OUT.	outlet
OVHD	overhead
P	power
PAR	parallel
PART	partition
PB	push button
PC	pull chain
PERM	permanent
PERP	perpendicular
PH	phase
PL	plate
PLAS	plastic
PLAT	platform
PLMG	plumbing
PLS	plaster
PLS BD	plaster board
PLY	plywood
PLYWD	plywood
PNL	panel
PREFAB	prefabricated
PRCST	precast
PROP	property
PT	part
PT	pressure-treated (lumber)
PTD	painted
PVC	polyvinyl chloride
PVMT	pavement
PWR	power
QTY	quality
QTY	quantity
R	radius
R	range
R	recessed
R	resistance
RAD	radiator
RAD	radius
RCPT	receptacle
RD	roof drain
RD	round
REBAR	reinforcing bar
REC	recessed
RECP	receptacle
REF	reference
REF	refrigerator
REFR	refrigerator
REINF	reinforced
REG	register
RET	return

RFG	roofing
RGH	rough
RH	right hand
RIS	riser
RM	room
RMC	rigid metal conduit
RNC	rigid nonmetallic conduit
RO	rough opening
ROW	right of way
S	south
S	switch
SAN	sanitary
SC	solid core
SCR	silicon-controlled rectifier
SCR	screw
SCRN	screen
SD	smoke detector
SD	storm drain
SDG	siding
SECT	section
SERV	service
SEW	sewer
SHTG	sheathing
SIM	similar
SL	sliding
SLT	skylight
SP	single pole
SPDT	single pole, double throw
SPECS	specifications
SPST	single pole, single throw
SST	stainless steel
STAT	thermostat
STO	storage
STL	steel
STR	structural
SUB	substitute
SUSB CLG	suspended ceiling
SW	switch
T	thermostat
T & G	tar and gravel
T & G	tongue and groove
TC	temperature control
TC	terra-cotta
TEL	telephone
TELE	telephone
TEMP	temperature
THK	thick
TR	tread
TS	terminal strip
TV	television

TW	top of wall
TUB	tubing
TYP	typical
TZ	terrazo
U	underground
UBC	Uniform Building Code
UF	underground feeder
UGND	underground
UL	Underwriters Labratories
UNFIN	unfinished
UR	urinal
USE	underground service entrance
UTIL	utility
UTY	utility
V	valve
V	volts or voltage
VA	volt-amp
VAN	vanity
VB	vapor barrier
VB	vinyl base
VD	voltage drop
VENT	ventilation
VIN	vinyl
VOL	volume
VP	vent pipe
VT	vinyl tile
W	watt
W	west
W/	with
W/O	with out
WC	water closet
WD	wood
WH	water heater
WH	weep hole
WIC	walk in closet
WM	washing machine
WP	waterproof
WS	waste stack
WT	weight
WTHPRF or WP	weatherproof
WV	wall vent
XP	explosion proof
YD	yard

PRINT LIST

NOTE: This section contains a list of all residential, commercial, and industrial prints supplied for use with the text.

Print Sets

1. Residential Print: Franklin Residence 6 prints)

RCS	Cover Sheet	C1	Site Plan
A1	First-Floor Plan	A2	Second-Floor Plan
A3	Elevations	A4	Sections

2. Commercial Print: IBEW Local 5 Office Building and Training Center (12 prints)

CCS	Cover Sheet
AT6.1	Training Center Reflected Ceiling Plan
M1.1	Symbols and Abbreviations
MO1.1	HVAC Duct Work Plan
E1.1	Electrical Site Plan
E1.2	Legend of Symbols and Panelboard
E1.3	Lighting Fixture Schedule
EO2.1	O.B. First-Level Lighting Plan
EO3.1	O.B. First-Level Power Plan
EO4.1	O.B. First-Level Systems Plan
E5.1	Electrical Riser Schedule
E5.4	Telecommunications Details

3. Industrial Print: NJATC Manufacturing Plant (20 prints)

ICS	Cover Sheet
A1	Floor Plan
A2	Basement and Mezzanine Plan
A3	Exterior Elevations
A6	Wall Sections and Details
M2	HVAC Plan Col. 1–3
M5	Piping Plan Col. 1–3
M8	Mechanical Room Piping
M10	Mechanical Room Isometric
M21	Control Diagrams
E1	Site Plan Electrical
E2	Plant Lighting Col. 1–3
E5	Plant Power Col. 1–3
E16	Legend and Lighting Schedule
E17	Panelboard Schedule
E18	Panelboard Schedule
E19	Panelboard Schedule
E20	Feeder Diagram
E21	Electric Details
E22	Electric Details

Addendum A change to prints or specifications before bids are opened.

Alphabet of lines System of defining the types of lines used in technical and construction drawings.

American National Standards Institute (ANSI) An organization that coordinates the development and use of voluntary consensus standards in the United States.

Area The number of square units it takes to cover a space.

As-built drawings A set of drawings that indicates exactly how the job was completed.

Auxiliary view A view of an object looking from an angle to show the actual size of a surface.

Ball note A symbol used to identify a section by number, page from which the section was taken, and page on which the section is drawn.

Balloon framing A method of framing in which the exterior wall studs run from the sill plate to the top plate of the second floor.

Beam The horizontal structural supporting member.

Change notes Notes that explain changes or revisions on prints.

Change order A change relating to a construction project that occurs after bids have been opened. Changes often cause an increase or decrease in job cost.

Computer-aided design (CAD) Using a computer, program, and plotter or printer to assist in the creation of drawings.

Concrete masonry unit A block of concrete that is mortared in place to create walls.

Curvimeter An instrument used for measuring distances on plans and maps. It has a wheel that rolls along the print, recording the distance traveled.

Demolition plan A drawing that indicates existing features of a building that will have to be demolished or moved.

Denominator The bottom number of a fraction.

Design build A project in which the electrical contractor designs and lays out all electrical for the job and performs the installation.

Detail A scaled drawing drawn to a larger scale to show exactly how a feature is made or where it is located.

Dimension lines Lines that indicate between which two points a measurement has been taken.

Direct job costs Any cost directly associated to a job, such as wages and materials.

Dormer A projection out of a sloped roof, typically containing a window that provides additional space in a room with a sloped ceiling.

Elevation detail A drawing in elevation view drawn to a larger scale to provide additional information.

Estimating The process of calculating what it will cost to build or complete a project.

Estimation sheets Sheets designed to aid in estimating a project.

Extension lines Lines on a blueprint that transfer a point on a plan out to get it away from the drawing, where it can be dimensioned.

Exterior elevation A scaled view of an exterior wall that often contains height and device locations.

Floor joist A horizontal framing member that supports a floor.

Floor plan A scaled view looking downward at one level or floor of a building.

Foundation plan A drawing that shows a plan view of foundation walls, footings, and load-bearing posts or columns.

Fraction A way of representing a part of a whole amount using a numerator and a denominator.

Freehand sketching Sketching a drawing without any drawing tools other than a pencil and paper.

Hourly labor cost The amount an employer has to pay for each hour an employee works, including wages, insurance, benefits, and so on.

Improper fraction A fraction that has a numerator that is larger than the denominator.

Indirect job costs Employers costs that are not directly associated with a specific job, such as office staff and shop expenses.

Interior elevation A scaled view of an interior wall that often contains height and device locations.

Isometric detail An isometric drawing drawn to a larger scale to provide additional information.

Isometric graph paper Paper with a grid of lines drawn 30° off the horizontal axis.

Isometric projection A pictorial view of an object that shows three views of an object simultaneously. All horizontal lines are drawn at 30° off the horizontal axis, and vertical lines are drawn vertically.

Let-in ribbon In balloon framing, a board notched into the framing members to support the floor joists of the second floor.

461

Lighting allowance A set amount of money set aside for a homeowner to purchase light fixtures.

Lighting plan A drawing in plan view that gives information on lighting.

Location dimensions Dimensions that define the distance a feature is from a known reference point.

Metal stud framing A framing method in which metal studs are used to create walls.

Mixed number A number that contains a whole number as well as a fraction.

Notations A few words or sentences used to convey additional information.

Numerator The top number of a fraction.

Oblique projection A pictorial drawing that shows three views of an object simultaneously. One side of the object is drawn true to size, as if looking straight on. The other sides are angled away, usually at 45°.

Orthographic paper Paper with a grid of lines drawn at 90° angles.

Orthographic projection Using two or more two-dimensional drawings to represent a three-dimensional object.

Overhead Costs associated with running a business, including indirect job costs.

Perspective view A pictorial view of an object showing three views simultaneously that is drawn as the human eye would view the object. All horizontal lines taper together as they get further away.

Plan detail A drawing in plan view drawn to a larger scale to provide additional information.

Platform framing Method of framing a wooden structure in which the wall studs are one floor in height. After one level has been completed, a platform is created that the framers can stand on to complete the next level.

Post The vertical structural supporting member.

Poured-in-place concrete Concrete mixed and poured at the job site.

Power plan A drawing in plan view that conveys where electrical devices and equipment are to be installed.

Precast concrete

Concrete panels poured at a plant and shipped to the construction site.

Profit The amount of money left from a project after all costs have been paid.

Reflective ceiling plan A drawing that indicates how the finished ceiling is to be laid out. It has information on the placement of ceiling tiles, lights, and any other device located in or on the ceiling.

Revision A dated change to blueprints or specifications.

Rise The amount a roof increases in height in a predetermined horizontal distance.

Riser diagram A diagram that shows the relationship of the electrical distribution system.

Roof pitch A representation of the slope or angle of a roof, typically written as a fraction of the rise over the run [such as 4/12, meaning a rise of 4 (inches) for every run of 12 (inches)].

Rough-in The stage in the construction process where electricians can install the boxes, cables, and raceways in the wall before the finished wall material is installed.

Run A horizontal distance used to determine the pitch of a roof.

Scale (1) An instrument, with evenly spaced graduations, that is used for measuring; (2) A mathematical size relationship between the actual object and a drawing of the object.

Scaling Using the appropriate scale to determine the actual size of a distance or feature.

Schedule A chart that contains detailed information about what size and type of materials are to be used or how a job is to be completed.

Section detail A section drawing drawn to a larger scale to provide additional information.

Section view A scaled drawing giving a view of an object or part of a building that has been cut away to see the inside features.

Sill plate A piece of lumber (attached to foundation walls and concrete floors) that is pressure treated to prevent rot from the moisture in the concrete.

Site plan A scaled drawing that shows the natural features of a property as well as any man-made features.

Size dimensions Dimensions that define the length, width, and height of a feature.

Specifications A list of detailed job requirements, under which all work must be performed.

Steel column-and-beam construction A type of construction in which large steel members are used for the horizontal and vertical framing members.

Systems plan A drawing in plan view that contains information on the various electrical systems of a premises. Examples would be fire alarm, communications, and security.

Title block An area of a blueprint dedicated to providing additional information about the drawing, designer, job, and so on.

Total job cost The amount a job costs, including direct costs, indirect costs, and profit.

Visualization Imagining what a finished product will look like while looking at a drawing.

Volume The amount of space occupied by a three-dimensional object or region of space expressed in cubic units.

Wood post-and-beam framing A type of construction in which large wooden members are used for the horizontal and vertical framing.

IMPORTANT! READ CAREFULLY: This End User License Agreement ("Agreement") sets forth the conditions by which Cengage Learning will make electronic access to the Cengage Learning-owned licensed content and associated media, software, documentation, printed materials, and electronic documentation contained in this package and/or made available to you via this product (the "Licensed Content"), available to you (the "End User"). BY CLICKING THE "I ACCEPT" BUTTON AND/OR OPENING THIS PACKAGE, YOU ACKNOWLEDGE THAT YOU HAVE READ ALL OF THE TERMS AND CONDITIONS, AND THAT YOU AGREE TO BE BOUND BY ITS TERMS, CONDITIONS, AND ALL APPLICABLE LAWS AND REGULATIONS GOVERNING THE USE OF THE LICENSED CONTENT.

1.0   SCOPE OF LICENSE

1.1   <u>Licensed Content.</u> The Licensed Content may contain portions of modifiable content ("Modifiable Content") and content which may not be modified or otherwise altered by the End User ("Non-Modifiable Content"). For purposes of this Agreement, Modifiable Content and Non-Modifiable Content may be collectively referred to herein as the "Licensed Content." All Licensed Content shall be considered Non-Modifiable Content, unless such Licensed Content is presented to the End User in a modifiable format and it is clearly indicated that modification of the Licensed Content is permitted.

1.2   Subject to the End User's compliance with the terms and conditions of this Agreement, Cengage Learning hereby grants the End User, a non-transferable, nonexclusive, limited right to access and view a single copy of the Licensed Content on a single personal computer system for non-commercial, internal, personal use only. The End User shall not (i) reproduce, copy, modify (except in the case of Modifiable Content), distribute, display, transfer, sublicense, prepare derivative work(s) based on, sell, exchange, barter or transfer, rent, lease, loan, resell, or in any other manner exploit the Licensed Content; (ii) remove, obscure, or alter any notice of Cengage Learning's intellectual property rights present on or in the Licensed Content, including, but not limited to, copyright, trademark, and/or patent notices; or (iii) disassemble, decompile, translate, reverse engineer, or otherwise reduce the Licensed Content.

2.0   TERMINATION

2.1   Cengage Learning may at any time (without prejudice to its other rights or remedies) immediately terminate this Agreement and/or suspend access to some or all of the Licensed Content, in the event that the End User does not comply with any of the terms and conditions of this Agreement. In the event of such termination by Cengage Learning, the End User shall immediately return any and all copies of the Licensed Content to Cengage Learning.

3.0   PROPRIETARY RIGHTS

3.1   The End User acknowledges that Cengage Learning owns all rights, title and interest, including, but not limited to all copyright rights therein, in and to the Licensed Content, and that the End User shall not take any action inconsistent with such ownership. The Licensed Content is protected by U.S., Canadian and other applicable copyright laws and by international treaties, including the Berne Convention and the Universal Copyright Convention. Nothing contained in this Agreement shall be construed as granting the End User any ownership rights in or to the Licensed Content.

3.2   Cengage Learning reserves the right at any time to withdraw from the Licensed Content any item or part of an item for which it no longer retains the right to publish, or which it has reasonable grounds to believe infringes copyright or is defamatory, unlawful, or otherwise objectionable.

4.0   PROTECTION AND SECURITY

4.1   The End User shall use its best efforts and take all reasonable steps to safeguard its copy of the Licensed Content to ensure that no unauthorized reproduction, publication, disclosure, modification, or distribution of the Licensed Content, in whole or in part, is made. To the extent that the End User becomes aware of any such unauthorized use of the Licensed Content, the End User shall immediately notify Cengage Learning. Notification of such violations may be made by sending an e-mail to delmarhelp@Cengage.com.

5.0   MISUSE OF THE LICENSED PRODUCT

5.1   In the event that the End User uses the Licensed Content in violation of this Agreement, Cengage Learning shall have the option of electing liquidated damages, which shall include all profits generated by the End User's use of the Licensed Content plus interest computed at the maximum rate permitted by law and all legal fees and other expenses incurred by Cengage Learning in enforcing its rights, plus penalties.

6.0   FEDERAL GOVERNMENT CLIENTS

6.1   Except as expressly authorized by Cengage Learning, Federal Government clients obtain only the rights specified in this Agreement and no other rights. The Government acknowledges that (i) all software and related documentation incorporated in the Licensed Content is existing commercial computer software within the meaning of FAR 27.405(b)(2); and (2) all other data delivered in whatever form, is limited rights data within the meaning of FAR 27.401. The restrictions in this section are acceptable as consistent with the Government's need for software and other data under this Agreement.

7.0   DISCLAIMER OF WARRANTIES AND LIABILITIES

7.1   Although Cengage Learning believes the Licensed Content to be reliable, Cengage Learning does not guarantee or warrant (i) any information or materials contained in or produced by the Licensed Content, (ii) the accuracy, completeness or reliability of the Licensed Content, or (iii) that the Licensed Content is free from errors or other material defects. THE LICENSED PRODUCT IS PROVIDED "AS IS," WITHOUT ANY WARRANTY OF ANY KIND AND CENGAGE LEARNING DISCLAIMS ANY AND ALL WARRANTIES, EXPRESSED OR IMPLIED, INCLUDING, WITHOUT LIMITATION, WARRANTIES OF MERCHANTABILITY OR FITNESS FOR A PARTICULAR PURPOSE. IN NO EVENT SHALL CENGAGE LEARNING BE LIABLE FOR: INDIRECT, SPECIAL, PUNITIVE OR CONSEQUENTIAL DAMAGES INCLUDING FOR LOST PROFITS, LOST DATA, OR OTHERWISE. IN NO EVENT SHALL CENGAGE LEARNING'S AGGREGATE LIABILITY HEREUNDER, WHETHER ARISING IN CONTRACT, TORT, STRICT LIABILITY OR OTHERWISE, EXCEED THE AMOUNT OF FEES PAID BY THE END USER HEREUNDER FOR THE LICENSE OF THE LICENSED CONTENT.

8.0   GENERAL

8.1   <u>Entire Agreement.</u> This Agreement shall constitute the entire Agreement between the Parties and supercedes all prior Agreements and understandings oral or written relating to the subject matter hereof.

8.2   <u>Enhancements/Modifications of Licensed Content.</u> From time to time, and in Cengage Learning's sole discretion, Cengage Learning may advise the End User of updates, upgrades, enhancements and/or improvements to the Licensed Content, and may permit the End User to access and use, subject to the terms and conditions of this Agreement, such modifications, upon payment of prices as may be established by Cengage Learning.

8.3   <u>No Export.</u> The End User shall use the Licensed Content solely in the United States and shall not transfer or export, directly or indirectly, the Licensed Content outside the United States.

8.4   <u>Severability.</u> If any provision of this Agreement is invalid, illegal, or unenforceable under any applicable statute or rule of law, the provision shall be deemed omitted to the extent that it is invalid, illegal, or unenforceable. In such a case, the remainder of the Agreement shall be construed in a manner as to give greatest effect to the original intention of the parties hereto.

8.5   <u>Waiver.</u> The waiver of any right or failure of either party to exercise in any respect any right provided in this Agreement in any instance shall not be deemed to be a waiver of such right in the future or a waiver of any other right under this Agreement.

8.6   <u>Choice of Law/Venue.</u> This Agreement shall be interpreted, construed, and governed by and in accordance with the laws of the State of New York, applicable to contracts executed and to be wholly preformed therein, without regard to its principles governing conflicts of law. Each party agrees that any proceeding arising out of or relating to this Agreement or the breach or threatened breach of this Agreement may be commenced and prosecuted in a court in the State and County of New York. Each party consents and submits to the nonexclusive personal jurisdiction of any court in the State and County of New York in respect of any such proceeding.

8.7   <u>Acknowledgment.</u> By opening this package and/or by accessing the Licensed Content on this Web site, THE END USER ACKNOWLEDGES THAT IT HAS READ THIS AGREEMENT, UNDERSTANDS IT, AND AGREES TO BE BOUND BY ITS TERMS AND CONDITIONS. IF YOU DO NOT ACCEPT THESE TERMS AND CONDITIONS, YOU MUST NOT ACCESS THE LICENSED CONTENT AND RETURN THE LICENSED PRODUCT TO CENGAGE LEARNING (WITHIN 30 CALENDAR DAYS OF THE END USER'S PURCHASE) WITH PROOF OF PAYMENT ACCEPTABLE TO CENGAGE LEARNING, FOR A CREDIT OR A REFUND. Should the End User have any questions/comments regarding this Agreement, please contact Cengage Learning at delmar.help@cengage.com.